CZECHOSLOVAK ACADEMY OF SCIENCES

Dualist Heresy in the Middle Ages

CZECHOSLOVAK ACADEMY OF SCIENCES

Dualist Heresy in the Middle Ages

Milan Loos

Springer-Science+Business Media, B.V.

1974

ISBN 978-94-015-1214-5 *ISBN 978-94-015-1212-1 (eBook)*
DOI 10.1007/978-94-015-1212-1

Ursprünglich erschienen bei Martinus Nijhoff N.V. The Hague in 1974
Softcover reprint of the hardcover 1st edition 1974

Contents

Foreword

There were two sectarian movements known to mediaeval Europe whose faith was founded on dualistic doctrines and which were very close. The history of the Byzantine Paulician sect of the seventh to tenth centuries is given briefly in Chapter II of the present work, summarizing the basic results of the author's previous studies of the subject. Further research can now rely on an excellent edition of the Greek sources, prepared by the Centre de Recherche d'Histoire et Civilisation byzantine in Paris (1970). After the historical work of Anna Comnena in the twelfth century, the sources tell us nothing of the Paulicians for a long time. Not until the sixteenth and seventeenth century is light thrown on the remnants of the sect surviving in Bulgaria for so many centuries, by the documents of the Catholic mission (studied in detail by *I.Dujčev*). The concluding chapter of this present work is devoted to this interesting epilogue in the story of the Paulicians. The main body of the book deals with the history of a later sectarian movement which spread into many countries and played an important part in the history of mediaeval Europe. This sect was given many names, but the most frequently heard in the Byzantine world was "the Bogomils", while in the West these heretics were usually known as Cathars or Patarenes.

A brief glance will suffice to show how far our knowledge of this vast subject has advanced since the end of the Second World War.

The pioneer work of *H.-C.Puech* (1945) has lost nothing of its value, throwing new light on the dualistic doctrines of the Bulgarian and Byzantine Bogomils by means of precise analysis of all the material and by comparative studies. The basic study by *E.Turdeanu* (1950), which does much to correct the views previously put forward by J.Ivanov, shows how the Bogomil myth was linked with the apocryphal tradition. Thanks to *I.Dujčev* we can now turn to a critical edition of the first evidence of the Bogomils - the letter from Patriarch Theophylact to Tsar Peter (1964). A new edition of the Sermon by Cosmas, prepared by *J.K. Begunov*, is now in press. While working on a new edition of the Byzantine Synodicon for the Sunday of Orthodoxy (1967), *J.Gouillard* discovered evidence of a Byzantine model for the sixteen anathemas of the Bulgarian Synodicon of Boril. It was identified by him in a letter of the Patriarch Cosmas of Constantinople, previously unknown (published in 1970).

The most important discoveries of new material are linked with the name of *A.Dondaine*. After the publication of the Cathar treatise "Liber de duobus principiis" and a fragment of Latin ritual (1939) he discovered the complete text of the anonymous work "De heresi catharorum in Lombardia" (published 1949) and the "Tractatus de hereticis", hitherto unknown, by the Italian Inquisitor Anselm of Alexandria (published 1950). The publication of these texts has made it impossible to doubt that the Cathar churches of western Europe had their origins in the eastern Bogomil movement; this is further confirmed by two texts which had long been known to scholars, the "Summa" of Rainerio Sacchoni and the "Acta" of the Cathar assembly in St.Félix de Caraman, which have been republished by A.Dondaine together with convincing arguments for their authenticity. Other important texts of an anti-Cathar character have been published by *T.Kaeppeli*, *Ilarino da Milano*, and *J.N.Garvin* together with *J.Corbett*.

While "De heresi" and the treatise by Anselm concern Lombardy, two works by Durand de Huesca refer to Languedoc; they were brought to light turn by *A.Dondaine*, and published by *Christine Thouzellier* er (Liber contra Manichaeos, 1964) and *K.Selge* (Liber Antiheresis, 1967). The latter made it possible to throw new light on the nature of the early Waldensian movement and its attitude to the Cathars. These questions are dealt with in the study accompanying K.Selge's edition of the text of the "Liber Antiheresis", and the book by Christine Thouzellier on the history of both these movements in Languedoc at the end of the twelfth and in the early thirteenth century (1966, 1969²). The early history of the Cathars in Languedoc is also dealt with by *E.Griffe* (1969, 1971), who pays great attention to the close links between the heretical movement and the interests of the landed aristocracy. Fundamental to the history of the Inquisition in what is now southern France is the work of *Y.Dossat*, who analyses the involved development of that institution in Toulouse from 1233 to 1273 (1959). Research into the archives of the Inquisition is constantly bringing to light more facts about the sect, its territorial distribution and social structure at different stages of development, and provides glimpses of the everyday life of the heretics. As far as the movement in Languedoc is concerned, recent studies in this field take up the earlier research of C.Douais, C.Molinier and J.Guiraud. Besides these French scholars, who have an important public platform in the Cahiers de Fanjeaux, the records of the Inquisition in Toulouse have been abundantly used by *G. Koch* for his work on the part played by women in the heretical movements of the Middle Ages (1962). Important light is thrown on the involved relations between the Cathar movement and the shifting political constellation in Languedoc by the recent study by *J.L.Biget,* which concentrates on the specific developments in the town of Albi (1971).

The text of Anselm's treatise on heretics has proved remarkably helpful in clarifying the nature of the mediaeval "Bosnian Church", which was the subject of much debate among Yugoslav historians. Against extreme scepticism voiced in various quarters, *D.Kniewald* and *A.Solovjev* have convincingly proved the authenticity of both Latin and Slavonic sources and thus confirmed again that the old conception of F.Rački was the right one, placing the "Bosnian Church" in the context of dualistic heresy in the Middle Ages. Texts newly re-edited or corrected provide a more reliable basis for historical research today. Nevertheless critics could easily point to weaknesses in Solovjev's arguments showing "Bogomil" elements in the text and illuminations of the Bosnian Gospels and in the strange Bosnian tombstones. Philological analysis and historical study of the art concerned has proved that such hypotheses are untenable, claiming too facile proofs of links between the Patarene church and the whole of mediaeval art and literature in Bosnia. This shows up all the more, on the other hand, the value of those native texts whose relation with the Bosnian Church is beyond doubt. Significant progress has been made in the interpretation of the Testament of *gost* Radin; new light has been thrown on this significant figure in the Patarene church by a document, hitherto unknown, published by *M.Šunjić* (1960). It is no longer possible to doubt that the doctrines of the Bosnian Church were dualistic. The most important opponent of Rački's view, *J.Šidak,* long ago recognised that it was fundamentally correct, and himself brought forcible arguments to confirm it. *M.Dinić* achieved valuable results by careful analysis of material from Dubrovnik (1948). The excellent synthesis of mediaeval Bosnian history elaborated by *S.Ćirković* (1964) gives us a good idea of the role of the Patarene church in the state of Bosnia, based on his own research. The work by *D.Mandić* (1962) gives a good survey of the source materials on the history of the Bosnian Church, even if the author's hypotheses are often very doubtful. So far *T.Okiç* has published only part of the results of his research into Turkish tax registers, which throw a surprising light on remnants of the *krst'jane* still to be found in Turkish times. The study of the Serbian Synodicon newly edited by *V.Mošin* (1960) has turned the attention of scholars to Bogomilism in Serbia itself. *A.Solovjev* has also drawn attention to some interesting Serbian texts.

The editions published and the research undertaken during the last quarter of a century have made a new attempt at synthesis possible. The author of the present work was led to essay it by prolonged study of the origins of mediaeval dualism and the continuity of its basic features throughout the varied history of the movement. The first attempt to cover the whole complex of these questions was that of *S. Runciman* (1947), and the need for a new synthesis is evident. On the history of the Bulgarian and Byzantine Bogomils we have the reliable works of *D.Obolensky* (1948) and *D.Angelov* (1969[3]) while for the West the book of *A.Borst* (1953) will long remain an essential reference book and guide through the unusual wealth of material. There are many questions, however, which can only be effectively dealt with in the context of the whole. The author is well aware of the weaknesses inherent in every such attempt, and will be grateful for critical suggestions and points for discussion based on real knowledge of the subject.

The author wishes to express his warmest thanks to Mrs. Iris Lewitová who undertook to translate this book into English. She carried out her task with an exceptional scholarly conscientiousness and fine sense of style.

A few notes on Slavonic orthography are necessary. In Slavonic languages using the Latin alphabet, the original spelling will be used. In languages using the Cyrillic alphabet (Russian, Bulgarian, Serbian) all bibliographical data will be transliterated in the Latin alphabet. The transliteration will follow the principles accepted in Slavic studies (see the journal Slavia 7 [1939] 317 - 320, and again, with certain modifications, 20 [1951] 158-161). In Slavonic expressions and sentences quoted in the English text, the same principles of transliteration will be followed, with the phonetical transcription adapted for English pronunciation in brackets. A phonetical transcription will also follow the names of persons and places. However if, in the text, various names with the same purely Slavonic graphic characters follow one another closely we consider it sufficient to give the phonetic transcription for the first of them only.

Prague, 4 May 1973 M. L.

Works referred to in abbreviated form

Acta Bosnae	*E. Fermendžin,* Acta Bosnae potissimum ecceleasiastica. Zagreb 1892.
Acta Bulg. Eccl.	*E. Fermendžin,* Acta Bulgariae ecclesiastica. (= Monum. Hist. Slav. Merid. XVIII) Zagreb 1887.
Acta St. Félix	*A. Dondaine,* Les actes du concile albigeois de Saint-Félix de Caraman, in: Studi e Testi 125. (= Miscell. G. Mercati 5) Città del Vaticano 1946. (Text: pp. 326-327).
Alexiad	Anne Comnène, Alexiade, ed. *B. Leib,* I - III. Paris 1937, 1943, 1945.
Alphandéry, Idées morales	*P. Alphandéry,* Les idées morales chez les hétérodoxes latins au début du XIIIe siècle. Paris 1903.
Angelov, Bogomilstvoto v Bălgarija	*D. Angelov,* Bogomilstvoto v Bălgarija, Sofia 1969³.
ASS	Acta Sanctorum, ed. Socii J. Bollandi, Antwerpen 1643 -
Beck, Kirche u. theol.Lit.	*H.-G. Beck,* Kirche und theologische Literatur im byzantinischen Reich. (= Byz. Hanbuch II, 1) München 1959.
Benac, Olovo	*A. Benac,* Olovo. Belgrade 1951.
Benac, Stećci	*O. Bihajli-Merin - A. Benac,* Stećci. Belgrade 1963.
Bern. Gui	Bernard Gui, Manuel de l'Inquisiteur, éd. *G. Mollat.*(= Les classiques de l'histoire de France au moyen âge 8 - 9) T. 1, Paris 1926.
Borst	*A. Borst,* Die Katharer. Stuttgart 1953.
Bouquet, Recueil	*M. Bouquet,* Recueil des historiens des Gaules et de la France, I - XXIV. Paris 1738 - 1904.
Brevis Summula	Brevis Summula contra errores haereticorum. Ed.: *C. Douais,* La Somme des autorités à l'usage des prédicateurs méridionaux au XIIIe siècle. Paris 1896.

7 - Sv.

Cathares en Languedoc	Cathares en Languedoc. (= Cahiers de Fanjeaux, fasc. 3) Toulouse 1968.
Chanson de la Croisade	La Chanson de la Croisade contre les Albigeois, ed. *P. Meyer* (Soc. de l'Hist. de France), t. I, II. Paris 1875, 1879.
Chênon, La Charité	*E. Chênon*, L'hérésie à la Charité-sur-Loire et les débuts de l'Inquisition monastique dans la France du Nord au XIIIe siècle, in: Nouv. rev. hist. de droit français et étranger 40 (1917) 299 - 345.
Cipolla, Verona	*C. Cipolla*, Il Patarenismo a Verona nel secolo XIII, in: Archivio Veneto 25 (1883) 64 - 86, 267 - 287.
Cosmas	See *Popruženko*, Cosmas.
Ćirković, Bosnische Kirche	*S. Ćirković*, Die Bosnische Kirche, in: Atti dei Convegni (Accad. Naz. dei Lincei) 361 (1964) 547 - 575.
Ćirković, Istorija	Id., Istorija srednjovekovne bosanske države. Belgrade 1964.
Ćirković, "Verna služba"	Id., "Verna služba" i vjera gospodska, in: Zbornik Filoz. fak. VI (Belgrade 1962) knj. 2, 95 - 112.
Dähnhardt, Natursagen I	*O. Dähnhardt*, Natursagen I. Sagen zum Alten Testament. Leipzig - Berlin 1907.
De heresi	De heresi Catharorum in Lombardia, ed. *Dondaine*, Hiérarchie I, 306 - 312.
Dinić, Državni sabor	*M. Dinić*, Državni sabor srednjovekovne Bosne, in: Srpska Akad. Nauka, Posebna izd., knj. CCXXXI (1955) 10 - 13.
Dinić, Prilog	Id., Jedan prilog za istoriju patarena u Bosni, in: Zbornik Filoz. fak. I (Belgrade 1948) 33 - 44.
Dinić, Trgovina robljem	Id., Iz dubrovačkog arhiva, knj. III. Belgrade 1967, section I: Gradja o trgovini robljem.
Dinić, Patareni	Ibid., section II: Patareni.
Dinić, Odluke veća	Ibid., section III: Odluke veća.
Dondaine, Durand de Huesca	*A. Dondaine*, Durand de Huesca et la polémique anti-cathare. AFP 29 (1959) 228 - 278.
Dondaine, Hiérarchie I, II	Id., La hiérarchie cathare en Italie. AFP 19 (1949) 280 - 312; 20 (1950) 234 - 324.

9 - 6357 - Sv.

Dondaine, Liber	Id., Un traité néo-manichéen du XIIIe siècle, le ‚Liber de duobus principiis' suivi d'un fragment de rituel cathare. Roma 1939.
Dondaine, Manuel	Id., Le Manuel de l'Inquisiteur (1230 - 1330). AFP 17 (1947) 85 - 194.
Dondaine, Nouvelles sources	Id., Nouvelles sources de l'histoire doctrinale du néo-manichéisme au moyen âge, in: Rev. des sciences philos. et théol. 28 (1939) 465 - 488.
Doresse	*J. Doresse*, The Secret Books of the Egyptian Gnostics. London 1960.
Dossat, Crises	*Y. Dossat*, Les crises de l'Inquisition toulousaine au XIIIe siècle (1233 - 1273). Bordeaux 1959.
DTC	Dictionnaire de Théologie Catholique. Paris 1903 -
Dujčev, Cattolicesimo	*I. Dujčev*, Il cattolicesimo in Bulgaria nel sec. XVII. Roma 1937.
Dujčev, Franciscanesimo	Id., Il franciscanesimo in Bulgaria nel secoli XIII e XIV, in: Medioevo I (see below), 395 - 424.
Dujčev, Rilskijat svetec	Id., Rilskijat svetec i negovata obitel. Sofia 1947.
Dupré - Theseider, Introduzione	*E. Dupré - Theseider*, Introduzione alle eresie medievali. Bologna 1953.
Duvernoy	*J. Duvernoy*, Le registre d'inquisition de Jacques Fournier, 1318 - 1325, t. I - III. Toulouse 1965.
Dvorník, The Slavs	*F. Dvorník*, The Slavs in European History and Civilization. New Brunswick 1962.
Ecb. Schon.	Ecbertus Schonaugiensis, Sermones XIII contra Catharos. PL 195, col. 11 - 102.
Euth. Acm., ed. *Ficker*	Euthymii monachi coenobii Peribleptae epistula invectiva contra Phundagiagitas sive Bogomilos haereticos, in: *Ficker*, Phundagiagiten (see below), 1 - 86.
Euth. Zig., ed. *Ficker*	Euthymii Zigabeni de haeresi Bogomilorum narratio, in: *Ficker*, Phundagiagiten (see below) 89 - 111.
Everw. Steinf.	Everwin Steinfeldensis, Epistola ad S. Bernardum. PL 182, col. 676 - 680.

Farlati	*D. Farlati,* Illyricum sacrum, I - VIII. Venice 1751 - 1819.
Ficker, Phundagiagiten	*G. Ficker,* Die Phundagiagiten. Leipzig 1908.
Ficker, Todesstrafe	*J. Ficker,* Die gesetzliche Einführung der Todesstrafe für die Ketzerei, in: Mitteilungen des Inst. f. Österreich. Geschichtsforsch. 1 (1880) 177 - 226.
Florence Ritual	Fragmentum ritualis Catharorum, ed. *Dondaine,* Liber, 151 - 165.
Förg	*L. Förg,* Die Ketzerverfolgung in Deutschland unter Gregor IX. (Historische Studien, H. 218) Berlin 1932.
Frédéricq	*P. Frédéricq,* Corpus documentorum inquisitionis haereticae pravitatis Neerlandicae I - V. Gent 1889 - 1906.
Fumi, Orvieto	*L. Fumi,* I Paterini in Orvieto, in: Archivio Stor. Ital., S.3, vol. 22 (1875) 52 - 81.
Georg. Disputatio	(Georgius). Disputatio inter Catholicum et Paterinum haereticum, ed. *Martène - Durand,* Thesaurus novus anecdotorum V, 1705-1753.
Gouillard, L'hérésie	*J. Gouillard,* L'hérésie dans l'empire byzantin des origines au XIIe siècle, in: Travaux et mémoires 1 (1965) 299 - 324.
Gouillard, Synodikon	Id., Le Synodikon de l'orthodoxie. Edition et commentaire, in: Travaux et mémoires 2 (1965) 1 - 316.
Griffe, Aventure	*É. Griffe,* Les débuts de l'aventure cathare en Languedoc (1140 - 1190). Paris 1969.
Griffe, Languedoc	Id., Le Languedoc cathare de 1190 à 1210. Paris 1971.
Grumel, Regestes	*V. Grumel,* Les regestes des actes du patriarcat de Constantinople, vol. I, fasc. I,II,III (1932, 1936, 1947), fasc. IV, ed. *V. Laurent* (1971).
Grundmann, Bewegungen	*H. Grundmann,* Religiöse Bewegungen im Mittelalter. (= Historische Studien, H. 267) Berlin 1935.
Grundmann, Eresie	Id., Eresie e nuovi ordini religiosi nel

sec. XII, in: Relazioni del X congresso internaz. di scienze storiche III. Firenze 1955, 357 - 402.

Guill. Podiol. — Guillelmus de Podio Laurensi, Chronica, in: *Bouquet*, Recueil (see above) XIX, 193 - 225; XX, 764 - 776. (*J. Beyssier's* edidion in: Université de Paris, Bibliothèque de la Faculté des Lettres 18 (1904) 85 - 175; 20 (1905) 233 ff. is not available to the author.)

Guiraud, Inquisition — *J. Guiraud*, Histoire de l'inquisition au moyen âge I, II. Paris 1935, 1938.

Harnack, Marcion — *A. Harnack*, Marcion. Ein Evangelium vom fremden Gott. Leipzig 1921.

Havet I, II — *J. Havet*, L'hérésie et le bras séculier au moyen âge jusqu'au treizième siècle, in: Bibliothèque de l'École des Chartes 41 (1880) 488 - 517; 570 - 607.

Hefele - Leclercq — *C. Hefele* - *H. Leclercq*, Histoire des conciles, I-X. Paris 1907 - 1938.

Hennecke - Schneemelcher — *E. Hennecke*, Neutestamentliche Apokryphen in deutscher Übersetzung, hrsg. v. W. *Schneemelcher*. I. Band: Evangelien. - II. Band: Apostolisches, Apokalypsen und verwandtes. Berlin 1961, 1966.

Higounet, Grand chapitre — *C. Higounet*, Un grand chapitre de l'histoire du XIIe siècle: la rivalité des maisons de Toulouse et de Barcelone pour la prépondérance méridionale, in: Mélanges Louis Halphen, Paris 1951, 313 - 322.

Hist. Sal. — Thomas archidiaconus, Historia Salonitana. (= Monum. hist. Slav. merid., XXVI. Scriptores, III). Ed. *F. Rački*, Zagreb 1894.

Interrog. Ioh. — Joannis et Apostoli et Evangelistae Interrogatio..., ed. *Ivanov*, Bog. knigi (see below), 73 - 87.

Ivanov, Bog. knigi — *J. Ivanov*, Bogomilski knigi i legendi. Sofia 1925.

Jireček, Heerstrasse — *K. Jireček*, Die Heerstrasse von Belgrad nach Konstantinopel und die Balkanpässe. Prag 1887.

Jireček - Radonić — Id., Istorija Srba, transl. by *J. Radonić*,

	I - IV. Belgrade 1922 - 1925.
Kniewald, Vjerodostojnost	*D. Kniewald*, Vjerodostojnost latinskih izvora o bosanskim krstjanima. Rad Jugosl. Akad. 270 (1949) 115 - 276.
Koch, Frauenfrage	*G. Koch*, Frauenfrage und Ketzergeschichte im Mittelalter. Berlin 1962.
Liber contra Manicheos	Une Somme anti-cathare le Liber contra Manicheos de Durand de Huesca, ed. *Christine Thouzellier*. Louvain 1964.
Ljubić. Listine	*Š. Ljubić*, Listine o odnošajih izmedju Južnoga Slavenstva i Mletačke Republike, I - X. Zagreb 1868 - 1891.
Luchaire, Albigeois	*A. Luchaire*, Innocent III, La croisade des Albigeois. Paris 1905.
Lyons Ritual	*L. Clédat*, Le Nouveau Testament, traduit au XIIIe siècle en langue provençale, suivi d'un rituel cathare.(= Bibliothèque de la Fac. des Lettres de Lyon, 4) Paris 1887 (Text: pp. IX - XXVI).
Maisonneuve	*H. Maisonneuve*, Études sur les origines de l'Inquisition. Paris 1960².
Mandić, Bogomilska crkva	*D. Mandić*, Bogomilska crkva bosanskih krstjana. Bosna i Hercegovina. Povjesno - kritička istraživanja, II. Chicago 1962.
Mansi	*J. D. Mansi*, Sacrorum conciliorum nova et amplissima collectio. Florentiis 1759.
Medioevo I	*I. Dujčev*, Medioevo bizantino - slavo, vol. I: Saggi di storia politica e culturale (= Storia e letteratura, Raccolta di studi e testi, 102) Roma 1965.
Meyendorff, Introduction	*J. Meyendorff*, Introduction à l'étude de Grégoire Palamas(= Patristica Sorbonensia 3). Paris 1959.
MGH SS	Monumenta Germaniae Historica, Scriptores.
Miklošić, Mon. Serb.	*F. Miklošić*, Monumenta Serbica. Wien 1858.
Miletić, "Krstjani"	*Maja Miletić*, I "Krstjani di Bosnia alla luce dei loro monumenti di pietra. (= Orientalia Christiana Analecta 149) Roma 1957.
MM I	*F. Miklosich - I. Müller*, Acta et diplomata Graeca medii aevi I. Vindobonae 1860.
Molinier, Église et société	*Ch. Molinier*, L'église et la société

I, II, III	cathares, in: Revue historique 94 (1907) 225 - 248; 95 (1907) 1 - 22, 263 - 291.
Moneta	Moneta Cremonensis, Adversus Catharos et Valdenses libri quinque, ed. *Th. A. Ricchini*. Roma 1743.
Mošin, Sinodik	*V. Mošin*, Serbskaja redakcija sinodika v nedelju pravoslavija. Analiz tekstov, in: Viz. Vremennik 16 (1959) 317 - 394. - Teksty, in: Viz. Vremennik 17 (1960) 278 - 353.
Muratori V	*L. A. Muratori*, Antiquitates Italicae medii aevi, t. V. Mediolani 1741.
Obolensky, Bogomils	*D. Obolensky*, The Bogomils. Cambridge 1948.
Ostrogorsky, Geschichte³	*G. Ostrogorsky*, Geschichte des byzantinischen Staates. (= Byz. Handbuch I) München 1963³.
Perrens	*F.-T. Perrens*, Histoire de Florence. Paris 1883.
Peter the Martyr	S. Petrus Martyr, Summa contra Patarenos, ed. *J. Kaeppelli*, Une Somme contre les hérétiques de S. Pierre Martyr (?). AFP 17 (1947) 295 - 335.
Peter Sarn.	Petri Vallium Sarnaii Monachi, Hystoria Albigensis, ed. *P. Guébin - E. Lyon*, t. 1 - 3. Paris 1926 - 1939.
PG	*J. P. Migne*, Patrologiae cursus completus, Series graeca. Parisiis 1857.
PL	*J. P. Migne*, Patrologiae cursus completus, Series latina. Parisiis 1844.
Popruženko, Cosmas	*M. G. Popruženko*, Kozma presviter, bolgarskij pisatel' X veka. (= Bălgarski starini XII) Sofia 1936.
Potthast	*A. Potthast*, Regesta Pontificum Romanorum I, II. Berlin 1874, 1875.
Pucić, Spomenici	*M. Pucić*, Spomenici srpski I,II. Belgrade 1858, 1862.
Puech, Catharisme médiéval	*H.-C. Puech*, Catharisme médiéval et bogomilisme, in: Atti dei Convegni (Accad. Naz. dei Lincei) XII (1956) Roma 1957, 56 - 84, 154 - 156.
Puech - Vaillant, Le traité	*H.-C. Puech - A. Vaillant*, Le traité contre les Bogomiles de Cosmas le Prêtre.

	(= Travaux publ. par l'Inst. d'Ét. slav. XXI) Paris 1945.
Račkí, Bogomili i patareni	*F. Račkí*, Bogomili i patareni, in: Srpska Akad., Posebna izd. knj. LXXXVII, Društv. i ist. spisi, knj. 38. Belgrade 1931, 337 - 599.
Rain. Sacch.	Rainerius Sacchoni, Summa de Catharis et Pauperibus de Lugduno, ed. *Dondaine*, Liber, 64 - 78.
Runciman, Medieval Manichee	*S. Runciman*, The Medieval Manichee. A Study of the Christian Dualist Heresy. Cambridge 1947.
Salvo Burci, Doell. II	Salvo Burci, Liber Supra Stella, ed. *I. Doellinger*, Beiträge zur Sektengeschichte des Mittelalters II. München 1890, 52 - 84.
Salvo Burci, Ilarino	Id. ed. *Ilarino da Milano*, Il "Liber Supra Stella" del piacentino Salvo Burci contro i Catari e altre correnti ereticali, in: Aevum 19 (1945) 309 - 341.
Schmidt	*Ch. Schmidt*, Histoire et doctrine de la secte des Cathares ou Albigeois I, II. Paris - Genève 1849.
Selge, Die ersten Waldenser	*K.-V. Selge*, Die ersten Waldenser I, II. Berlin 1967.
Serb. Sinodik	See *Mošin*, Sinodik (texts).
Sinodik of Boril (Palauz., Drin.)	*M. G. Popruženko*, Sinodik carja Borila. (= Bălgarski starini VIII) Sofia 1928 (Palauzov recension, Drinov recension).
Smičiklas, CD	*T. Smičiklas*, Diplomatički zbornik (Codex diplomaticus), I - XII. Zagreb 1904 - 1934.
Solovjev, Autour des Bogomiles	*A. Soloviev*, Autour des Bogomiles, in: Byzantion 22 (1952) 93 - 104.
Solovjev, Doctrine	Id., La doctrine de l'église de Bosnie, in: Bull. Acad. Belg., Cl. des Lettres 5ᵉ série, XXXI (1948) 481 - 534.
Solovjev, Gost Radin	Id., Gost Radin i njegov testament, in: Pregled II (Sarajevo 1947) Nº 7, 310 - 318
Solovjev, Svedočanstva	Id., Svedočanstva pravoslavnih izvora o bogomilstvu na Balkanu, in: Godišnjak Ist. društva B. i H. (1953) 1 - 103.
Stojanović, Povelje i pisma	*Lj. Stojanović*, Stare srpske povelje i pisma I 1,2. Belgrade 1929, 1934.

Synodikon — See *Gouillard*, Synodikon.

Šidak, Današnje stanje — *J. Šidak*, Današnje stanje pitanja "Crkve bosanske" u novijoj literaturi, in: Hist. zbornik VII (1957) 129 - 142.

Šidak, "Dubia" — Id., Franjevačka "Dubia" iz g. 1372/3 kao izvor za povijest Bosne, in: Ist. časopis V - 20 (1955) 207 - 231.

Šidak, Heretički pokret — Id., Heretički pokret i odjek husitizma na slovenskom jugu, in: Zbornik Matice Srp. za društv. nauka 31 (1962) 5 - 24.

Šidak, Misija dominikanaca — Id., Ecclesia Sclavoniae i misija dominikanaca u Bosni, in: Zbornik filoz. fak. u Zagrebu III (1955) 11 - 40.

Šidak, O pitanju — Id., O pitanju heretičkog "pape" u Bosni 1223 i 1245, in: Razprave V (= Hauptmannov zbornik) (Ljubljana 1966) 145 - 160.

Thallóczy, Studien — *L. Thallóczy*, Studien zur Geschichte Bosniens und Serbiens im Mittelalter. München-Leipzig 1914.

Theiner, Mon. Hung. — *A. Theiner*, Vetera monumenta historica Hungariam sacram illustrantia I,II. Roma 1859 - 1862.

Theiner, Mon. Slav. I — Id., Vetera monumenta Slavorum Meridionalium historiam illustrantia I. Roma 1863.

Theloe — *H. Theloe*, Die Ketzerverfolgungen im 11. und 12. Jahrhundert. Berlin-Leipzig 1913.

Thouzellier, Languedoc — *Christine Thouzellier*, Catharisme et valdéisme en Languedoc à la fin du XIIe et au début du XIIIe siècle. Paris 1966.

Thouzellier, Un traité cathare — Ead., Un traité cathare inédit du début du XIIIe siècle, d'après le "Liber contra Manicheos" de Durand de Huesca. (= Bibl. de la Rev. d'Hist. eccl., 37) Louvain - Paris 1961.

Tollius — *J. Tollius*, Insignia itinerarii Italici. Traiecti ad Rhenum 1696.

Tract. her. — Tractatus de hereticis, ed. *Dondaine*, Hiérarchie II, 308 - 324.

Truhelka, Testamenat — *C. Truhelka*, Testamenat gosta Radina, in: Glasnik Zem. Muz. XXIII (Sarajevo 1911) 355 - 375.

Turdeanu, Apocryphes bogomiles I, II	*E. Turdeanu*, Apocryphes bogomiles et apocryphes pseudo-bogomiles, in: Rev. de l'Hist. des Relig. 138 (1950) 22 - 52, 176 - 218.
Vaissète - Molinier	*C. Devic - J. Vaissète*, Histoire générale de Languedoc, ed. *A. Molinier*. Toulouse 1872 - 1892.
Vasiliev, Anecdota	*A. Vasiliev*, Anecdota Graeco-Byzantina. Mosquae 1893.
Vidal, Derniers ministres	*J. M. Vidal*, Les derniers ministres de l'Albigéisme en Languedoc. Leurs doctrines, in: Rev. des questions hist., N. S. 35 (1906) 57 - 107.
Vidal, Doctrine I, II	*J. M. Vidal*, Doctrine et morale des derniers ministres albigeois, in: Rev. des questions hist., N. S. 41 (1909) 357 - 409; 42 (1909) 5 - 48.
Visconti	*A. Visconti*, Storia di Milano. Milano 1937.
Volpe, Movimenti	*C. Volpe*, Movimenti religiosi e sette ereticali nella società medievale italiana (secoli XI - XIV). Firenze 1926.
Vrana, Književna nastojanja	*V. Vrana*, Književna nastojanja u srednovječnoj Bosni, in: Poviest hrvatskih zemalja Bosne i Hercegovine. Sarajevo 1942, 794 - 822.
Wadding	*L. Wadding*, Annales Minorum I-XX. Firenze 1931 - 1948.
Zlatarski, Istorija	*V. N. Zlatarski*, Istorija na Bălgarskata Dăržava prez srednite vekove I 1, 2, II, III. Sofia 1918, 1927, 1934, 1940.

AFP	Archivum Fratrum Praedicatorum
AfSlPh	Archiv für Slavische Philologie
ByzSlav	Byzantinoslavica
BZ	Byzantinische Zeitschrift
ΕΕΒΣ	'Επετηρὶς 'Εταιρείας Βυζαντινῶν Σπουδῶν
REB	Revue des Études Byzantines
RESl	Revue des Études Slaves
RHE	Revue d'Histoire Ecclésiastique

Chapter I

THE FOUNDATIONS OF MEDIAEVAL DUALISM IN THE PERIOD OF LATE ANTIQUITY

The startling discovery of a library of Gnostic writings in Upper Egypt in 1945 or 1946[1] did not reveal to the eyes of scholars a world hitherto unknown. The works of the Church Fathers had already provided sufficient evidence of certain fundamental features to be discerned in the apparently hopeless confusion of myths which reflected to a more or less perfect degree the ideas of the different Gnostic schools and groups in the first centuries of our era. Behind all these heterogeneous beliefs there lay a negative attitude to the world which was something unknown to the minds of classical antiquity.[2] The world had changed, and men now felt lost and perplexed, living as they did in a vast Empire whose burdensome administrative system exhausted their material resources and prevented their minds from imaginative flight.[3] Escapist trends appeared in the philosophy of the time, another aspect of this development being the revival of mysteries.[4] Even the New Testament, and especially the writings of Paul and John, show the Christians' rejection of the "cosmos", in which they lived only because they were forced to, while their thoughts were fixed on the "world to come".[5] Nowhere, however, was the feeling of alienation so strong as in Gnostic thought. The poetry of the Gnostic sects echoes with the feelings of dread aroused by the idea of eternity, of time inexorably repeating the cycles of the heavenly bodies.[6] Instead of resignation to "Heimarmene" - to Fate ineluctably ruling the world and the lives of men, open rebellion now arose, boldly challenging many of the accepted values. The ancient astral gods of Babylonian myth became demonaic forces, encircling and imprisoning the world. The Jewish Jehovah, the God of the Old Testament, found himself on the same footing with them and indeed merged with the greatest of them. The visible world, the Greek cosmos which had been the object of aesthetic admiration, now became the domain of the powers of Evil, a place of despair where not a ray of light could penetrate.

The human soul is imprisoned in this dungeon of the world, confined in the darkness of matter; but it originated in a very different place, in a timeless world, in the bright abode of another God of whom the earthly world and its rulers have no conception. The soul partakes of the very substance of this Unknown God, but in the material world it has lost consciousness of its identity. The world has confused it with its "clamour", poisoned it with its "venom", "benumbed" it and "put it to sleep", until the soul has forgotten whence it came. Only *gnosis* can awaken it, knowledge which comes from above and is really only the reviving memory of the soul's origin. The Unknown God comes to earth in the person of the Saviour; he passes through the circles of the planets and brings the soul this knowledge - the soul is part of His substance, and thus He is liberating Himself. This "story of the soul" which escapes from the bonds of time through the saving knowledge of its eternal essence brings Gnostic thought close to that of the mysteriosophies.[7] From the moment the soul has received this supernatural enlightenment it longs ceaselessly for its lost home. The image of this unknown but better world becomes the only reality - as the moving hymns of the Mesopotamian Mandaean sect put so effectively. The realization that his soul is a part of God's being places a man far above the rest of the world, however humble his position in that world may be for the moment. It was for this pride that the Neoplatonist Plotinus attacked the Gnostics so sharply; he believed that however imperfect the material world might be, it was nevertheless an emanation of Eternal Being.[8] Gnosticism shared with Neoplatonism the idea of a vertical descent from the perfect to the less perfect, but in Gnostic thought the descent was the result of catastrophic disturbance, and therefore necessarily takes on the character of a myth.[9] The Gnostics created not a philosophical system, but an artificial myth which always offered what was essentially the same answer to the fundamental problems of human existence. This gave Gnostic thought a configuration not to be mistaken for anything else, although the individual elements used to build up this Gnostic myth were taken from the treasure-house of late classical syncretism (with a marked predominance of the Jewish component). This accounts for the great variety of forms found. Within the different Gnostic schools individual speculation was unfettered, and the most varied solutions were reached. Nor did the Gnostic schools shut themselves off from outside influences. Indeed the documents discovered at Nag Hamadi in Egypt provide plentiful illustration of how one group borrowed from another without hesitation, and each assimilated heterogeneous mythical ideas. One or another sect could possess Gnostic writings of widely different character; the most frequent form was that of the apocalypse, visions attributed to individuals whose names are sometimes a mystery to us, but for the most part well known, like Zoroaster (Zarathustra); Adam, Seth and other Old Testament figures; or the Apostles. The discovery of these Gnostic texts has also helped to clarify points of agreement between Gnostic teachings and Manichaean thought, and not least to show how some Gnostic writings were used among the Manichees.

Manichaeism is an independent religion; like Buddhism or Christianity. Mani, its founder (216-277), early parted ways with the Elchasaites,[10] the Judeao-Christian Baptist sect to which he originally belonged in southern Babylonia, but he retained their belief that Christ had returned to earth over and over again, each time in a new form. Mani believed himself to be the latest incarnation of Christ, and was convinced that he possessed the complete and ultimate revelation of the truth. Elements of different religious doctrines can in fact be traced in Manichaeism, but its essential characteristics derive from Gnostic thought in late antiquity.[11]

The "story of the soul" is the Ariadne's thread that guides us through the maze of Manichaean myth, crowded with figures and events, embracing images from very different sources and of very different aesthetic value, the whole varying considerably according to the environment. The Gnostic drama, however, took on an original character in Manichaean teaching, in a bold fusion with Iranian Mazdaist beliefs on the conflict between two eternal forces. The Manichaean view of the world was based on the symmetrical image of two principles, two everlasting dominions: that of the spirit and that of matter, of good and of evil - or of darkness and light, in the spiritual sense. The drama of the world follows a three-fold rhythm: at first these two realms existed side by side, but *separate;* then the moment came when the powers of evil, hitherto embroiled in internal conflict, turned against the realm of Light. The first battle ended with the demons of Darkness swallowing some of the particles of Light, and this was the beginning of the second epoch, during which *part of the Light is mingled with darkness*. This means that the Spirit is imprisoned in matter, in ignorance, in oblivion. The forces of the realm of Light, however, were aroused to new efforts, fighting for the gradual liberation of the imprisoned particles of Light. The Creation of the world was part of this magnificent intention. The liberation of the particles of Light imprisoned in matter takes and has taken place in various ways, and the world will continue until all the particles have returned to their everlasting source. Then, once more, the two spheres will be *separated, as they were in the beginning.*

It is an important point that in the Manichaean view the world was not created by the forces of Evil, although they intervened in it with dramatic effect, to hinder the liberation of the particles of light. In this way man was also created,[12] his soul being no other than light, imprisoned in the darkness of matter. Although the first man - Adam -

was aroused from ignorance to clear awareness of his nature, through the good offices of the forces of Good, human beings nevertheless continued to fall into sin, and sin kept them bound to matter. Messengers of Truth, the greatest of whom was Mani himself, were therefore sent to instruct men in their true nature and show them how to conduct their lives in order to liberate their souls, not to have to remain prisoners of matter in further incarnations. It is necessary to avoid as far as possible everything that increases the suffering of the divine substance in the world - suffering caused not only when man kills living creatures or destroys plants, but even simply by moving in the air or the water. The class of the "Elect" formed in the Manichaean church were expected to renounce sexual intercourse, and to restrict the taking of food to the minimum required for the maintenance of life. The Gnostic idea of liberation of the soul thus led the Manichees to abnormal austerity and an unnatural degree of asceticism. On the other hand, it was a powerful source of inspiration for poetry, song, music and painting; we are still enchanted today by the beauty of the miniatures and the fine calligraphy of their books, and paintings adorned the walls of their churches, too.

To some extent Manichaeanism drew on the heritage of the great Gnostic schools whose teachings spread in the Christian communities of Egypt and Syria, but we can find many common mythical themes and ideas in texts deriving from the broad stream of earlier Gnostic tradition, originally untouched by Christian influence. Some Gnostic texts are completely free from Christian influence, while others show clearly how Christian elements penetrated the Gnostic myth at a later stage. In a book said to record secrets revealed to St.John the Apostle the figure of the "Mother" was later replaced as the vehicle of salvation by that of Christ, and the two versions clumsily combined. Irenaeus, the great Christian theologian of the second century, undoubtedly had this Gnostic text in front of him when he wrote his great polemic against the Gnostics. Certainly the teachings of the Gnostic sect described in Irenaeus, Book I, Chapter 29, show a remarkable similarity to those found in the "Apocryphon of St.John".[13] The teachings of another group described in the following chapter also belong to the same type, and further variants can be seen in the recently discovered Coptic texts, "The Hypostasis of the Archons" and "The Origin of the World".[14] Everywhere we find Sophia, the heavenly being whose error plunges the cosmos into the catastrophe which leads to the creation of the world.

This was the tradition which inspired the teachings of the great Gnostic Valentine,[15] who began preaching in Alexandria and was active in Rome from 135 to 165. The teachings presented by these Coptic texts embrace, often in a somewhat chaotic form, a large number of mythical episodes; the school of Valentine, on the other hand, elaborated a clear system informed by a simple mental process projecting psychological phenomena into the supernatural. They distinguished three substances: spiritual (pneumatic), psychic and material (hylic), a division which was not typical of this school alone; other Gnostics too, like Goethe's Faust, distinguished two souls in the human being. The first, usually called the *pneuma*, was the fundamental element of being, the immortal core; as we have seen, it was part of the divine substance, a spark of divinity which had come to earth in its wanderings.

Then came the sum of qualities acquired in this world, wrapped round the immortal core and forming the second soul, usually called the *psyche*, which bound man to the world in which he lived. Finally there was the outer layer, the material body. Matter, polymorphous, fluid, subject to change and decay, was identified with moral evil in Gnostic thought; this was a feature common to all the Gnostic schools, and one which was present in Manichaeism, too. In the latter, of course, it was accomodated to Iranian religious ideas, so that the principle of Evil, of Darkness, which was identified with matter in accordance with the Gnostic conception, was believed to have existed from all eternity and to be an active force, attacking the realm of Light. Even in Gnostic thought itself we often find the idea that matter has always existed, although only as a passive element. Some forms of the Gnostic myth, however, do not assume that matter has existed from all eternity, but that it came into being as the tragic result of the Fall which took place in the sphere of eternal spiritual powers. This is the idea to be found in the "Hypostasis of the Archons" and the "Origin of the World", which in this respect derive from a common source. Sophia, one of the eons, an emanation from the Supreme God, deranged the eternal order when she attempted an individual creative act. The result was a veil dividing the universe into the two separate regions of light and shadow; the shadow thus cast became matter.

In the system of the Valentinian school, as perfected by Valentine's disciple Ptolemy, the figure of Sophia has become two. "The first Sophia" is the last in the series of eons which make up the fullness of God *(pleroma)* and which at the same time, by their descending order, represent the decline from perfection to imperfection. The fatal disturbance occurs at the lowest level when Sophia, the youngest of the eons, upsets the balance of the *pleroma* by her excessive longing to penetrate the eternal mystery. Her passion, cast out into outer emptiness *(kenoma)*, is finally given form in a new Sophia. In her longing for lost divine light this second Sophia is tormented by a similar combination of suffering as was her mother before her; her spiritual states, later taken from her by her Saviour, acquired an independent existence as *psychic* and *hylic* (material) substance.

The creator of the visible world also belongs to the psychic element. The Demiurge of Valentine is not demonic, unlike the figure called up by the Gnostic texts we have already quoted. On the contrary, albeit unwittingly, he is the instrument of the divine forces in Sophia, and the world he created is in the Platonic sense an imperfect reflection of a higher spiritual world.

Gnosis in the teachings of Valentine again shows the profound difference between the conception of salvation among the Gnostics and that of the Christians. It is not faith, but gnosis, that ensures salvation. In Ptolemy's system gnosis comes first to the eons, disturbed by the tragedy of the first Sophia. A new eon, Christ, imparts to them the knowledge of unattainable divinity, and thus restores the pleroma to peace and harmony. The pleroma then gives forth a new emanation - the Saviour (Jesus), who brings gnosis to the second Sophia and thus assuages her sorrows. The sight of the angels in the train of the Saviour causes the second Sophia to bring forth *spiritual seed* which is placed in the visible world. This seed is the souls of the elect *(pneumatikoi)*, the female counterparts to the angels just as Sophia herself is the bride of the Saviour, and just as the pleroma is made up of "pairs" of eons. The gnosis received by the Elect in the school of Valentine led the spiritual seed to perfection, so that after death, freed from the psychic and the hylic layers, these souls could at last enter the pleroma. There they would be united with their angels, as their mother Sophia would celebrate her wedding with the Saviour.

Even the second category of men *(psychikoi)* could achieve a form of salvation, by means of faith and good works; this of course was only a pale reflection of the true bliss of the Elect, just as the Demiurge and his world were only dull images of supernatural reality. These "psychics" were the simple Christians among whom the Valentinians lived.

Men of the third category, identified with the heathen, were destined to utter destruction. They were the *hylikoi*, irretrievably entangled with matter which itself was predestined to destruction.

The revelation of gnosis as supernatural knowledge was attributed by Gnostic groups to various prophets,[16] but the Valentinians linked it with the *appearence of Jesus in the world.* In their teachings not only Christ or the Saviour, from the abstract sphere of the eons, but the historical Jesus whose story is told in the Gospels also had a place.[17] Nevertheless the Gnostic view of matter would not allow the Valentinians to accept his incarnation, and they were bound to resort to the idea that the birth and death of Christ were only apparent (docetic Christology). The Gnostic teachings could not go more than half-way towards those of Christianity.

The great Gnostics like Basilides, Valentine and their disciples were originally members of Christian communities who tried to remain within the fold, and who declared that their gnosis was the true meaning of Christianity. In fact, however, they clothed in Christian form ideas which were essentially alien to it. In the end they were cast out from the Christian community as it became more tightly organized and realized the need for a firm foundation of dogma.

For Irenaeus, writing at the end of the second century, Gnosticism was a very real threat. In his own diocese of Lyons he had had to deal with "false" doctrines which had found supporters especially in high society, and the first generation of Valentine's disciples were the living contemporaries of this Church Father. Irenaeus directed his polemics against other teachers of heresy, too, outstanding among them being Marcion,[18] the son of a bishop of Sinope. Marcion had been in Rome about the same time as Valentine. Disagreeing with the Christian church in Rome, he founded one of his own, a powerful church which was regarded for years as the most dangerous, indeed the only really serious threat to the rest of Christianity. Marcion carried the teachings of St.Paul the Apostle to their logical extreme; the words "old things are passed away; behold, all things are become new", were taken to mean that the Gospel had replaced the Old Testament. Marcion could not avoid Gnostic dualism, however;[19] he carried the contrast between the new religion of love taught by Christ and the stern justice of the Old Testament into the sphere of metaphysics. There is no reason to doubt the traditional view that in this Marcion was influenced by Cerdon, the Syrian Gnostic to whom Irenaeus attributes this same dualist belief, distinguishing the God of the Old Testament from the God preached by Christ. Evil matter as a principle, eternal and not created, has its place in Marcion's teachings, but the main emphasis is on the contrast between the two Gods. He did not distinguish the three substances, pneumatic, psychic and hylic; the God of this world did not resemble the Demiurge of Platonic myth nor had he created this world to the image of a higher one. Marcion gave no explanation of the origin of the Creator; both he and the Unknown God of the invisible world existed independent of each other, and the abyss between them was simply impassable. The God of this world, according to Marcion was *not* the demon of evil, the father of the evil archons of the planetary spheres and at the same time the highest of them all - a being presented by many Gnostic sects in the shape of the grim figures of Iranian myths and identified by them with the Jewish God the Creator. For Marcion He was *only* the God of the Old Testament, while the author of evil - the devil - was merely His creature. The fundamental characteristic of the Creator was justice in the Old Testament sense, the merciless justice of a despot, often turning to sheer malignity. The weakness and imperfections of this God were clearly revealed, and he was responsible for the misfortunes of all human life. Man was at the mercy of his Creator, who had formed his body and breathed a soul into it. Marcion thus rejected the basic Gnostic idea that the soul is a particle of the divine Being and that God liberates Himself in the soul. Marcion's God is not only the "Unknown", but the "Stranger" as well; he has nothing in common with man, and redeems him from durance merely out of His charity. Only the soul of man will be saved, however; the body, formed of matter which is evil, will decay. Here again the ideas of Marcion come back to those of the Gnostics.

Like other early Christian teachers, Marcion was a modalist; he regarded Christ as a manifestation of God. He could not accept the incarnation of Christ as real, for God could not inhabit unworthy matter. Nor did he turn to the explanation that the birth of Jesus was apparent. According to Marcion Jesus did not appear until the fifteenth year of Tiberius' reign; he preached the true God, of whose existence not only mankind, but even the Creator Himself as yet knew nothing.

On the other hand, the Unknown God had known from the very beginning that the Creator existed, as an incomparably inferior God, bound to the visible world, which was made in His image. Thus the struggle between the two is an unequal one, and the Unknown God has no difficulty in reaching the earth, in the form of Christ. Since His essence is love, He does not take advantage of his superiority over the Creator to take mankind from him by force, but literally redeems mankind by His death on the cross. The sufferings and death of Christ were real, although his body was mere illusion. This inconsistent docetism was not at all unusual in Marcion's day. The early Christian belief that Christ descended into Hell was also very much alive in his teachings; in this way Christ brought salvation to those previous generations which had known only the Creator's Law. Not all were ready to receive salvation, though; the heathen and sinners chastised by the God of the Jews took refuge with the victorious Saviour, while the "just men" of the Old Testament remained faithful to the Creator who had rewarded them. The paradoxical result was that Cain and not Abel would be saved!

The only way of salvation for the living was faith in the Unknown God, the Stranger, foretold by Christ. Belief in this loving God before whom man need not tremble in fear would carry the soul up to the Supernatural heights where no sin could reach and where the commandments of the stern Old Testament fade away as something obvious which there is no need to pay special attention to. The "Strange God", however, was the God of the world to come, when this world and apparently its Ruler as well would have passed away. In this world mankind is entirely at the mercy of the Creator's wrath. In this world, too, false Jewish apostles have obscured the teachings of Christ, falsified the Gospels and the Epistles of St.Paul. Marcion took it upon himself to purge the Scriptures, and did so with remarkable consistency, not flinching before the most forced interpretations. After his death Marcion's disciples saw him as seated with Paul by the side of Christ, but he himself said he was but a modest disciple of Paul.

The organization of Marcion's church resembled that of the other Christian communities of his day, and was headed by bishops and deacons. In ritual and in church order, however, it seems to have been simpler, nor was there such a barrier between the clergy and the laity. A strict way of life was maintained, with insistence on absolute asceticism. Marcion felt the same revulsion against indulgence of the flesh, derived from matter which was evil, as he did against the shedding of blood, of which so much was written in the Old Testament. Everything that served to strengthen the physical side of human life should be restricted to the absolute minimum, and it was therefore desirable to refrain from eating meat or drinking wine, and to keep strict fasts. Marcion's rejection of marriage shows the extreme abhorrence in which he held the whole of man's physical nature. At the same time it was a clear protest directed against the Creator, for it was meant to prevent the expansion of His domain by the multiplication of the human race.

Baptism and the Eucharist were administered in the Marcionite church more or less as in other Christian communities, and it is likely that Sunday worship also followed the same lines, so that a Christian arriving in a strange town might easily stray into a Marcionite church by mistake. The church could not survive in the towns for long, however; from the time of Constantine the Great, Roman law had condemned heretics as violaters of the religious unity of the Empire.[20] Christianity had been made the official religion under Theodosius the Great, and from then on heresy was considered a public crime, included therefore in the term "lese-majesty" (law of Arcadius, Honorius and Theodosius II, 22.II.407). Although the Marcionites were never explicitly named in the imperial constitutions, there can be no doubt that they were included in the persecution. They were thus gradually driven into the country districts, where they were better protected. In out-of-the-way rural areas the remnants of what had been a great church lingered on for a very long time. This explains the surprising renaissance of Marcionism somewhere in the middle of the seventh century, in the form of the new sect of *Paulicians*.

[1] See *H.-C.Puech*, Les nouveaux écrits gnostiques découverts en Haute-Egypte (Premier inventaire et essai d'identification). In: Coptic Studies in Honor of Walter Ewing Crum (Boston, Mass., 1950), 91 - 154; *Doresse*, Secret Books; *H.Jonas*, Gnosis und spätantiker Geist I. Göttingen 1964³, 377 ff.. Bibliographic information on texts as published is given regularly by Byzantinische Zeitschrift (4 B: Apokryphen).

[2] *Jonas*, op.cit., 140 ff.

[3] See *Simone Pétrement*, Le dualisme chez Platon, les gnostiques et les manichéens, Paris 1947, 158.

[4] See A.D.Nock's review of the first edition of Jonas in Gnomon 12 (1936) 605-612. Cf: *E.v.Ivánka*, Religion, Philosophie und Gnosis: Grenzfälle und Pseudomorphosen in der Spätantike, in: Le origini dello gnosticismo, Colloquio di Messina 13 - 18th April 1966, Leiden 1967, 319.

[5] The references are given by H.Sasse, in: *G.Kittel*, Theol. Wörterbuch zum Neuen Testament III, 884 ff.. Cf. *Pétrement*, op.cit., 208 ff.

[6] *Jonas*, op.cit., 94 ff.

[7] See *U.Bianchi*, Le problème des origini du gnosticisme, in: Colloquio di Messina, 9 ff.

[8] See *E.v.Ivánka*, Plato christianus, Einsiedeln 1964, 70 - 82.

[9] See *H.Jonas*, Delimitation of the gnostic phenomenon - typological and historical, in: Colloquio di Messina, 92 ff.

[10] See *A.Heinrichs - L.Koenen*, Ein griechischer Mani-Codex, in: Zeitschrift f. Papyrol. u. Epigraphik 5 (1970) H.2, 97 - 216 + 3 Taf.

[11] The best treatment of Manichaeanism is so far to be found in the following basic studies: *H.J.Polotsky*, Abriss des manichäischen Systems, Stuttgart 1935, special reprint from *Pauly-Wissowa*, Realenzyklopädie der klassischen Altertumswissenschaft, Bd. VI, Sp.241 - 272; and *H.-C.Puech*, Le Manichéisme. Son fondateur - sa doctrine, Paris 1949.

[12] See *K.Rudolph*, Gnosis und Manichäismus nach den koptischen Quellen. In: Koptologische Studien in der DDR, in: Wissenschaftl. Zeitschr. d. Martin-Luther-Univ. Halle-Wittenberg (1965) 156 - 190.

[13] *F.M.Sagnard*, La gnose valentinienne et le témoignage de Saint Irénée Paris 1947, 445.

[14] "The Hypostasis of the Archons", German translation by *H.-M.Schenke*, in: *Leipoldt Schenke*, Koptisch-gnostische Schriften aus den Papyrus-Codices von Nag-Hamadi, Hamburg-Bergstedt 1960, 69 - 78. "The Origin of the World", published by *A. Böhlig*, - *P.Labib*, Die koptisch-gnostische Schrift aus Codex II von Nag Hammadi (DAW-Inst. f. Orientforsch. 58) Berlin 1962.

[15] For Valentinian Gnosis see the fundamental work by *F.M.Sagnard* (N. 13 above).

[16] See *Doresse*, Secret Books, 146 ff.

[17] See *H.I.Marrou*, La théologie de l'histoire dans la gnose valentinienne, in: Colloquio di Messina, 215 - 226, esp. 221 ff., and the discussion on 225 f.

[18] The best monographical work on Marcionism is still *Harnack*, Marcion. Note should be taken, however, of the criticisms of Harnack's view of Marcion expressed particularly by *H.v.Soden*, in: ZKG 40 (1922) 191 - 206. Cf. *E.C.Blackman*, Marcion and his influence. London 1948.

[19] Gnostic dualism had a more profound influence on Marcion's teaching than Harnack admitted. See *Soden*, l.c., 204. For the contrast between the God of the New Testament and the God of the Old Testament in Marcion and the Gnostics, see *Sagnard*, op.cit., 75.

[20] See *J.Cronz*, La lutte contre l'hérésie en Orient jusqu'au IXe siècle. Pères, conciles, empereurs. Paris 1933; Maisonneuve, 30 - 34.

Chapter II

THE PAULICIAN MOVEMENT IN BYZANTIUM

The name Paulician[1] is of Armenian origin, and from the outset the sect was clearly linked with the Armenian ethnic environment. Throughout the first century of its existence there were two centres round which the life of the sect gathered. It was born in the Byzantine province of the Fourth Armenia, near the town of Arsamosata (Asmosata);[2] the other refuge of the sect was the Pontic region, where the Armenian element was also strong. This early period in the history of Paulicians is known to us only from their own semi-legendary tradition, first recorded by Byzantine authors as late as the ninth century. At first the Paulicians existed secretly in out-of-the-way villages, and at the end of the seventh century the sect was almost wiped out. The constant threat of persecution drove the Paulicians to seek refuge once more in their original home, in the time of Leo III (717 - 741); at that time the power of Byzantium did not reach so far, for the Fourth Armenia had been securely in Arab hands since the beginning of the eighth century.

It was not until the second half of the eighth century, when the Paulicians again appeared on Byzantine soil, that the movement gained in strength and in importance, and came out into the light of history. The Byzantine chroniclers mentioned them, and the Church indulged in more and more frequent polemics against this dangerous heresy. Militant and enterprising, the sect took root in the towns and even reached Constantinople itself. The times were unusually favourable. From the late twenties of the eighth century the ecclesiastical policy of the Byzantine rulers had been determined by the powerful iconoclastic movement which originated in the eastern provinces of the Empire. There can be no doubt that one reason lay in the constant threat from the Arabs, forcing the emperors of the Syrian dynasty to pay careful attention to what the people of the eastern themes of Asia Minor thought; the region was of vital importance for the defence of the Empire. The iconoclastic movement reached its peak under Constantine V Copronymus (741 - 775), during whose reign the Paulicians again appeared in Byzantium. The walls of the churches were painted over with secular pictures; icon worship, so deep-rooted in Byzantine religious life, had its strongest support in the monasteries, and here resistance was brutally overcome. The emperor himself was sympathetic to the extremists who even attacked saint worship and the cult of the Virgin Mary. It was an atmosphere in which the Paulicians, who hated the monks and refused to show respect to pictures of the saints and of the Virgin, could expect to be tolerated,[3] and could more readily make converts to their own beliefs; these of course went much further than the most radical iconoclastic ideas.

The Paulicians aimed not at reform of the Church, but at its abolition and replacement by primitive Christianity. There were no ordained priests in their religious community, but only the body of "Christians", among whom certain individuals had been blessed with the gift of the Spirit, giving them authority in matters of the faith while others were entrusted with administrative matters. Essentially this was a return to the conditions of apostolic times, but there were very curious features to the Paulician "reconstruction" of early Christianity. Their "churches" had neither bishops, deacons nor presbyters, which was probably the ultimate outcome of their rejection of the official Church whose hierarchy used these old titles. The leaders of the sect called themselves "pastors and teachers", after Paul's Epistle to the Ephesians (4.11). Their assistants were also charismatic; they were known as *synekdemoi*, a term used in II.Corinthians 8.19 and in the Acts of the Apostles 19.29. The Paulician "pastors and teachers" believed themselves to be the bearers of the true faith preached in the early days of Christianity by the "apostles, prophets and evangelists". They really lived their roles as disciples of the Apostle Paul, taking upon themselves the names of Paul's followers and giving names from his Epistles to the churches they founded. The fragments of epistles which have survived from the pen of the greatest of the Paulician teachers, Sergius (d.834/835), are permeated by the ideas and phrases of St.Paul. The narrower canon of the New Testament accepted by the Paulicians consisted of the four Gospels and the Epistles of St.Paul, which was very nearly that of the Marcionites' Bible. It is also significant that they laid the greatest emphasis on the Gospel according to St.Luke. It is known that Marcion took St.Luke's Gospel as his basis for the reconstruction of the "true" gospel of Christ. The broader Paulician canon included the Acts of the Apostles and the other Epistles, with the exception of those by Peter. It was a fundamental difference between the Paulicians and the Marcionites, however, that the former did not change the wording of the New Testament, being content to interpret it in the spirit of their anti-ecclesiastical teachings.

The polemical writings of the Byzantine ecclesiastical authorities show very clearly the general Byzantine tendency to classify new ideas under old headings. Often on very superficial grounds they would liken "newly appeared" heretical teachings to the classical heresies against which the great Fathers of the Church had fought. In the case of the

Paulicians this tendency stands out very clearly. In the eyes of orthodox writers the Paulicians were simply the followers of the famous heretical teacher, Paul of Samosata. The ecclesiastical polemists confused this well-known Bishop of Antioch with the Paul who really, according to to Paulician tradition, revived the sect which had practically died out at the end of the seventh century. Confusion was certainly all the easier because Samosata in Syria was often confused with Arsamosata, close to which the original Paulicians had arisen. The name *pavlikeank* (Greek: Παυλικιάνοι), used in Armenia to denote the Nestorians, as heirs to the adoptionist doctrine of Paul of Samosata, was thus easily transferred to the new sect, erroneously believed to derive from this third century heretical teacher. Soon, however, Paulicianism was also declared an offshoot of *Manichaeanism;* the combination of these two versions produced the well-known legend of how the sect arose and got its name. The superficial assumption that the Paulicians "confess two principles" sufficed the orthodox polemists, who identified the new movement with the supreme dualist heresy, in their eyes, that of Manichaeanism.

The truth lies elsewhere. Paulician teaching is not based on the Iranian dualism of Darkness and Light, which merged in Manichaeanism with the Gnostic dualism of Spirit and Matter, but preserved in a remarkably pure form the dualism of God and the World preached by Marcion.

It is well known that some of Marcion's disciples developed his teachings in various directions not envisaged by the Master, who devoted himself entirely to reform of the Biblical text and did not construct a system of thought. The mythologising trends which developed sometimes took his metaphysical dualism nearer to Manichaeanism, while other teachers toned it down in the Platonic spirit. Compared with these later variants of Marcionism, Paulician teachings show indubitable archaic features which could have survived in isolated Marcionite groups in rural areas.

The fundamental idea of the two Gods remained as untouched by Gnostic myths in Paulicianism as in Marcionism. The supreme God does not display the fulness of His being *(pleroma)* through a descending line of eons, but simply rules an invisible world which the Paulicians endowed with a specific form - that of Heaven inhabited by angels. On the other hand the "ruler of this world" (ἄρχων τοῦ κόσμου τούτου) for them, as for the Marcionites, was simply the God of the Old Testament, rejected by the Paulician teachers as totally as by Marcion. There is remarkable agreement in the anthropological attitudes of the two sects; for the Paulicians, too, the human being as a *whole* was the work of the Old Testament God the Creator. Both assign the same motive to divine redemption of man: out of pure love for suffering mankind God decided to "take man unto Himself". The modalist Christology of the Marcionites, however, is replaced among the Paulicians by a form of adoptionism: Christ was one of the angels, and only after his work on earth had been completed could he be called the Son of God. The whole story of the salvation of mankind was part of the secret teaching revealed only to the Paulician initiates. In itself this esoterism is evidence of a development away from Marcionism, whose founder rejected the idea of mystery. The content of this esoteric teaching also reveals a tendency to myth which was alien to Marcion himself although it flourished among his disciples.

The soteriological teachings of the Paulicians leave us with the impression that unlike Marcion they regarded Christ's death as a sacrifice but by no means as the price of redemption, paid for the salvation of mankind. The Paulicians made individual salvation dependent solely on faith in Christ's teaching of the true God.

Their view of Christ's humanity shows clear continuity with that of Marcion. The sources say nothing of the Paulicians' attitude towards *matter*, but agree that they did not believe in the physical incarnation of the Son of God.

Christ's life on earth, for them, began with His birth; nevertheless they did not consider it real, and Mary was regarded as an ordinary woman. The Saviour's sufferings and death upon the cross, however, were accepted as reality, just as they were by Marcion. The "cursed instrument", as they called the cross, seems to have meant it as the instrument of Christ's suffering, although the phrase itself is an echo of the well-known words of St.Paul in his Epistle to the Galatians (3.13), quoting Deuteronomy 21,23.

Just as they did not believe in the physical body of Christ, so the Paulicians rejected the bread and wine of the Eucharist and baptism by water. They interpreted these words in a spiritual sense only, maintaining that what Christ had in mind was His word, and not bread and wine. Baptism, too, meant for them word of the gospel only, for did not Jesus say "I am the living water"? Indeed, an allegorical interpretation helped them out of their difficulties whenever they were forced to conform to the orthodox articles of faith. They would declare their faith in the terms required of them, while assigning to the words a different meaning, in accordance with the teachings of their sect.

As we have seen, the Paulicians stripped their religion of the basic forms of Christian worship, baptism and the Eucharist, thus going much further than Marcion. On the other hand, there is no evidence among the Paulicians of the asceticism which was such a characteristic feature of the Marcionite church, and indeed there are facts which argue against the sect either preaching or practising asceticism. Nor do the polemists ever accuse the Paulicians of criticizing the existing order of society.

In the eyes of the Paulicians themselves their dualist faith, rejecting this world and its God, marked them off quite distinctly from orthodox believers. When they were living in safety on Arab territory, and free to speak openly of their faith, they made no secret of it. While the Byzantine Church believed in one God, the Creator and Ruler of the world, they - who called themselves Christians - knew very well that there were two Gods, one who created and ruled the visible world, while the other had no power in this world but in the world to come. It was of this other God that the words of the Gospel spoke: "Ye have neither heard his voice at any time, nor seen his shape". Thus for the Paulicians dualism was not mere theological speculation, but the metaphysical explanation of the impassable rift between them and the established church.

A movement which thus consistently discarded the whole system of dogma, worship and organization of the church was bound to come into conflict with the Byzantine state, which was too closely involved with the Orthodox church. No doubt when iconoclasm was at its height, under Constantine V. the persecution of the Paulicians slackened. The evidence also suggests that at the beginning of the ninth century the emperor Nicephorus I (802 - 811), a realistic statesman who had the interests of the state at heart, was tolerant of both the iconoclastic movement, still strong, and the Paulicians who had been growing more powerful. During this time the sect had spread throughout the empire. During the short reign of Nicephorus's successor Michael I (811 - 813), however, the laws against the Manichees were applied in all their severity to the Paulicians. The laws instituted by Justinian had made it the duty of all in authority to seek out the followers of Manichaeanism, to whom very severe punishment was meted out, including the death penalty.[4] The inexorable application of these measures cleared the Byzantine Empire of Manichaeanism.[5] The same measures were again prescribed, however, in the Eclogue of the Syrian emperors (726), where the method of execution was laid down as death by the sword. This was the law now applied to the Paulicians and also to the Athingan sect, widespread in Phrygia and Lycaonia. (Similarly the Eclogue prescribed death for the Phrygian sect of the Montanists, as well as for the Manichees.)[6]

These drastic measures instituted by the Emperor Michael I were instigated by the leaders of the Byzantine church, headed by the patriarch Nicephorus; they were prepared to go to all lengths to fight the growing influence of heresy. The radical monastic trend, however, headed by Theodore of Studios, was opposed to such severity, remaining firm in their belief that violence and the shedding of blood went against the laws of the Church. Contemporary sources allow us to assume that for a time Theodore influenced the Emperor and the Patriarch to moderation in the matter of death for heresy.[7] The imperial edict[8] remained in force, however, and not even the official revival of iconoclasm since 815 brought the respite hoped-for by the Paulicians. The sect was centred at that time in the north-eastern regions of Asia Minor, where the population was mainly Armenian. From a laconic comment by Peter of Sicily it appears that the Metropolitan of Neocaesarea and the exarch of the region (in charge of supervision of the monasteries) carried out the imperial edict in their persecution of the Paulicians. In his paraphrase of Peter's text the patriarch Photius gives a detailed description of the way the edict was put into execution.[9] He was evidently not concerned with the historical facts, but with a model application of the regular canonical procedure in such cases. The judicial investigation of the accusations was in the hands of the Church, as was the power of absolution and the pronouncement of ecclesiastical disciplinary measures. The secular organs served only as auxiliaries; they sought out the heretics and brought them before the ecclesiastical court, and ultimately carried out the punishment prescribed for unrepentant heretics, who were handed over to the secular court.[10]

In this region the persecution of the Paulicians was repaid with violence, and after taking their revenge large numbers of the sect took refuge over the border, in the territory of the Emir of Melitene. The Arabs used this influx of refugees to found new military colonies on their borders, and the Paulicians then took part in the frequent Arab raids on Byzantium. This went on in spite of the express disapproval of their "teacher" Sergius, who even as a refugee went on living like the Apostles by the work of his hands. By this time, of course, not even the most authoritative spiritual leadership could deflect the inevitable flow of developments. Already in the reign of Theophilus (829-842), the last of the iconoclastic emperors, the Paulicians had their own military forces in the border region of the Caliphate. Their military leader became Carbeas, who had held high rank in the Byzantine army Some of the commanders of the army in the eastern themes were also secretly in touch with them. It was a grave situation for the Empire. Heresy had become so widespread in the eastern provinces that the Empress Theodora, who restored icon worship (843), tried to suppress it by force of arms. Thousands of Paulicians perished in the terrible massacre, and their goods were confiscated. A later chronicler, looking back, realized that this display of violence had been the cause of "much evil" that befell the Empire. Streams of refugees now swelled the ranks of the Paulicians living under Arab protection. Argaun and Amara, the towns given over to them, could not absorb such numbers, and a new town had to be established. Carbeas chose the site of this centre, Tephrice (now Divrighi), to be easily accessible to further refugees from Byzantium while withdrawing as far as possible from the reach of the Emir of Melitene. Thus a new half-independent domain grew up on the western frontiers of the Caliphate, its rulers little different from the Arab emirs of the region. Carbeas (d. 863) and his successor Chrysochir (Chrysocheris) were undoubtedly Paulicians in belief, but there is no doubt that they were out to build up their own power. In the ranks of their army there were certainly many who had nothing to do with the faith of the sect. The names of these two rulers of Tephrice occurred later in the Byzantine epic of Digenis Akritas. From Tephrice Chrysochir ruled a considerable stretch of territory, and his forces carried their raids into the most westerly parts of Asia Minor.

The bold exploits of the Paulicians were all the easier because the Byzantine armies were tied up in the fighting against the Arabs in the west, who were threatening the last strongholds of Byzantium in Sicily and penetrating southern Italy. The Emperor Basil I (867-886) therefore tried to negotiate with the Paulician leader Chrysochir and secure his eastern frontiers. The attempt at peace was wrecked by Chrysochir's obstinacy; his successes had given him confidence in his own powers. Greek and Arab sources allow us to date these negotiations about the year 870. It is known that Peter of Sicily was in Tephrice at this time (869-870); he writes that he was there to arrange for the exchange of Byzantine officers who had been taken prisoner. Peter made good use of his nine months among the Paulicians to inquire into the life of the sect. His work is our main source today for knowledge of the history of Paulicianism. The diplomatic mission with which Peter of Sicily was entrusted was concerned with the military leaders of Tephrice, who had been able to organize such a menace to Byzantium on Arab soil. No less dangerous, however, was the spiritual power of this sect, driven out of the Empire but now flourishing under the protection of the rulers of Tephrice. Peter saw with his own eyes the fanatical and unwavering faith of the Paulician teachers, who moved people of all social groups to follow them. The theological discussions Peter held with the Paulicians in Tephrice are clearly reflected in the homilies appended to his Historia Manichaeorum. These disputations seem to have been part of his mission to Tephrice, for Byzantium had still not given up hope of peaceful conversion for the Paulicians. There is no evidence that Peter achieved any success in this direction, however. Indeed, he became convinced of the dangerous threat of the Paulician spiritual leaders, who were not only carrying on their work very successfully on Arab territory after their expulsion from Byzantium, but also planning a bold mission to convert far-off Bulgaria. A land so recently converted to Christianity seemed to offer fruitful ground for the energetic and skilful prosyletizing of the Paulicians.

No doubt this report that the Paulicians of Tephrice had designs on Bulgaria aroused considerable anxiety in Constantinople. Bulgaria was then in the forefront of Byzantine interests. The year that Peter returned home from his mission in the east (870), this Balkan country had finally come within the Byzantine ecclesiastical sphere, and the rulers of the Empire had every interest to see that their political influence would be furthered by the church there. Now the threat loomed large that the cancer of heresy might disrupt this design at the outset, and become a menace to the Empire on its western borders as well. Peter of Sicily was probably acting on orders from above when he introduced his history-polemic, completed after his return from Tephrice, with a message of warning addressed to the Archbishop of Bulgaria.

[1] See the author's Le mouvement paulicien à Byzance, in: ByzSlav XXIV (1963) 258-286; XXV (1964) 52-68; and his critical survey: Où en est la question du mouvement paulicien? In: Izv. na Inst. za Ist. BAN 14-15 (Sofia 1964) 357-371. To the literature given there there must now be added: *Nina G. Garsoïan*, The Paulician Heresy, The Hague - Paris 1967. - Les sources grecques pour l'histoire des Pauliciens d'Asie Mineure. Texte critique et traduction par *C. Astruc, W. Conus-Wolska, J. Gouillard, P. Lemerle, D. Papachryssanthou, J. Paramelle*, in: Travaux et Mémoires du Centre de Recherche d'Hist. et Civilis. byz. 4 (1970) 1-227. - *Nina G. Garsoïan*, Byzantine Heresy, A Reinterpretation. Dumbarton Oaks Papers 25 (1971) 85-113. - *K. N. Juzbašjan* K proischoždeniju imeni "pavlikiane", in: Vizantijskije očerki, Moscow 1971, 213-235.

[2] Arsamosata is now localized at Şimşat Kale near Haraba, about 55 km. east of Elâziğ. See Anatolian Studies XXI (1971) 45-46.

[3] On the settlement in Thrace of Armenians and Syrians from the regions of Theodosiopolis, Melitene and Claudia, under Constantine V., see our study in ByzSlav XXIV (1963) 267, n.46. All references in the sources are quoted there; Theophanes is the only writer to connect this with the spread of Paulician heresy, and even from what he says it is not absolutely clear that these exiles brought Paulician beliefs with them from their original homes in the East. In so far as the account given by Theophanes corresponds to reality, it can be understood as meaning that Paulician propaganda had its effect on Armenian immigrants in Thrace, whence the heresy spread still further.

[4] CJ I v 11, 12, 16, 18.

[5] See H.-C.Puech, in: *Puech-Vaillant,* Le traité, 304.

[6] Ecloga XVII, 52. C.E.Zachariae v. Lingenthal, Collectio Juris Graeco-Romani, Lipsiae 1852, 51.

[7] See *Grumel,* Regestes, NN. 383, 384.

[8] πρόσταγμα - see Peter of Sicily, Ἱστορία χρειώδης, ed. *D.Papachryssanthou-J.Gouillard,* in: Travaux et Mémoires 4 (1970) 65,13-19.

[9] Photius, Διήγησις, ed. *W.Conus-Wolska - J.Paramelle,* ibid., 167,11-29.

[10] See *J.Scharf,* Zur Echtheitsfrage der Manichäerbücher des Photios, BZ 44 (1951) 491.

Additional note:

The important work by *P.Lemerle,* based on the new edition of Greek sources for the Paulicians (see n.1), appeared when this book was already in press. See *P.Lemerle,* L'histoire des Pauliciens d'Asie Mineure d'après les sources grecques, in: Travaux et Mémoires 5 (1973) 1 - 144, with 2 plates and 1 map. - Cf. the present author's review in ByzSlav XXXV (1974) (in printing).

Chapter III

THE RISE OF A HERETICAL MOVEMENT IN THE FIRST BULGARIAN EMPIRE

We do not know, now, whether the Tephrice Paulicians did in fact send missionaries to Bulgaria. We can only assume that the life and death struggle which broke out between them and the Byzantine Empire just at this time obliged them to postpone their plan, and that it ultimately came to nothing as the war ended in disaster (872). The mission might in any case have come to grief before reaching its goal. It was of course very feasible for the Paulicians to have penetrated into Bulgaria from places much nearer than eastern Asia Minor. We know that Paulicianism was widespread throughout the Byzantine world, and it cannot be discounted that some followers of the heresy may have taken refuge from persecution by crossing into Bulgaria.

The confused situation in Bulgaria after Khan Boris (Michael) accepted baptism in 864 or 865 may indeed have appeared propitious to the spread of various heretical doctrines. For some years Bulgaria wavered between the ecclesiastical jurisdiction of Constantinople and that of Rome. Missions were despatched to Bulgaria from all sides, offering each their own form of Christianity, and there were even impostors claiming to be Christian priests. It would appear that the instability of religious and ecclesiastical conditions in the country offered an excellent opportunity for Paulician propaganda, which was skilful and persevering - assuming that the sect already had missionaries in the country.[1]

It can be questioned, however, whether the period immediately following the conversion of Boris presented the right spiritual atmosphere for the reception of the Paulician faith. The world of Byzantium, where the life of society was permeated by Christian culture and the church already an institution of long standing, might offer fertile ground for Paulician seeds of criticism directed against the reprehensible aspects of ecclesiastical life, and for the original attempt of the Paulicians to revive primitive Christianity, shorn of all later accretions. In the favourable conditions created by the iconoclastic movement in Byzantium, the Paulicians were able to fight very effectively against icon-worship, likening the holy pictures to the idols of the heathen. They also ingeniously called into question the worship of the saints, which distracted the faithful from the pure truth of Christ's gospel. It is difficult to imagine this line of argument succeeding in the atmosphere of Bulgaria, still not entirely free of the thralls of paganism. Bulgaria was indeed not yet ready for conflict between orthodoxy and heresy; the field of religion remained dominated for a long time by the opposition between Christian and pagan ideas. While in earlier times the rulers of Bulgaria had persecuted the Christian faith which was finding its way into the country from neighbouring Byzantium by devious ways, the same vehemence was now turned against the native pagan religion. It was no doubt political necessity that made Prince Boris decide to accept baptism, but he then proceeded to introduce Christian ways with great thoroughness, and ruthlessly swept away whatever stood in his way. Decades later the legend of the miracle of St.George still echoed memories of the violent upheavals of those years. It speaks of Boris-Michael the chosen of God by the strength of Christ overcoming the stiff-necked and proud Bulgarian people", turning them aside from "dark and false sacrifices" and destroying their heathen altars. These words can perhaps be taken as referring to the bloodshed that shocked Bulgaria shortly after the conversion of Boris, when he crushed those of the boyars who took an open stand against Christianity, massacring them and their families. The well-known Responsa of Pope Nicholas I, answering the questions put to him by a Bulgarian mission despatched to Rome in 866, contain a lively reaction to this inhumanity. The surviving boyars hid their seething antagonism, not daring to rebel openly again; it was not until later, under the ephemeral rule of Boris's son Vladimir (889 - 893), that there was a short-lived return of the old ways. The enforcement of Christianity was also bitterly resented by the common people of Bulgaria, who were obstinately attached to their old pagan customs and worship. The papal Responsa also throw very interesting light on this question, and provide invaluable information about the life of the pagan Proto-Bulgars and Slavs. Even a century later, however, the Testament of St.John of Rila (d.946) and the Sermon of Cosmas the Presbyter prove that remnants of pagan beliefs still lingered in Bulgaria.[2] It took a long time for a deeper understanding of the Christian faith to take root, and it was no easy task to present it in terms intelligible to people still in the thrall of primitive religious ideas.

We can surely assume that the God of the Paulicians would have remained alien and inaccessible to people living thus, their doctrines being too abstract and too thoroughly purged of material symbolism. In contrast the orthodox church could attract believers through just those aspects which so powerfully affect the human senses: the magnificence of its churches, their colourful paintings, the beauty of liturgical hymns and the splendour of the religious ceremonies themselves.

At the time Peter of Sicily addressed his warning to the Bulgarians no more than the foundations of church organization had as yet been laid. Following a decision by the council held in Constantinople in 869/870, a Greek archbishop was sent to Bulgaria, accompanied by a certain number of bishops. For the conversion of the land to be really effective large numbers of the lower clergy were necessary, and even so Christianity would have remained superficial so long as the liturgical language (Greek) was unintelligible to the people. The turning-point was the arrival in Bulgaria of Slav teachers driven out of Moravia after catastrophe had overtaken the Slavonic church there. After the death of Methodius (885), three of the great archbishop's disciples took refuge in Bulgaria; Angelarius died soon after arrival, from the deprivations of the journey. Clement and Naum were later joined by others who had been redeemed from slavery in Venice.

Boris was anxious to see the work of the mission embracing not only the heart of his Empire, in the north-east, between the Danube and the Balkan range, but also the great expanse of Macedonia recently annexed by Bulgaria. The spread of Christianity was to bind this new region of the south-west more firmly into the fabric of his state.[3] In north-east Bulgaria the Slav "teachers" began their work in the immediate ambiance of the Prince's court; Naum taught there for some years,[4] as did Constantine, later to become the Bishop of Preslav, and some others. With the active help of Boris's brother Duks, a monk, they soon created an important literary centre. Among the young Bulgarian scholars here the learned John the Exarch soon stood out.[5]

The sources yield very little factual information about Naum, a trick of fate which has meant that this outstanding figure is overshadowed by that of Clement, whose work has been recorded in a considerable literary tradition. The scene of Clement's missionary activity was far-off Macedonia, where the local governor gave him every assistance, on the express orders of the ruler.[6] When Boris's son Symeon came to the throne in 893 Clement (d.916) became the first "Slavonic" bishop in Bulgaria. Naum then joined him in Macedonia, to carry on his work as the "teacher". Clement's diocese[7] was not far from where he had been working hitherto, and so the work of the two loyal companions remained associated with the lovely country round Lake Ohrid. Today the remains of two churches bear witness to them; they are strikingly alike in their trefoil ground-plan, Naum's on the eastern shores of the lake and Clement's in Ohrid itself, on the site of a Christian church ruined long before. Now only the written accounts of their contemporaries can call up before our eyes the world they built by tireless effort, long ago disappeared. The full Greek life of St.Clement was written by Theophylact, Archbishop of Ohrid, according to a Slavonic text now lost.[8] The account by an anonymous disciple of Clement gives a vivid impression of the outstanding ability of the teacher, and of the talents also evident in his surviving literary works. Clement wrote liturgical hymns which were profoundly lyrical, and he also wrote simple homilies for holy days which were clearly meant to help the future Slav preachers. He aimed at bringing both the doctrinal and the ethical message of Christianity within the understanding of the simplest minds. He did not underestimate the power of visual impressions, and we read how he had churches built, to attract the people to divine service by the beauty of their architecture. Yet he was not blind to the material poverty of his flock, and cared devotedly for the poor and homeless of his diocese.

The continual exertions of Clement, Naum and their fellows soon produced fruit. We learn from Theophylact that many young disciples trained by Clement were appointed to the churches then being built all over Bulgaria, to serve as "četci" (tchettsi = "readers"), subdeacons, deacons and priests. After Symeon's accession to the throne bishops of Slav origin were gradually appointed.[9] This entire generation was the product of the school of Clement and Naum, excellently equipped to spread and establish Christianity in their country. The Bulgarian disciples of these teachers from Moravia were able not only to conduct divine service in the Slavonic tongue, but also to explain the Christian faith to their congregations in a way they could understand. At home among the people, they influenced them not only by their preaching but by the exemplary Christian lives they led. Here there would be little chance of heretical teachings taking root.

Nor, indeed, do we hear of heresy in Bulgaria under Boris, or even in Symeon's reign. By this time the Slavonic learning brought here from Moravia was already yielding good fruit. The East Bulgarian literary school centred in the vicinity of Prince (later Tsar) Symeon's new court, Preslav, was stamped with the personality of the new monarch, purposeful and active, quite weaned from the primitive way of life of the semi-barbarian environment. It was his aim to bring into the country as much as possible of the mature civilization of Byzantium. The Preslav school remained in touch with that of Macedonia, which kept more closely to the Moravian-Pannonian literary tradition.[10]

Literary activity did not die out in Bulgaria even in the second half of Symeon's reign, when fierce struggles against Byzantium destroyed the peace (913-927). This was no ordinary war of conquest; it was Symeon's ambition to achieve universal dominion, which in the minds of his contemporaries was linked with Constantinople. The Empire was in a temporarily weakened state, which seemed to promise easy success; Symeon did in fact win great military victories, and took the title of Tsar. The accession of an energetic ruler to the Byzantine throne, Romanus Lecapenus (920-944), marked the beginning of the consolidation of the imperial forces. In the last years of Symeon's reign the Bulgarian state wore itself out with long campaigns while the goal of Symeon's ambitions seemed further and further away. Even during his lifetime there was great discontent in the country, culminating in a rebellion and the flight of large numbers of Bulgarians to Byzantium. The peace signed immediately after Symeon's death (927) did nothing to strengthen Bulgaria. Indeed the departure of this ruler of strong personality caused great upheaval in the country, and Byzantium was not slow to turn the situation to its own advantage and intervene in Bulgarian affairs. In particular, when the young Tsar Peter (927-969) on his accession married the granddaughter of Romanus Lecapenus Byzantium had every opportunity to influence the Bulgarian court. Bulgaria became very dependent on the neighbouring Empire. The sharp decline in state power was accompanied by a drastic sharpening of social contrasts, the result of Symeon's endless wars.[11]

The Hexaemeron of John the Exarch, which was written during the earlier, peaceful part of Symeon's reign, contains a passage which shows up clearly the sharp contrast between rulers and ruled in the Bulgarian Empire. The writer is describing the magnificence of the churches and palaces of Preslav and the splendid attire of the Prince and his boyars. This gorgeous picture is presented to us through the eyes of a villager, coming from the countryside with its straw-thatched hovels.[12] We can feel the enormous gap between the simple Bulgarian people and those who ruled over them. Some of the boyars lived at the Prince's court while others were "outside", "among the people". They ruled vast estates with large herds, and could not only secure their households against bad harvests, but put the produce of their estates to profitable use in trade with Byzantium; from the outset this trade had brought in great profits to the landowners of Bulgaria.[13] In contrast the peasants, dependent on the modest returns of small holdings, were far worse off. In times of bad harvest or natural catastrophe the people of whole regions were reduced to starvation. Symeon's long wars undoubtedly made the life of the peasants much worse, for quite apart from their enforced military service it was the peasants who bore the main brunt of taxation and the other burdens connected with the fighting. Even under Peter the internal conflicts of the country must have been a source of distress to the peasants, and so undoubtedly were the raids of Magyar horsemen. The boyars, on the other hand, enriched themselves with booty during the fighting against Byzantium, in the course of Symeon's successful raids on the territory of the Empire.

The social gap between the ruling class and the ordinary people was made even more apparent by the fashion for Byzantine ways. Together with the title of Tsar the court ceremonial of Byzantium was certainly also adopted by Symeon. The marriage of Peter to Princess Maria (Irene) undoubtedly made the Preslav court even more Byzantine in character, and it does not need much imagination to see the life of the upper classes in Bulgaria imitating more and more the manners of their Greek model.[14]

Since the baptism of Boris the whole system of church organization had been introduced from Byzantium. Monasteries had also been set up, with as significant a place in Bulgarian society as that of their counterparts in the Byzantine world. Even the scant historical sources which have survived relating to this period are sufficient to give us a good idea of how close the ruling house of Bulgaria felt itself to the monastic environment. The foundation of monasteries was of course accompanied by the gift of lands and serfs, and although no legal documents of these transactions have survived, there can be no doubt that such donations were very frequent. When the First Bulgarian Empire was no more than a faint memory the Bulgarian church still remembered Peter as the Tsar who heaped favours "on the monks and servants of the Church of God".[15]

There can be no doubt that the rich donations to church institutions, and the fact that the monasteries were closely linked with the interests of secular society, presented no small moral danger. The Testament of John of Rila, written in 941, earnestly exhorted the monks to beware of the evil influence of gold and silver and the fettering effect of dependence on secular rulers. This holy man enjoined on his monks not only poverty and abstinence, but fervour in the reading of holy books and in work among the people, who were still very pagan in these out-of-the-way parts.[16]

Thus among the Rila mountains a new and important centre of missionary activity arose, as well as a nucleus of literary life which did not weaken even under Byzantine domination.[17] The spirit of Clement, Naum and their fellow-workers continued to live in Peter's times, but was no longer the over-ruling influence throughout Bulgarian ecclesiastical life. Material interests were coming forward, over-riding the selfless devotion of the earlier generation. A church of the Byzantine type had by now become firmly rooted in Bulgarian soil, and its less desirable aspects began to make themselves felt. The first suggestion of heresy in Bulgaria was now heard.

The earliest evidence of the existence of a heretical movement in Bulgaria is provided by the letter of Theophylact, Patriarch of Constantinople (933-956), to the Tsar Peter.[18] The Patriarch was a son of the emperor Romanus Lecapenus and thus the uncle of Peter's wife Maria. The letter which has come down to us gives some idea of the previous correspondence about this "newly appeared" heresy. It appears that the first reply sent by Theophylact answered Peter's questions in the typical Byzantine rhetoric, formulated somewhat vaguely and not covering everything that Peter wanted to know. To some degree it was a repetition of what happened when Photius wrote to Boris.[19] Tsar Peter, not satisfied with the Patriarch's answer, wrote again, asking for everything to be explained "more fully and clearly", keeping to the "bare facts" and written in plain language. This time Peter gave Theophylact fuller information about the Bulgarian heresy, so that the Patriarch was able to determine its nature more accurately. He called it "Manichaeanism mixed with Paulianism". The latter was the heresy preached by Paul of Samosata, who was generally assumed by ecclesiastical writers to have been the founder of the Paulician sect, although this did not prevent them from declaring at the same time that the sect was an offshoot of Manichaeanism. "Manichaeanism mixed with Paulianism" thus means simply Paulicianism. According to this classification the Patriarch applied the 19th canon of the Council of Nicaea to this "newly appeared" heresy; this canon laid down the conditions for the re-acceptance into the Church of followers of Paul of Samosata. Theophylact introduced a significant distinction, however: it was only leaders and teachers of the heresy who were to be treated according to the letter of the 19th canon, which demanded rebaptism for the "Paulians" and that heretical priests should have their orders declared null. Ordinary people who had been led astray were to be treated more leniently, while the greatest charity was to be shown to those who had joined the sect out of ignorance but never accepted the doctrines or adopted the practices of the heretics. These three categories to be distinguished both among the laity and the priests concerned those who were willing to abjure heresy and return to the true Church. Unrepentant heretics were abandoned by the Church and subject to its anathemas. Concerning the secular laws, about which the Tsar had enquired, the Patriarch pointed out that the penalty was death, especially when the "evil" was spreading rapidly. While not excluding the extreme penalty, he was nevertheless against it in principle, urging the Tsar in the spirit of the tenets of the Church to strive continually to save these souls so obstinately persisting in their errors.

A great deal of Theophylact's letter is taken up with the formula to be used by the Bulgarian "Paulicians" in abjuring their heresy by word of mouth or in writing, and which was therefore an important part of the rite of re-acceptance into the Church. Texts of this kind must always be treated with caution, for they usually combine elements of quite different heretical teachings, brought together in an arbitrary manner. Realizing that it was common practice to identify Paulicianism with Manichaeanism, we are not surprised by the strange collection of names in Theophylact's formula. Besides names authentically found in Paulician history we find traditional names from the history of Manichaeanism. Nor is it surprising to find among typical Paulician "errors" some tenets of the Manichees. There is another Greek abjuration formula known to us, also for the use of repentant Paulicians, in which this compilatory technique is even more marked.[20]

Since Theophylact's anathemas seem to have been drawn up on the basis of the usual anti-Paulician material, they cannot provide us with concrete information about this tenth century Bulgarian heresy.

What is shown quite clearly by this letter, however, is the fact that in Peter's reign there was such a strong heretical movement in the country that the authorities were disturbed, and the Tsar felt it necessary to turn to Constantinople for advice on how to deal with this sudden danger. But it is only from later sources that we can determine whether the heresy then rife in Bulgaria was in fact Paulicianism.

[1] On the situation in Bulgaria after the country was officially converted to Christianity, see: *I.Dujčev*, Au lendemain de la conversion bulgare, in: Medioevo I, 107-123; Id., Die Responsa Nicolai I. Papae ad consulta Bulgarorum als Quelle für die bulgarische Geschichte, ibid., 125-148.

[2] Testament of St.John of Rila, published by *I.Gošev*, in: Godišnik na Duch. Akad. IV (XXX) 12 (Sofia 1954 - 1955) 447 Cf. *I.Dujčev*, Rilskijat svetec, 155. Cosmas, 86,9.

[3] The traditional view (see *Zlatarski*, Istorija I 2,219 ff.) is that Boris planned the "Slavonization" of the Bulgarian Church from the outset, to get it out of Greek hands. It was not until after the accession of Symeon, however, that Slav bishops were appointed, and this could only be carried out gradually. Under Boris the immediate task was to complete the conversion of the country to Christianity as fast as possible. The use of Slavonic in the liturgy was in no way contrary to the policy of the Byzantine Church at that time. (See *D.Zakythinos*, Byzance et les peuples de l'Europe du Sud-Est. La synthèse byzantine, in: Actes du Ier Congr.Intern.

des Et.Balkaniques et Sud-Est-Europ. III, Sofia 1969, 20 ff.). There does not seem to be any justification, therefore, for the suggestion that Clement was sent to work in Macedonia, far from the centre of government, in order to keep him out of sight of the Greek ecclesiastical authorities assumed to be opposed to the Slavonic liturgy. It should not be overlooked that Naum, about whom we have far less information, remained for the early years of his mission in north-east Bulgaria, where the administration and the ecclesiastical life of the country was centred. There can be no doubt he worked there in the same manner as Clement in Macedonia.

4 For the latest study see *I.Dujčev,* Prostranno grăcko žitie i služba na Naum Ochridski, in: Konstantin Kiril Filosof, Sofia 1969, 261-279.

5 For biographical data on John the Exarch see I.Dujčev, in: *Cv.Kristanov - I.Dujčev,* Estestvoznanieto v srednovekovna Bălgarija, Sofia 1954, 54 ff.

6 See *F.Dvornik,* Byzantine missions among the Slavs, New Brunswick - New Jersey 1970, 410 f., n.52.

7 For the localization of this diocese see: *I.Snegarov,* Po voprosu o mestonachoždenii eparchii Klimenta Ochridskogo, in: Actes du XIIe Congr. Intern. d'Et.Byz. 1961, II, Belgrade 1964, 188-195. - *P.Gautier,* Clément d'Ochrida, évêque de Dragvista, in: REB 22 (1964) 199-214. - *P.Karlin-Hayter,* Clément d'Ochrid, la guerre bulgare et la prise de Thessalonique en 904, in: Byzantion XXXV (1965) 606-611.

8 See *I.Dujčev,* Kliment Ochridski i negovoto delo v naučnata knižnina. Kritikobibliografski pregled, in: Kliment Ochridski. Sbornik ot statii po slučaj 1050 godini ot smărtta mu, Sofia 1966, 416-424. The Life of St.Clement by Theophylact was published by *N.L.Tunickij,* Grečeskoje prostrannoje žitije sv.Klimenta slovenskogo, Sergijev Posad 1918. - *A.Milev,* Teofilakt Ochridski, Žitie na Kliment Ochridski. Sofia 1955.

9 *Zlatarski,* Istorija I 2, 233 ff., 276 f.

10 *B.Koneski,* Ochridska književna škola, in: Slovenska pismenost, Ochrid 1966, 57-62.

11 Zlatarski, Istorija I 2, 356-515, 535 ff.

12 See *J.Trifonov,* Svedenija iz starobălgarskija život v Šestodneva na Ioana Ekzarcha, in: Spisanie na Bălg.akad. na naukite, klon ist.-filol. 35/19 (1926). Cf. *D.Stričević,* La rénovation du type basilical dans l'architecture ecclésiastique des pays centrales (sic!) dans les Balkans aux IX-XIe siècles, in: Actes du XIIe Congr. Intern. d'Ét.Byz. (10-16 Septembre, Ohride 1961), I, Belgrade 1963, 209 ff.

13 *N.A.Săkăzov,* Bulgarische Wirtschaftsgeschichte, Berlin-Leipzig 1929, 49 ff.

14 *Zlatarski,* Istorija I 2, 541 ff.

15 Second liturgical service in memory of Tsar Peter, ed. by *J.Ivanov,* Bălgarski starini iz Makedonija, Sofia 1931, 393.

16 *I.Gošev,* loc.cit., 441, 446, 447; *I.Dujčev,* Rilskijat svetec, 150 ff., 154 ff.

17 *I.Dujčev,* Rilskijat svetec, 161 ff.

18 Newly edited by *I.Dujčev,* in: Mélanges Eugène Tisserant II (Studi e Testi 232), Città del Vaticano 1964, 63-91. (Reprinted: Medioevo I, 283-315). See the present author's review, ByzSlav XXVII (1966) 373-376.

19 *I.Dujčev,* ed.cit., 75.

20 ByzSlav XXVII (1966) 375.

Chapter IV

THE BULGARIAN BOGOMIL HERESY AS PORTRAYED BY COSMAS THE PRESBYTER IN HIS SERMON AGAINST THE HERETICS

The times of the "orthodox Tsar Peter" were already over when a sermon attacking heresy was written by one Cosmas Presbyter. The language clearly points to East Bulgaria;[1] the text dates from the time when Preslav was still the seat of the Tsar. There is internal evidence suggesting the date of this polemical work: power is still in the hands of the "Tsar and the boyars", in other words the country was not yet ruled by a foreign power; nevertheless God had already sent down upon it the "sufferings of war". The monasteries provided no safe refuge from the devastation of war, some indeed were reduced to ruins and their monks wandered homeless about the countryside. Tsar Peter was no longer among the living, but older men could still remember John the Exarch, a high dignitary of the Bulgarian church and a learned writer, who lived in Symeon's reign.

All these circumstances referred to in the Sermon of Cosmas point to the conclusion that it was written after the death of Tsar Peter but before the tsars of Preslav had disappeared. This dates the Sermon in the short period between 969 and 972. The sufferings of war referred to could thus be those caused by the invasion of Bulgaria by the Grand Prince of Kiev, Svjatoslav (Svyatoslav), who took Preslav in the winter of 969-970, or the invasion of the spring of 972, when the Byzantine emperor John Tzimisces captured the country from the Russians.[2]

The Sermon, whole or in part, was widely copied for centuries, not only among Bulgarians but also in Serbia and especially in Russia,[3] and thus there has come down to us one of the most remarkable written works of the first Bulgarian Empire. It is an original work, and not a translation or paraphrase of a Byzantine text. Cosmas had neither the literary ability of Clement nor the learning of John the Exarch; he was decidedly no practised stylist nor had he mastered the art of literary composition.[4] In his arguments against heretical ideas he followed no logical plan, but allowed himself to be carried along by a free association of ideas. In this way his polemic against heresy slips quite naturally into a criticism of the evils rife in orthodox society, particularly those seen in the clergy and among monks. The whole work has thus as it were two distinct aspects, a fact which led later copyists to distinguish in the title of the work between the polemic against the heretics and "counsel from holy books". In reality all the polemical attacks, the observations and the counsel Cosmas put down in his Sermon form a single flow of eloquence directed to a single aim: the fight against a "newly appeared heresy" which a priest by the name of Bogomil had begun to preach in Bulgaria in the reign of Tsar Peter.

The Sermon itself reveals how well-read its author was, and also the limits of his learning. He was thoroughly grounded in the Scriptures, but his knowledge of patristic writings was not remarkable, nor was he particularly proficient in church history. It must be remembered, too, that the wisdom he culled from books only enabled him to use foreign words to express what he himself thought. Naturally he could not fail to be influenced by Greek habits of expression, which left a clear mark on his language. Even so the Sermon, far more than other Slavonic works of the period, has preserved much of the freshness and the natural quality of the spoken language.[5]

Cosmas's treatise is very important for our knowledge of the life of Bulgaria in the tenth century. In spite of individual points which are not clear,[6] the Sermon gives a convincing picture of the misery of the common people. They complain that the heavy burden of duties leaves no time for prayer during the day, if they are to provide for their families. They have a hard job to get their daily bread, and abject poverty sometimes drives them to seek refuge in monasteries, leaving their children to starve.[7] These observations which Cosmas lets drop in the course of his discussion of monastic life are telling enough. They are signs of profound social crisis rooted, as we know, deep in the past.

While secular life is only touched on here and there in the Sermon, there are whole passages devoted to the life of the Church. The main features and the problems of monastic life in Bulgaria could not be very different from those of Byzantium, whence the institution had been adopted. Cosmas describes things which are already known from Byzantine monasteries.[8] In Bulgaria there were many who assumed the monk's habit without due thought, made pilgrimages to Jerusalem or Rome, and regretted it when they came back home. Some made themselves ridiculous by returning to their wives. Others wandered about from monastery to monastery, boasting of what they had seen in far-off lands. Cosmas was stern in his censure of those monks who left their monasteries in order to found new ones run in accordance with their own wishes; they did so for worldly glory, and made a great show of magnificence in their dress and their companies of horsemen. "But who will tell of the injustices wrought by them against the weak?" "Their tables are loaded with all kinds of food ...", "disgusting talk" and quarrelling is rife among them, and when they are not quarrelling they are indulging in drink.[9]

Nevertheless there were some who sincerely longed for a life of asceticism, following the example of such great and holy men as St. Anthony, St.Sabbas and others. Cosmas advised them to follow the only way to the Christian ideal, life in a monastic community, obedient to the higumen, and devoted to the prescribed regular prayers. Hermits living on their own, not attached to a monastery, lived on the faithful and fattened "like pigs", and in addition were swollen with pride, taking unto themselves the right to give absolution and to advise people on how to run their lives and their businesses.[10] How rare, in contrast, were those holy men, living in twos and threes in the mountains or in caves, and rarely appearing among the people.[11] The Sermon of Cosmas, like the earlier Testament of John of Rila,[12] vehemently praised the cenobitic monastic life, a trend which had become very strong in Byzantium after the iconoclastic period, although it could never become general there.[13] Cosmas was particularly anxious to convince the monks not to leave their house once they had entered it; if armed bands destroyed their peaceful home they should find another monastery without delay.[14] His exhortations seem to echo faintly the call to his scattered flock sent out by Theodore of Studios in the time of persecution.

Cosmas himself was neither monk nor higumen; he was a member of the secular clergy. In the title of the Sermon he is called a presbyter[15] but his text suggests convincingly that he already held the office of bishop.[16] He addresses the Bulgarian bishops as one of their number. He was profoundly disturbed by the moral degradation he observed among them, and therefore spoke to them as "the dog spoke to the people of Sodom".[17]

The bishops in Bulgaria were indeed far from meeting the requirements set by the apostle Paul for a Christian bishop. These "lawmakers ignorant of the law" often acquired office by outright corruption; they were "shepherds who milked and sheared their flocks but took no care of them". It is noteworthy that Cosmas does not refute the accusations made against the clergy by the heretics; he quotes their reproach that the priests do not live according to the commandments of Christ and His apostles, but live in drunkenness and theft, committing other sins as well in secret, and there is no-one who interferes in their sinful lives. Heretics can be heard to say: "The bishops ... do not live as soberly as we do; they prevent us from praying, but do not hinder the priests in their sins". The priests either sinned in secret or lived openly "according to the flesh and not according to the spirit, as we do". All that Cosmas can say in answer to this is that "though a priest may be a sinner, he is not a heretic". He cannot deny that the "orthodox priests live in idleness", but stresses that, after all, they do not blaspheme God as the heretics do.[18]

It is probably of members of the clergy that Cosmas is speaking when he says there are people who have the Old and the New Testament at home, as well as other religious works, but use them only to boast of their learning and to prevail over others in battles of words. What they should do is willingly lend the books to all who would like to read or copy them.[19] It was one result of insufficient instruction in the true faith that there were still many who had not given up their pagan customs and ideas.[20] And there was only the laziness and ignorance of the clergy, and especially of the bishops, to blame for the spread of heresies, running wild like "wolves and wicked dogs", attacking and scattering the flocks of the faithful. It was the primary duty of the clergy to instruct their flocks ceaselessly. The example Cosmas places before the Bulgarian bishops in this respect is the great figure of John the Exarch. It was not so very long ago that he lived,[21] and there was therefore no excuse for saying that nowadays no-one could live up to the standards of the great Church Fathers.

There were few priests and bishops in Bulgaria, however, who took to heart the words once spoken by the monk Duks to John the Exarch: "What other business has the priest, except to teach?" And yet Christianity had taken firm enough root in the country for many people to show sincere interest in questions of Christian doctrine and practice. The Church at this time gave them little in the way of real instruction and living example, and it was all the easier for heretical sects to win them over.

For the author of the Sermon heresy was a very real evil, one of the sins for which the country was troubled with wars as a divine punishment. He was little concerned to place the new sect in the context of earlier heresies, and did not turn to the heresiological catalogues to find a well-tried name for it. His "historical" introduction is brief indeed: various heretics such as Arius and Sabellius and Macedonius proclaimed false doctrines of the Holy Trinity, after which other heresies appeared in various places, concerning "the creation of God". One of these heresies was first taught in Bulgaria by a priest named Bogomil, in the reign of Tsar Peter.[22] It is clear that Cosmas regarded the contemporary Bulgarian heresy as a *dualist doctrine*. He mentions none of them by name, however, although he may have known Manichaeanism from the Hexaemeron of John the Exarch or from the Catecheses of Cyril of Jerusalem, which had been translated in Bulgaria as early as under Tsar Symeon.

It is surprising that there is no hint in the Sermon that Cosmas had heard of the letter of the Patriarch Theophylact, which dealt with the heretics of Bulgaria. This letter must have soon fallen into oblivion, which gives the work by Cosmas the added significance of being the second and quite independent evidence that a dualist heresy had appeared in Bulgaria in the reign of Tsar Peter. Unlike the Byzantine Patriarch, the Bulgarian bishop presents not a stereotyped bookish outline, but a real account based on his own experience with the heretics.

It was not his purpose to give a systematic account of the heresy such as those presented by Byzantine heresiologists in encyclopaedical or monographic works. Cosmas was polemizing with the heretics, and therefore concentrated on the individual points on which they differed from the dogma of the Orthodox Church. His choice of facts, of course, reflects his own attitude. In spite of its fragmentary character, the evidence supplied by Cosmas shows clearly that Theophylact was right in his assessment of the fundamental nature of the Bulgarian heresy. There can be no doubt that the doctrines discussed in the Sermon were closely connected with *Paulicianism*.

The Bulgarian sect, too, was violently opposed to the Church, its hierarchy and worship; the priests were regarded as "blind Pharisees",[23] the churches as the "corners of the streets" where hypocrites pray, and the liturgy and other prayers recited in church as no more than the "vain repetitions" that Jesus warned his followers against, when he taught them to say only the Lord's Prayer. The liturgy did not go back to the Apostles, but only to St.John Chrysostom. People who bow before icons are like the heathen, and the words of St.Paul about idols refer equally to icons. What was even more serious, the Bulgarian heretics rejected baptism and the eucharist. Like the Paulicians they took the body and blood of Christ to mean His teachings, the body being the four Gospels and the blood the Acts of the Apostles.[24] The followers of Bogomil interpreted the five loaves with which Jesus fed the multitude as these five New Testament books, as well. The other miracles were interpreted in a similar spiritual way; the blind man whose sight was restored, the sick man healed, the man raised from the dead, - all these miracles were merely allegories of man cured of sin.[25] They did not believe that Christ could come into contact with the material world, and thus could not attribute to Him a material body; they considered His birth imaginary. Cosmas does not say so explicitly, and he refuses to repeat the "insolent words" of the heretics about the Mother of God. All we can deduce from what he says is that they saw in her no more than an ordinary sinful woman.[26]

Cosmas reveals an interesting detail in the attitude of these Bulgarian heretics to the cross: they would cut down crosses and make tools out of them. Hatred of the cross is more decided than it was among the Paulicians: "How can we bow to it, for on it the Jews crucified the Son of God?". Sometimes they explained their attitude by a parable: "if someone were to kill the son of a king with a piece of wood, is it possible that this piece of wood could be dear to the king?"[27] This would seem to suggest that they regarded the sufferings and death of Christ as real, as did the Paulicians.

Certainly the cross was not the symbol of salvation for this Bulgarian dualist sect; Cosmas tells us explicitly that they believed Christ was crucified "not of His own will, nor for the salvation of mankind, but because he was forced to".[28] This clearly meant that the death of Christ was not the means of redemption, and that man's salvation could come only through His teachings. It is equally certain that the Bulgarian heretics regarded the teachings of Christ, set forth in the Gospels, as directly opposed to the Old Testament; as they put it, they listened neither to David nor to the prophets, living not by the law of Moses but by that of the Apostles. This also shows that they made no exceptions in their rejection of the books of the Old Testament, being no more kindly disposed to the Psalms of David or to the prophets, and indeed the latter were scorned and said not to be worthy of salvation; they prophesied of their own accord and not because they were moved by the Holy Ghost. John the Baptist was then quite clearly the "forerunner of Antichrist".[29]

The Old Testament thus had nothing to do with the God of whom Christ preached. It follows essentially from this that the Bulgarian heretics did not identify the Old Testament God the Creator with the God of the Gospels. The Sermon also tells us explicitly that they believed the Evil One - the Devil - was the Creator and Ruler of the world. This was a basic heresy against which Cosmas fulminated with all his eloquence. How could the magnificent work of the Creation be attributed to the Devil! The heretics believed that everything we see around us has come about by the Devil's will - sun and stars, air and earth, man, churches, crosses - in short, all living and non-living things in the world. He is the Creator and master of worldly things, and his dominion over the whole world is proved, in their eyes, by the temptation of Christ, when the Devil offered him "all the kingdoms of the world and their glory". Like the Paulicians, the Bogomils based this belief on the New Testament, where it speaks of "the prince of this world".[30]

In accordance with the polemical nature of his treatise Cosmas dealt only with this negative aspect of dualist metaphysics, and said nothing of the heretics' ideas on the supernatural world and its God. In spite of this one-sided view we cannot doubt that the dualist foundation of the Bulgarian heresy was the same as that on which Paulicianism was based. There was an important difference between them, however, in that the Bulgarian dualist faith had more profound ethical implications and that, at least in Cosmas's day, it affected social attitudes much more than did Paulicianism.

We read in the Sermon that the heretics turned away from baptised children in disgust,[31] a curious habit that would be difficult to understand without the evidence of later sources. It derived from the belief that from conception onwards man is entirely in the power of the evil creator of the material world; baptism by water cannot change this, so that even after baptism children are still "the children of the Devil". Cosmas does not tell us whether the sect already had a rite of spiritual baptism in his day, such a baptism offering the only way to save man from the power of the "prince of this world". It is certain that in the tenth century the sect consistently turned away from everything that binds human beings to the material world and thus keeps them in the clutches of its ruler. To eat meat and drink wine were the Devil's commands, and so was marriage; people who "married and lived in the world" were called "servants of the Devil" by the heretics. They considered themselves those "who live in Heaven",[32] no doubt in the sense of St.Paul's words to the Philippians (Phil. 3, 20). We have already seen how they contrasted their ascetic way of life with the immorality of the Orthodox bishops and priests, and indeed their very appearance was different, modest as was their behaviour, and their faces were pale from frequent fasting.[33]

Relinquishing life in the world, the Bogomils gave up all wealth too, and were fierce in their condemnation of all who held secular power or possessions. Cosmas specifically accused them of teaching "their own people" not to obey the authorities, reviling the wealthy, hating the Tsar, ridiculing their superiors, insulting the boyars, believing that God abhors those who serve the Tsar, and admonishing servants not to obey their masters.[34]

The strictest and most ascetic community, however, can only exist with the help of people outside it, who live in the every-day world. Cosmas sharply condemns the heretics for wrongly interpreting Christ's words, "Take no thought for your life, what ye shall eat; neither for the body, what ye shall put on ...", they wander about not working, but living on the people they have deceived.[35] We do not know who these people were who provided the apostles of heresy with a refuge and food and drink. There may have been people of good social standing among them, but the sect, calling for disobedience to the authorities, appealed primarily to the poor and oppressed, and it was among them that it apparently made its greatest following. The faith could only be spread in secret, for persecution was relentless; Cosmas speaks of chains and prison to be endured. The teachers of the sect themselves exhorted their followers to caution: "If people hear about our prayers and our actions, the whole of our work will come to nought." The heretics therefore kept up an appearance of orthodoxy, going to church, kissing the cross and the icons there, and even uttering the five daily prayers at home as prescribed, with open doors. On the other hand they showed themselves willing to suffer for their beliefs, and gained the reputation of martyrs worthy of reward in Heaven.[36] Their very humility and abstemious lives gained them adherents, who looked to them with the respect found in the Christian East for ascetics and monks.

Indeed the ascetic teachings of the Bogomils were in remarkable harmony with an atmosphere which existed in the country independently of the sect. We have touched on various reasons given by Cosmas for the flow of men into the monasteries. Some were moved by material interest, but others were tormented by their need for salvation, kept out of their reach by the troubles and anxieties of daily life, or at the other end of the scale, by excess of wealth and luxury. They turned to the epistle of John: "Love not the world, nor the things that are in the world".[37] Yet there were also voices urging the need for monastic asceticism not in order to attain a truly Christian life, but to reject on principle the practice of marriage and life in the world. Cosmas knew full well that this came dangerously near the views of various heretics at intervals in the past, who had declared the institution of marriage as something impure and prohibited the eating of meat and the drinking of wine - which used in moderation are not in any way contrary to the Christian ethics. He was aware of the canons of the Council of Gangra which had condemned this trend in the middle of the fourth century, when it had flourished in the monasteries of Asia Minor.[38]

Cosmas called to his aid not only canon law, but the Epistles of St.Paul as well, to refute on all sides both this erroneous conception of asceticism and the dualist heresy for which tendencies of this kind undoubtedly prepared the way in Bulgaria.

It would give a distorted picture of the sect, however, if we were to look only at the ascetic aspect and ignore the intense interest in dogma which was evident even at this early stage, in the constant meditation on the texts of the New Testament in search of their "true" hidden meaning. Besides the allegorical interpretation of the miracles Cosmas recorded some curious explanations of parables from the Gospels, in which simple events of everyday life were transposed to the transcendental sphere. The prodigal son, for instance, who left his father's house and indulged in riotous living, represented the Devil in the eyes of the Bogomils, while the elder son who lived an orderly life in his father's house stood for Christ.[39] Other parables from the Gospel according to St.Luke were also seen as referring to the Devil, hidden for example in the figure of the unjust steward (Luke 16, 1-13). In this text the term "unrighteous mammon" occurs, as well as the oft-quoted: "Ye cannot serve God and mammon". The word "mammon", sometimes used in Greek patristic literature in the sense of "devil",[40] was used remarkably frequently by these Bulgarian dualists. It became a mere synonym for the Devil, as can be seen from the phrase "servants of Mammon" applied to people who got married and lived the life of the world. In his Sermon Cosmas even quoted the strange term of abuse *mamoničišta* (mamonitchishta = children of Mammon), hurled at baptised children; it simply meant "children of the Devil".[41]

Individual ingenuity had free play in their interpretation of the biblical texts, and Cosmas declared that "their words ... do not agree with one another, and fall apart like a piece of rotten cloth". The heretics vied with each other in using their wits, and Cosmas said sarcastically that each "found a name for his father and teacher" (i.e. the Devil) on his own. For some he was the "unjust steward", for others a "fallen angel".[42] The Sermon does not refer to the names Samael or Satanael, which the sect must have adopted for the Devil later, from apocryphal literature. There is no evidence that they had at this time a connected myth about the fall of the first of the angels, and the creation of the world.

It is unfortunate for us that Cosmas tells us very little about the organization of the sect and their manner of worship. All we can say for sure is that in his day the Bogomils used only the Lord's Prayer, repeating it four times during the day and four times during the night. Otherwise Cosmas tells us only that the heretics confessed their sins to each other and gave each other absolution, and that not only men, but women, too, did so.[43] This active role played by women in the religious life of the sect was common in the history of religious movements. We cannot doubt that in the tenth century the Bogomil sect had some sort of organization, but we can form no idea of what it was like.

The growth of heresy in Bulgaria is often explained by the need for the native Slav environment to defend itself against Byzantine influence, which was particularly disastrous in church life - and against Byzantinism in general, an alien spirit artificially encouraged in the land by the governing class.[44] In reality ***the heresy itself was one further proof of Byzantine influence;*** this influence was not confined to the official ecclesiastical channels, but came at lower levels, too, in popular movements. The very active Byzantine sects were not to be held by the frontiers of an easily accessible adjacent country where favourable conditions for the growth of heresy had already existed under Tsar Peter. The dualist doctrine deriving from Paulicianism could now rouse lively interest in Bulgaria. People had begun to think deeply, and Cosmas the Presbyter found many people asking: "Why does God permit the Devil to work against men?"[45] The more widespread suffering to be seen in the land, the more urgent this question of the origin of evil became. A doctrine which attributed the very existence of the world to the principle of evil offered a surprising answer to many burning doubts, and one which was particularly suited to the time.

The form in which Cosmas found the dualist heresy in Bulgaria and presented in his Sermon, *was still very close to the Paulician doctrine.* The "archaic" character can be seen clearly in the Bogomils' complete rejection of the Mother of God, of all the books of the Old Testament, all the figures of Old Testament history, and the direct precursor of Christ, John the Baptist. On the other hand, even at this early stage we can trace new features in their doctrines which remained characteristic of the sect in Bulgaria and in other lands. They include the frequent repetition of the Lord's Prayer, the identification of the evil ruler of the world with the fallen angel, and the allegorical motif of the "unjust steward". *The combination of dualist metaphysics with a strict ascetic life* became the hall-mark of the sect.

In tenth century Bulgaria this negation of the world took a very powerful form, fully in keeping with the general atmosphere. Yet the dualist doctrine may have assumed this radical form even earlier, in Byzantium, perhaps in some branch of the Paulician sect as yet unknown to us. It could be thought that what we find in Bulgaria is a specific form of Paulicianism. It should be remembered, however, that the Bulgarian sect did not follow the Paulicians' emphasis on the teachings of St.Paul, which indeed formed the starting-point of their beliefs. The Bogomils, as Cosmas knew them, recognised the whole of the New Testament and did not attach greater importance to the Epistles of St. Paul nor reject those of St.Peter. They showed no signs of the exclusive cult of St.Paul propagated by the Paulicians, whose teachers claimed to be his disciples. The evidence supplied by Cosmas reveals a significant point of development, when *the dualism derived from Gnosticism is still alive in the new sect, but Paulinism is dying out in its beliefs.*

The old Paulician sect preserved its individuality for centuries; even after the fall of Tephrice the Paulicians in the armies of the eastern themes represented a constant threat to the Byzantine Empire.

They were particularly strong in the themes of Chaldaea and Armeniakon. The Emperor John Tzimisces (969-976) followed the example of some of his predecessors who had settled Armenians from the eastern themes in Thrace. Shortly after Cosmas had written his polemic against the new dualist heresy in Bulgaria, Paulicians from the east were brought to settle in the region round Plovdiv (Phipippopolis). Later writers attribute a double motive to this measure of the Emperor's: to remove the Paulicians from the forts and strongholds where they ruled as they wished, and to make them reliable guards of the Balkan frontiers of the Empire, exposed to constant threat from nomads in the north.[46] While traces of the Paulicians are lost in Asia Minor in the course of the eleventh century, in the Balkans they remained a distinct group among the natives for centuries.

They never merged with the related dualist sect which took its name from the Bulgarian priest Bogomil.

Cosmas himself did not use the name *Bogomils,* which appeared in Byzantine sources in the eleventh century. Nor does he tell us anything about the person of Bogomil himself.[47] We are led to conclude that he was a country priest, living as simple and poverty-stricken a life as his flock. In his Sermon Cosmas makes reference elsewhere, too, to "popes" who were active in the heretical movement. Basing their conclusions on the evidence supplied by Cosmas, historians have assumed that Bohomil the "pope" founded the mediaeval dualist sect which spread both in Eastern and Western Europe. All that the author of the Sermon says, however, is that Bogomil was the first to preach this heresy *in Bulgaria.* This does not in itself necessarily mean that the new sect, clearly distinct from that of the Paulicians, was born in Bulgaria. Until there is sufficient evidence we should not exclude the possibility that it arose earlier, in Byzantium.[48]

It should not be forgotten that the development of Byzantine Paulicianism was certainly more varied and involved than would seem from the existing sources, which mainly date from the ninth century. The Paulicians' own traditions, noted down by Peter of Sicily, show that the frequent arguments about primacy which broke out among their teachers were also reflected in doctrinal differences. The doctrines of the sect did not remain without change for long, as can be seen from the attested existence of different canons of the New Testament. In Byzantine Asia Minor, the age-old breeding-ground of sects, the Paulicians may have come into touch with remnants of older sects and merged with them in various forms. In the middle of the eighth century, returning once more to Byzantium, the Paulicians settled in Pisidia, which had been a hotbed of sects in the fourth and fifth centuries. Among those attested for that period in epigraphical and literary sources, the Novatians were outstanding; they called themselves "pure", *katharoi* in Greek. There were still members of this sect in Asia Minor in the following century. Some ecclesiastical writers identify it with the sect of the *Athingani,* recorded in Phrygia and Lycaonia in the ninth century. The habit of the latter, holding themselves apart from those who did not share their faith, is not without analogy with the views of the earlier Cathars.

In his Chronicle Theophanes, writing 810/811 - 814/815, mentions the Athingani as one of the heretical groups in Asia Minor who were carried along with the wave of iconoclasm. At several points he lists them along with the Paulicians, and both sects suffered the same fate in the twists and turns of the iconoclastic period. After enjoying a period of toleration under Nicephorus I they were the victims of a relentless extermination campaign, yet in the first half of the eleventh century the "Paulicians and Athingani" were still in existence, named together in a letter by the Metropolitan Demetrius of Cyzicus. A degree of symbiosis can be assumed, but the sources give us no concrete facts about their influence on each other. It is in any case worth noting that in the eleventh century the new Bogomil heresy had taken root firmly in Phrygia, which in the ninth century was the home of the Athingani and undoubtedly within the sphere of influence of the Paulicians settled in neighbouring Pisidia. Nor should it be forgotten that in Western Europe the Greek word *katharoi* was used from the outset for the dualist sect of eastern (Bogomil) origin.

The Paulician movement was widespread throughout the vast Byzantine Empire, and conditions were not favourable for the preservation of its unity. The persecution which came down heavily in the twenties of the nineth century forced the leaders of the sect to take refuge in Arab territory, from where it was difficult for them to keep up regular contacts with their members. Nor was it easy for the scattered Paulician communities to keep in touch with each other at this time. It is more than likely that they often developed very much along their own lines.

A Byzantine euchologion of the year 1027 contains six articles directed at a dualist heresy,[49] the last of which lays a curse on Paul of Samosata and a number of other names which mean nothing to us. These articles are appended to the well-known abjuration formula for Paulicians, which already contains the name of Chrysochir who died in 872, and therefore cannot have come into existence as a complete text before the last quarter of the ninth century.

The six articles concerned can be assumed to refer to an independent branch of the Paulicians, or to a related dualist sect. In any case the views listed in them could as well be Bogomil as Paulician beliefs.

In the first half of the tenth century we find interesting evidence in the south-western regions of Asia Minor, round the city of Miletus and in the Cibyrrhaeot theme. On Mount Latros (later Latmos) near Miletus, there lived St.Paul the Younger (d.955)[50], the founder of one of the many monasteries in the place. An anonymous biographer praises his ardent fight against the "Manichees" to be found in the region. He managed to have the most dangerous of these heretics moved far away. At the time of which he was writing the term "Manichaean" certainly referred to heresy of the Paulician type. The question remains open, however, to what degree the heretics of this part of Asia Minor still preserved characteristic features of Paulicianism, and particularly how far they were aware of their connection with the sect. It is certain that there is clear proof of the existence of Bogomil heretics in the Cibyrrhaeot theme in the eleventh century. Here and elsewhere in Asia Minor the new dualist sect undoubtedly took up the still living tradition of earlier heresies, unless indeed it grew directly out of it.

[1] A.Vaillant, in: *Puech - Vaillant,* Le traité, 37 ff.

[2] A bibliography of works on the Sermon of Cosmas, and a summary of theories about its date, are given by *D.Angelov,* Bogomilstvoto v Bălgarija, 41 ff., n.4.

[3] For the text of the Sermon see *Popruženko,* Cosmas, IV ff.; Vaillant, in: *Puech - Vaillant,* Le traité, 10 ff. For the time being Popruženko's 1936 edition has to be used, until J.K.Begunov's new edition, now in press, appears; it is based on much fuller MS material, and the editor has already published a number of preparatory studies. All the complete copies of the Sermon are from Russia, and date from the 15th to the 19th centuries. In addition extracts and compilations taken from the work have also been preserved, from both Russia and Serbia. The fact that the Sermon was so widely known in Russia cannot be taken to mean that the Bogomil heresy had penetrated to the eastern Slavs. On the Bogomil heresy in Russia see the exhaustive study by *D.A.Kazačkova,* Kăm văprosa za bogomilskata eres v drevna Rusija prez XI v. Ist. Pregled XIII (1957) No.4, 45-78. In spite of all that has been written on the subject, there is not a single piece of reliable evidence proving that the Bogomil sect existed in Russia.

[4] Vaillant, in: *Puech - Vaillant,* Le traité, 40 f.

[5] Ibid., 37 ff.

[6] The words властеле and владꙑкꙑ embrace all who are endowed with power (власть). They are used in the same general sense in Cosmas's quotations from Titus 3, 1-2 and Romans 13, 1-4 (Cosmas, 37). The words властеле and владꙑкꙑ

are used there for the Greek ἀρχαί and ἐξουσίαι, which in these passages from St. Paul mean "the authorities" in the broad sense (see *G.Knittel*,Theol. Wörterbuch zum Neuen Testament I, 481; II, 562. властеле (владыкы) земьнии means "temporal powers", as D.Angelov rightly comments (Prezviter Kozma i besedata mu protiv bogomilite, Sofia 1948, 24). The words старѣи, старѣишина (Cosmas, 35, 11-14; 44, 2) express superiority in standing; of the many examples which could be given, see Cosmas himself: послоушаите гласа старѣишины пастоушска (75, 13; similarly 76, 1-2; cf. p.76, quoting from I. Peter 5, 2-4: началникоу пастоухѡ). In the given context, however, we cannot assess the precise meaning of the word.

7 Cosmas, 43,21 - 44,2; 47,12 - 48,6.

8 See *J.M.Hussey*, Church and Learning in the Byzantine Empire (867-1185), Oxford 1937, chapters IX, X.

9 Cosmas, 42,22 - 43,10; 52,4-12; 53,1-11.

10 Ibid., 55,11 - 56,4.

11 Ibid., 56,21 - 57,16.

12 Ed. *Gošev*, loc. cit., 443 f. (see chap.III, n.2); cf. *Dujčev*, Rilskijat svetec, 153 ff.

13 *Beck*, Kirche u. theol. Lit, 126 ff. Cf. *F.Dvorník*, Les légendes de Constantin et Méthode vues de Byzance, Prague 1933, 117 ff.

14 Cosmas, 51,4 - 52,3.

15 According to A.Vaillant, "Cosmas, comme Jean l'Exarque, fait partie du clergé instruit, des riches du siècle présent (29,14): ce sont des presviteri." The word *pop*, on the other hand, referred to the "clergé pauvre et ignorant". See A.Vaillant, in: *Puech - Vaillant*, Le traité, 29. It is a fact that Cosmas himself never uses *presviter* to refer to the common order of priests, for whom he employs the terms попъ, иереи, жрьць (Complete list of examples in Popruženko's edition under the relevant words). I.Gošev gives examples from Constantine of Preslav and from the Sinai Euchologion. See *I.Gošev*, Starobălgarskata liturgija, in: Godišnik na Sof. univ., Bogosl. fak. IX (1931) kn. 2,9. The title of *presviter* as used by Cosmas, John the Exarch and Constantine of Preslav (who used it until he was ordained bishop), certainly implies that they were members of the higher clergy, and we therefore think that Vaillant is closest to the truth.

16 This has been convincingly shown by *J.Gagov*, Theologia antibogomilistica Cosmae presbyteri bulgari (saec. X), Rome 1942, 27 ff.

17 74,23 - 79,21.

18 Ibid., 12 - 13; 15,6.

19 Ibid., 72,3 - 73,2.

20 Ibid., 67,19 - 23; 74,13 - 14,18 - 21. Referring to Cosmas and his mention of the "games" and "diabolic songs", *I.Dujčev* gives the evidence for these customs among the ancient Slavs, Slavjanobolgarskije drevnosti IX-go veka, ByzSlav XI (1950) 23.) An almost literal parallel to the passage 74,13 - 14 (мнози бо от чловѣкъ паче на игры текоут неже въ црькви, и кощоуны и блади любать паче книгъ) can be found in the Izmaragd, edited by *A.S.Archangel'skij*, Tvorenija otcov cerkvi v drevnerusskoj pišmennosti IV, Kazan 1890, 64. A.Vaillant points out this coincidence in: *Puech - Vaillant*, Le traité, 51. He gives further evidence which suggests that Cosmas "a emprunté une partie de ces citations à quelque recueil comme le Zlatostruj, qui est vieux-slave, ou à quelque florilège analogue au Margarit et à l'Izmaragd, qui paraissent postérieurs et russes, mais dont l'origine n'est pas exactement connue..." (p.52). On the other hand, whole passages of the Sermon were

included in anthologies of the kind as moral instruction (see *A.I.Ponomarev*, Pamjatniki drevnerusskoj cerkovno-učitel'noj literatury III, SPb 1897). There are frequent descriptions of pagan customs and heathen ways still common among the people, in the Izmaragd (see *Archangel'skij*, op. cit., 29 ff.), and expressions similar to those used in the same context by Cosmas (74,19) are also found there (ibid., esp. pp. 31, 37). For the opposition of *igry* and *cerkvi* see also: Poučenije o vedre i kažnjach božjich which was included in the Russian chronicle Povest' vremennych let, ed. *V.P.Adrianova-Peretc*, t. I, 1950, 114, and also: Slovo istolkovano mudrostiju ot sv. Apostol. Both texts are quoted by *Kazačkova*, art. cit. (see n. 3 above) 57.

It can justifiably be assumed that quite apart from whether the formulation is his own or not, in these passages Cosmas described real survivals of pagan customs, such as lived on for a long time in Bulgarian, as in Russian life.

21 Cosmas, 79,3 - 15. It is not very likely that John the Exarch was still alive a few years after the death of Tsar Peter (as Vaillant thinks, op. cit. 39). Nevertheless it follows from what Cosmas says (и не гл҃ите не можеть такъ быти в си лѣта) that the "times" of his contemporaries were still the "times" of John the Exarch. The whole passage is a typical example of how the author of the Sermon used borrowed formulas to express his own observation. His call for priests to follow the example of the Church Fathers echoes the Pseudo-Chrysostom, Sermo de pseudoprophetis, PG 59, col. 560 (see Vaillant, in: *Puech - Vaillant*, Le traité, 50). Cosmas immediately proceeds to expand his own idea there is no need to go back to St. John Chrysostom, the example of "the younger presbyter John" (ивана прозвитера новаго) is closer to hand. Apart from the famous author of the Preslav school the Bulgarian sources mention no other John as "presbyter and exarch in the land of Bulgaria". John the Exarch, one of the members of the literary ambience of Symeon, is called thus in the titles of his works. The evidence is given by *K.M.Kuev*, Kăm văprosa za titlata "ekzarch" v starobălgarskata literatura, in: Izv. na Inst. za Ist. BAN 14 - 15 (1964) 326. See also *I.Iufu*, Sbornicul lui Gherman (1359), Ortodoxia XII (1960) No 2, 260. This Bulgarian collection dating from the reign of Tsar Ivan Alexander contains a copy of a panegyric on "the holy apostle and evangelist John the Divine", composed иωаномъ, екзархомъ блъгаръскымъ. Cf. *Dora Ivanova - Mirčeva*, Ivan Ekzarch Bălgarski, Slova, Sofia 1971, 16 ff. Dujčev's identification of the "new priest John" with St. John of Rila is based on the assumption that at this time the term "exarch" was still current for the office of higumen (Slav: načalnik). See *I.Dujčev*, Rilskijat svetec, 385 ff., n.87; and most recently: Zur Datierung der Homilie des Kosmas, in: Die Welt der Slaven 8 (1963), 319 ff. In fact the testimony of Cyril of Scythopolis, on which Dujčev relies, shows that even in sixth century Palestine the title of exarch had a broader significance; it was given to two archimandrites appointed to supervise all coenobitic convents and all anchorets in the district of Jerusalem. Similarly in Constantinople the abbot of the monastery of Dalmatos, as archimandrite, was at the same time exarch for all the convents in the capital. (See *Beck*, Kirche u. theol. Lit. 137, cf. 138, n. 1). For the existence of exarchs charged with supervision of monastic establishments see Just. Nov. CXXXIII, cap. IV, ed. *Schoell - Kroll* III, 671:... ὁ δὲ τῶν μοναστηρίων ἔξαρχος (εἴγε τις εἴη, κατὰ τὸν τόπον ὥσπερ ἐπὶ ταύτης ἐστὶ τῆς εὐδαίμονος πόλεως) ταῦτα περιεργαζέσθω ... (cf. *Kuev*, l.c., 339). There is proof from the 9th and 10th centuries of the institution of exarchs charged with supervision of monasteries under the direct jurisdiction of the Patriarch of Constantinople (stauro-

pegial houses). See *Grumel*, Regestes, NN.451, 474, 764, 791. For the use of the title *exarchos* to designate the Metropolitans of large regions in the middle and late Byzantine period, see *Beck*, Kirche u. theol. Lit., 69. The context of the relevant passage in Cosmas suggests that John the Exarch was high in the ecclesiastical hierarchy of Bulgaria. We believe it most probable that he was the Bulgarian archbishop.

[22] Cosmas, 2,5 - 8.

[23] Ibid., 12 - 13.

[24] Churches: 34, 15; liturgy: 11,12 - 16; the prayers of the Bogomils: 11,14; 32,19 - 23; icons:18 - 20; relics of the saints: 5,16 - 26; the eucharist: 8,24 - 11,11; baptism: see n.31 below.

[25] Ibid., 32,11 - 18; 56.

[26] Ibid., 17,20 - 22; 33,15 - 16.

[27] Ibid., 5,13 - 14; 6,23 - 7,4. (Cf. Patriarch Germanos II, In exaltationem venerandae crucis et contrà Bogomilos, PG 140, col. 632B.)

[28] Ibid., 7,18 - 19.

[29] Old Testament and the prophets: 16,5 - 8; 17,9 - 11; 20,10 - 12. John the Baptist: 17,12 - 14.

[30] Ibid., 23,5 - 6; 22,19 - 24,16; 26,9 - 10.

[31] Ibid., 31,4 - 32,10.

[32] Ibid., 26.

[33] Ibid., 3,1 - 12.

[34] Ibid., 35,11 - 14.

[35] Ibid., 34,22 - 35,5. работати means "to serve". Cf. Cosmas 53,14: serve God and Mammon, Luke 16, 13; and id.: Savina kniga, 1.63, ed. *V.Ščepkin*, 48.

[36] Cosmas, 22,15 - 17.

[37] Ibid., 45,5 - 8.

[38] Ibid., 58. Cf. A.Vaillant, in: *Puech - Vaillant*, Le traitê, 44 ff.

[39] Cosmas, 26,6 - 9. See Byzantinobulgarica III, 28 ff.

[40] Greg. Nyss., De oratione dominica, orat. V, PG 44, col. 1192; Epiphanios, Adv. haer., lib. II, t.II, PG 42, col. 1373. For the etymology of the term μαμωνᾶς see Kittel, Theol. Wört. zum N. Test. IV, 390 ff.

[41] Cosmas, 31,22 - 25.

[42] Ibid., 22,19 - 22. Cf. H.-C.Puech, in: *Puech - Vaillant*, Le traitê, 188 ff.

[43] The Lord's Prayer:Cosmas, 32,20 - 23; confession: 39,13 - 21.

[44] On the penetration of Byzantine influences into Bulgaria see *P.Mutafčiev*, Der Byzantinismus im mittelalterlichen Bulgarien, BZ 30 (1929-1930) 387-394.

[45] Cosmas, 24,6 - 7.

[46] Alexiad XIV, t. III, 179,31 - 180,25; Kedrenos, Historiarum compendium,Bonn, II, 382; Zonaras, Epitome historiarum, ed. L.Dindorf, IV, Lipsiae 1871, 251, 3-6. Cf. ByzSlav XXV, 66.

[47] Scholars who denied the historical existence of a priest Bogomil pointed out that no such Bulgarian priest was named in any Byzantine sources (See *V.S.Kiselkov*, Săštestvuval li e pop Bogomil? in: Ist. Pregled XIV (1958) kn.2, 57 - 67). This argument broke down, of course, after Gouillard's discovery of Byzantine references to a Bulgarian priest Bogomil (see Chapters VIII and XVI. The attempt by G.Wild to explain the name Bogomil itself as an expression of the idea "Gottesfreundschaft" is not convincing. See *G.Wild*, Bogu mili als Ausdruck des Selbstverständnisses der mittelalterlichen Sektenkirche, in: Kirche im Osten 6 (1963) 16 - 33.Cf. *E.Werner*,

Θεόφιλος - Bogumil, Balkan Studies 7 (1966) 49-60. The whole learned argument put forward by Wild can be confronted with the simple fact that the Slavonic name Bogomil (a calque of the Greek Θεόφιλος) was already a common personal name by that time. This is proved by the feminine form Bogomila found among the Bulgarian names listed in the Cividale Gospel of the second half of the ninth century. See *I.Dujčev*, I Bogomili nei paesi slavi e la loro storia, in: Atti del Convegno internaz. Accad. Naz. dei Lincei 62 (1964) 626 ff. (= Medioevo I, 261 ff.).

48 On this point see our study La question de l'origine du bogomilisme (Bulgarie ou Byzance?) in: Actes du Ier Congr. Intern. des Et. Balk. et Sud-Est Eur. III Sofia 1969, 270.

49 Text: Synodikon, ll. 250-276. See *Gouillard*, L'hérésie, 313 f. The words of the first article in this euchologion differ significantly from the parallel text in the Synodicon: instead of ἐπεισακτον τινὰ ἄγγελον 'Αμὴν λεγόμενον τὸν Υἱόν, (Synodikon 1.252) we read: ἐπεισακτόν τινα ἄγγελον ὁμολογούμενον τὸν Υἱόν, (Euchologion, Coisl. 213, ed. *A.Dmitrijevskij*, Opisanije liturgičeskich rukopisej... II, Kiev 1901, 1025). Note: There are twenty-three anathematizing articles added to the Synodicon for the Sunday of Orthodoxy in the earliest known copy, dating from the eleventh century (*Ma* in *Gouillard*). See the text and commentary, *J.Gouillard*, Synodikon, 304-313. Nothing definite can be said either of the origin of these articles, or of the purpose for which the whole series was put together.

50 Anal. Boll. XI (1892) 156.

Additional note:

Begunov's new edition of the Sermon appeared when the present book was already printing. See *J.K.Begunov*, Kozma Presviter v slavjanskich literaturach, Sofia 1973.

Chapter V

THE BOGOMIL HERESY AND BYZANTINE MONASTICISM

The first Byzantine writer to give reliable evidence of the existence of the Bogomil heresy in Byzantium itself was Euthymius of Acmonia.[1] He seems to have already been living for some time as a monk in the Peribleptos monastery in Constantinople, which had been founded in 1031 by the Emperor Romanus III Argyrus[2], who died in 1034. Euthymius refers to him as deceased, but gives no other pointer to a more exact date for his work.

It takes the form of an epistle[3] addressed to the people of his home region, in the Opsikion theme of Asia Minor, in the diocese of Acmonia. Another text of which a fragment has been preserved probably represents part of an earlier epistle, to which Euthymius refers in the one discussed here.[4] In his monastic retreat in the capital he heard with anxiety reports of the way heresy was spreading in his home, growing fast since the time of his youth. The purpose of both his epistles was to reveal the diabolic power of the heresy in "its true light".

Euthymius of Acmonia is the first writer to use the term Bogomils.[5] The Bulgarian origin of the name is suggested by his comment that the heretics are so called "in the West",[6] i.e. in the western areas of the Byzantine Empire, in the Balkan peninsula; the name was also well-known in the Ciberrhaeot theme and "elsewhere". In the writer's own home, however, the old region of Phrygia, the heretics were known only as Phundagiagitae.[7] This name, which is still a mystery to us, occurs in other Byzantine texts as well,[8] and it is not impossible that it was the original name of the sect.

In the region of Acmonia the heresy was not a new development. Euthymius had met the leader of the sect there, John Tzurillas, when he himself was still a boy and came to the town of Acmonia with his mother, in connection with a lawsuit. This was in the reign of Basil II (976 - 1025), and Romanus Argyrus, later to become Emperor, was at that time judge of the theme of Opsikion. This incident occurred some time in the first quarter of the eleventh century, and Tzurillas already had the reputation of a heretic then.[9] The recipients of the epistle would remember that for three years Tzurillas had preached his heresy in the neighbouring Thracesian theme, winning over "whole towns" in the region of Smyrna and elsewhere.[10] The village in which he now lived was well known in the diocese of Acmonia. Euthymius himself could still remember the time when all the people there were orthodox but now there remained no more than ten who still held the true faith, for which they were forced to suffer ridicule and humiliation from the heretics around them.[11]

This provincial focus of heresy was not isolated, but the heretics obviously kept in contact with fellow-believers elsewhere in the Empire Euthymius himself came across a group of them in Constantinople; they knew Tzurillas and his companion Racheas, and were preparing to set out to spend the winter with them, when a strange coincidence brought them to the Peribleptos monastery in Constantinople. There, in the absence of Euthymius, the heretics almost succeeded in "corrupting" his young "disciple".[12]

It was by no means unique that heretics penetrated into a monastery. Euthymius himself was continually disturbed by the number of monks and clerics won over for heresy.[13] When he had seen John Tzurillas years before, it was in the guise of a monk. He tells us that the heretic had compelled his wife to wear monastic garb too, and that his disciples followed his example. They openly preached the belief that only a man who leaves his wife, and a woman who leaves her husband, can attain salvation.[14]

This conception of monastic asceticism could of course exist independently of the dualist sect, but in any case it prepared very favourable ground for it to work on. We have already seen this in Bulgaria, but in Byzantium the apostles of the Bogomil heresy appear as *monks*. This fact explains much of its success among the people, even in the upper ranks of society, for we know the great influence and authority exerted by monks in Byzantine society. It is clear from the description given us by Euthymius[15] that "father" Tzurillas (Greek: *papas*) was deeply respected by the people of Acmonia; it is equally clear that our author is over-anxious to present the heretic to them in the worst possible light. The scandalous affair he describes carries little conviction, while his determination to give as black a picture as possible of the lives of the heretical monks is clear from other places in the text, too.

It can be said that the epistle of Euthymius of Acmonia is not outstanding for its moral concern, the serious undertone which runs through the work of the Bulgarian presbyter Cosmas however urgent his polemic ardour. The Byzantine monk was writing sheer invective, inspired by malicious spite and the determination to present those who practised the Bogomil heresy as worshippers of the Devil. He was out to achieve horrific effect, to turn the simple minds of his readers away from all thought of communion with the heretics.

This naive tendentiousness is obvious throughout, so that it is not at all difficult to recognise the facts beneath the horrifying disguise. This allows us to form a valuable idea of the life and teachings of the sect, which the author knew from his own experience. He reveals, too, that he asked the heretical teachers he met with in the Peribleptos monastery to explain the whole of their doctrines to him, pretending deceitfully that he wanted to join the sect.[16]

It is no good for Euthymius to assure us that whenever the heretics pray earnestly: "Our Father, which art in Heaven", they are in reality calling on the Devil.[17] He promptly convicts himself of lying when he goes on to tell us how the heretics explained the world. They believed that there were seven invisible heavens, ruled over by God, while the visible heaven is the domain of the evil principle, called by the Phundagiagitae of Euthymius, as by the Paulicians, the *prince of this world*, according to the New Testament term (ὁ ἄρχων τοῦ κόσμου τούτου). He exercises power over the whole of the world he has created, and over the paradise with which he rewards those who please him, and the place of damnation to which he consigns his enemies. He and God have nothing in common; the sphere of each is clearly marked, one governing worldly things and the other the things of Heaven.[18]

No other text gives us such convincing proof of the Marcionite origin of the dualist doctrines of the Middle Ages. In this case the abyss between God and the evil creator stands out particularly clearly. Nevertheless we can distinguish a clear incipient tendency to adapt the original Marcionite–Paulician dualism to the normal Christian doctrine of the origin of evil. For Marcion and the Paulicians the creator and ruler of this world has always existed, quite independently. The teachings of the Phundagiagitae, however, suggest clearly that there was once a bond between him and God. Not until God was angered and cast him out "from before His face", did the archon leave the heavenly kingdom. We are not told the reason for his ejection, nor is there explicit mention of an angel who became evil in consequence of his fall from God's grace. Nor is the name Satanael yet given, although the contemporary of Euthymius, the learned Michael Psellus, found it in use among the heretics of this sect. The influence of the Christian tradition of the fall of Satan is quite apparent, however, and we know that already some of the Bogomils of whom Cosmas wrote used the term "fallen angel" for the Evil One who rules this world.

The visible heaven and the world were not created until after the archon's fall; everything in our world is his work, except for two things which have come from God: the sun and the human soul. The prince of this world simply stole them as he was driven out from God's heaven.[19] No less primitive was the explanation of how the soul came to be bound up with the material body, which was created by the evil archon. His first efforts to keep the soul imprisoned in the body were fruitless, and for three hundred years Adam's body lay lifeless. Then a mess of many unclean beasts fouled the soul, held for a moment in the body; the soul thus received evil qualities and was bound to the body for ever.[20]

This is where Euthymius stops in his presentation of the dualist myth. We learn no more about what happened to the human race, or about its salvation.

Their eschatology is disposed in a negative sense: they do not believe in the resurrection of the body, nor in the Second Coming of Christ and the Last Judgment.[21]

In spite of its gaps, the information given by Euthymius offers a unique glimpse into the minds of the heretics, whose primitive imaginations had created this bizarre myth with all its naturalistic details. Their wild fantasy provoked the monkish writer to loud and scornful laughter. But even within the sect, they did not dare to reveal the myth to new adherents straight away. This secret learning about the origin of the world was communicated to them by degrees, and not until after the believer had been initiated by a special rite which signified his rescue from the power of the Evil principle and the descent of the Holy Spirit on the new member of the sect.[22]

Euthymius, of course, deliberately distorts the sense of the Bogomil initiation ceremony, in his anxiety to show that this secret rite put the unfortunate victim in the power of the Devil for ever more. In spite of his distortions, however, we can distinguish the individual elements of the ceremony as he describes it.[23] We are helped in this by a later synodical tomos which clearly derives from Euthymius but also draws on other sources.[24] Before initiation into the sect, the candidate had to go through a preparatory period of stern asceticism and constant prayer.[25] The initiation ceremony itself began with an original form of exorcism and the symbolical washing away of all trace of church baptism.[26] There is clear evidence that the book of the Gospels was laid on the initiate's head,[27] while the culmination, the laying-on of hands, can only be guessed at under the words "seal of the devil" (τὴν σφραγῖδα τοῦ διαβόλου). [28] The "Satanic incantation" (ἐπῳδὴν σατανικήν) [29] clearly refers to the epiclesis, calling upon the Holy Spirit.[30]

There can be no doubt that the initiation rite of this mediaeval dualist sect was not elaborated without being influenced by the liturgy of the Church. There are remarkable parallels to it in ancient Christian baptismal ritual and also in the ceremony of taking monastic vows; the most ancient forms of this rite were clearly modelled on the rite of baptism, stressing the significance of the monastic initiation as "a second baptism".[31] The synodical tomos mentioned above provides valuable evidence that a *monk's habit* was assumed as part of the Bogomil initiation rite.[32]

It would thus appear that the mediaeval dualist sect which was closely bound up with monastic life in Byzantium had itself grown up under strong monastic influences. Its real members, later known in the west as the "perfect", or Cathars, formed something of a monastic group, and in Byzantium were undistinguishable from wandering orthodox monks in their appearance as well. The difference between them was that our heretics did not think they were simply living a more Christian way of life, but believed that they were the only true *Christians;*[33] the initiation they received was not a second baptism in their eyes, but the only true baptism.

The Byzantine monk Euthymius of Acmonia was the first to introduce us to this secret rite, so reminiscent of the ancient ritual of taking monastic vows in the Christianity of the East. He also witnessed the *common prayers* of the sect[34] which had a set form, such as that known in the sect later in western Europe and in Bosnia.

Euthymius succeeded in penetrating into the hidden life of the sect, which kept its ceremonies secret. Here, too, we find the forced hypocrisy with which presbyter Cosmas also reproached the heretics. In Byzantium, too, the Bogomils attended divine service in the churches, to all appearances even accepting the sacraments, which they thought "did neither harm nor good". They would secretly spit out the bread and wine of the Eucharist, and tread it under foot; if there was no help for it, they would swallow it like ordinary food and drink, and did not fast beforehand.[35] They had their children baptized, but at home they washed them with dirty water to remove all trace of baptism.[36] The monks and priests who were members of the sect in secret would carry out all the duties of their cloth. As Euthymius describes their conduct, of course, it takes on monstrously distorted proportions; he tries to convince his readers that these false servants of God paint icons, set up crosses and build churches only in order to be able to blaspheme and desecrate them.[37] Among the tales he tells is an incredible one of a monk who was also a priest, who found a most original way of desecrating the church he himself had built.[38]

Leaving aside these exaggerated stories, it cannot be denied that Euthymius understood very well the danger that a heresy skilfully hidden beneath a monk's habit meant for the world of Byzantium. In this he rightly distinguished between his Phundagiagitae-Bogomils and the Paulicians, heretics whom he also knew enough about. Byzantine heresiology soon found a convenient formula to cover the "newly appeared" heresy. The connection between this sect and the monks almost inevitably meant that it was identified with the much older Messalian heresy, appearing in Syria in the fourth century as a typically monastic heresy. The Syrian name "Messalian", the equivalent of the Greek "Euchitae" , came from the ceaseless prayers which these mystics saw as the only way to free each soul from the power of the demon and the only path to spiritual perfection. The Messalian doctrines were the extreme expression of the longing to comprehend mystical revelation through sensual experience. This strange movement was essentially alien to the cosmological dualism of the Bogomils.[39] A few external similarities sufficed for the Byzantines to confuse one with the other,[40] and the Bogomils were often simply called "Messalians" or "Euchitae". The contemporary of Euthymius, Michael Psellus,[41] already uses the name, although Euthymius himself does not. For him the Bogomil heresy was simply the "last and greatest error" heralding the coming of the Antichrist.[42]

The teachers of the sect, who called themselves "apostles", covered the whole area of the eastern Christian Empire, wrote Euthymius, in their tireless proselytizing,[43] and like the apostles of Christ, drew lots to decide where each should work.[44] As we have already seen, Euthymius gave quite definite information about the localities infected with the heresy, which had overrun the western regions of Asia Minor - the Opsikion, Thracesion and Cibyrrhaeot themes, penetrated into the capital and spread through the Balkan peninsula. Our author says nothing, however, of the sect's activities among the southern Slavs, where nothing had been heard of them since the time of Cosmas the Presbyter.

[1] See *J.Darrouzès*, Euthyme moine de la Péribleptos (XI siècle). Dictionnaire d'Hist. et de Géogr. Ecclés. XVI (1964) col.63-64. (I did not have access to this article). The monk Euthymius with whom we are concerned has sometimes been identified with the later Byzantine ecclesiastical writer, Euthymius Zigabenus. See e.g. *F.Cumont*, La date et le lieu de la naissance d'Euthymios Zigabénos, BZ 12 (1903) 582-584. Other scholars have convincingly refuted this, see *M.Jugie*, Phoundagiagites et Bogomiles, Échos d'Orient 12 (1909) 257-262.

[2] *R.Janin*, Le monastère de là Théotokos Péribleptos à Constantinople. Bul.Sect. Hist. Acad. Roumaine 26 (1945) 192-201.

[3] The text was published by *G.Ficker*, from two MSS in Vienna: Cod. Vindob. Theol. Graec. 307 and 193. See *G.Ficker*, Die Phundagiagiten. Leipzig 1908, 1-86 (pp. 1 - 9 are printed in parallel, giving the variant text of a MS in the Utrecht University library: Cod. Graec. 3 univ. Rheno-Traiectinae.)

[4] Printed in PG 131, col. 48B - 57C, from the Vatican MS (Cod. Vatic. Graec. 840). On this text see *Ficker*, op.cit., 150 ff. and 157 ff. Besides these two epis-

tles attacking the Bogomils Euthymius was probably the author of the "first invective against the Armenians", attributed until recently to the Armenian Catholicus Isaac. See *V.Grumel*. Les invectives contre les Arméniens du "Catholicos Isaac". REB. 14 (1956) 174 - 194.

⁵ Euth. Acm., ed. *Ficker*, 34; 4,19; 42,21; 62,13.

⁶ ... εἰς τὴν Δύσιν ... Ibid., 62,12; PG 131, col.48C. See *Ficker's* commentary, op. cit., 249 f.

⁷ Euth. Acm., ed. *Ficker*, 62,10-13.

⁸ Besides the form Φουνδαγιαγῖται, found in Ficker's text, shorter forms are also attested. The Vatican MS quoted above (Cod. Vatic. Gr. 840) gives the genitive Φουνδαγιατῶν (PG 131, col.48C). The form Φουνδαῖται or Φουδαῖται is found in two texts connected with the tract of Euthymius, in particular the synodical tomos of the Cod. theol. gr. 40 (fol. 250v - 251v) of Vienna, ed. *Tollius*, 106 - 125 = PG 131, col. 40C - 48B (*Grumel*, Regestes, No 989), where we find the genetive Φουνδαϊτῶν in the title (Φουνδαϊατῶν in the text).

Closely connected with this text is the form prescribed for re-accepting Bogomils into the Church; the title gives the form Φουδαιτῶν (see text in *Ficker*, op. cit., 172) while the form φουνδαΐαττῶν is used in the text of Thallóczy's formula (cf. *Ficker*, op. cit., 175).

In the liturgical commentary written by the bishop Theodore of Andida in the twelfth or perhaps already in the eleventh century, the name of the sect is given (in the accusative) as Φουνδαΐτας. See text in *Migne*, PG 140, col. 461B. (For Theodore of Andida see *H.G.Beck*, Kirche u. theol. Lit., 645).

The name also occurs in the Greek version of the Palaea historica, ed. *Vasiliev*, Anecdota I, 191: Τοῖς δὲ ... Φουνδαΐταις. It is worth noting that among the names of supposed disciples of Mani Euthymius of Acmonia gives the name Φουνδᾶς, apparently as a variant of Βουνδᾶς i.e. Buddha (see text in Ficker, op. cit., 42,10). It is possible that the shorter form Φουνδαῖται or Φουδαῖται was first derived from the name Φουνδᾶς, followed by the longer form Φουνδαγιαγῖται. *M.Jugie* (Phoundagiagites et Bogomiles, l.c., 260) and H.-C.Puech (*Puech - Vaillant*, Le traité, 281, n.3) support the theory that the name Phundagiagitae was derived from the name Phundas. Yet another etymological explanation, based on the word φούνδα (= scrip) and linking the name Φουνδαγιαγῖται with the name of the Christian sect of the Σακκοφόροι, was rightly rejected by *Ficker* (op. cit., 193).

⁹ Apart from this epistle of Euthymius of Acmonia (ed. *Ficker*, 66,24 - 67,1) there is no other Byzantine source which refers to Romanus Argyrus holding the office of judge in the Opsikion theme. All we have on record is that under Basil II he was one of the judges of the hippodrome (Πεῖρα, ed. *Zachariä v. Lingenthal*, in: Jus Graeco-Romanum I,223; cf. *N.Skabalanovič*, Vizantijskoje gosudarstvo i cerkov' v XI v., St Petersburg 1884, 13. Both these functions (i.e. judge of a theme and κριτὴς τοῦ βήλου καὶ ἐπὶ τοῦ ἱπποδρόμου) could be held simultaneously, as we know from evidence on seals (See *V.Laurent*, Collection Orghidan, N° 201, 221, 241). Romanus Argyrus was sixty when he married Zoë, the daughter of Constantine VIII (12. XI. 1028 (See Zonaras, Epistome historiarum, ed. *L.Dindorf*, IV, Lipsiae 1871,133, 19 - 20; cf. *Skabalanovič*, op.cit., 13, 14). He was therefore born about 968. The time during which he could have been judge of the Opsikion theme must have been approximately 990 - 1025. There is another point to be borne in mind, however; when Euthymius wrote his epistles he had already been a monk in the Peribleptos monastery for a long time, and the monastery was founded in 1031. The pilgrimage to

Jerusalem of which he speaks had been undertaken years before, and already then he was not only a monk of that house, but a senior, experienced monk to whom the spiritual care of a younger monk had been entrusted. He was thus advanced in age when he wrote the epistles. Many years had passed since the time when, a youngster, he had seen John Tzurillas; the latter was then a grown man, certainly much older than Euthymius. Since in his invective against the sect he speaks of Tzurillas as being still alive, (Euth. Acm., ed. *Ficker*, op. cit.,66), it will certainly be nearer the truth to place his meeting with John Tzurillas in the first quarter of the eleventh century.

The expressions ἐν τῷ ὑφ' ἡμᾶς θέματι (ed. *Ficker*, 66,25), ἐν τῇ ὑφ' ἡμᾶς ἐπισκοπῇ (ibid., 67,2 - 3) cannot have any other meaning in the given context than the theme and bishopric from which the author originally came. This is how *Ficker* correctly interprets the text (op. cit., 179) while *F.Cumont* (art. cit., 583 f.) does not arrive at a satisfactory conclusion. In the sentence: Καὶ μέμνημαι ἐλθόντος τούτου ἐν τῇ ὑφ' ἡμᾶς ἐπισκοπῇ 'Ακμονείᾳ καὶ ἦλθον κἀγὼ μετὰ τῆς μητρός μου ἔχοντος πρὸς ἄλλον τινὰ δίκην (ed. *Ficker*, 67,2 - 4) an error has clearly occurred, since the form ἔχοντος is incompatible with μητρός. This passage is taken from the Cod. Vindob. Theol. Graec. 307 (MS A in Ficker's edition), while the MS Cod. Vindob. Theol. Graec. 193 (B) gives ἔχοντες (see apparatus below the text, p. 67). This nominative plural fits into the construction of the sentence (it refers to κἀγὼ μετὰ τῆς μητρός μου) and there can be no doubt that MS B gives the correct reading here. There is thus no need for the forced interpretation attempted by S.Pétridès who thought that the words ἐλθόντος τούτου and ἔχοντος πρὸς ἄλλον τινὰ δίκην referred to John Tzurillas while καὶ ἦλθον κἀγὼ μετὰ τῆς μητρός μου was an insertion.(*S.Pétridès*, reviewing the article by Cumont mentioned above, in Échos d'Orient (1904) 63 f.) This interpretation was also accepted by *V.Grumel*, Les invectives..., l.c., 183, n.6. It is clear, however, that ἐλθόντος τούτου ἐν τῇ ὑφ' ἡμᾶς ἐπισκοπῇ 'Ακμονείᾳ here refers to Romanus Argyrus, mentioned in the previous sentence as being a judge in the theme of Opsikion. The text describes his arrival in the diocese of Acmonia with some interesting details. He was accompanied by the *strategos* of the theme of Opsikion (here given the title of *archistrategos*). We learn that subordinates (οἱ τάξαντες) put up the judge's tent (καλύβην) close to (that of) the archistrategos (πλησίον τοῦ ἀρχιστρατήγου). The term οἱ τάξαντες here seems to be used for the well-known term οἱ ταξεῶται, meaning police employees in the services of the judge or of the strategos of the theme. (For examples of οἱ ταξεῶται in this sense see *Hélène Glykatzi-Ahrweiler*, Recherches sur l'administration de l'Empire byzantin aux IXe - XIe siècles. in: Bull. de corresp. hell. 84 (1960) 71).

While Euthymius and his mother put their case before the judge of the Opsikion theme (we are not told what it was about), John Tzurillas was brought in to answer a charge of rape.(Ed. *Ficker*, 68 f.) A bishop who was present, 'Οβαρνακουμέων, allowed himself a bitter sarcastic comment on a monk who wore the "holy" cloth to snow off on such an occasion before so large a lay audience (ἐπὶ τοσούτου λαοῦ κοσμικοῦ). It appears to have been the bishop of Acmonia, although the name may give rise to various suggestions - see *Ficker*, op. cit. 68, n. 16. The bishop's derision turned to amazed indignation when he was told by Galenos, the archdeacon and *oikonomos* of the Acmonian diocese, that the man before him was not a true monk but the leader of the heretics (πρῶτον τῶν αἱρετικῶν). Surprisingly enough, Euthymius does not tell us what the judge's verdict was. It is of prime importance for us, however, to see that in the first quarter of the eleventh century Tzurillas was already openly known as a heretic, which gives us the right to assume that the Phundagiagite heresy had already taken root in this region of Asia Minor in the previous century.

V.Grumel gives the year 1015 as the latest possible date for the birth of Euthymius of Acmonia. He starts from the assumption that he would have to reach the age of 16 in order to enter the monastery of Peribleptos, and that he did so when the monastery was first founded in 1031. Grumel admits that this limit should probably be moved several years backwards, since in the last year of the reign of Basil II (1025), the latest at which Euthymius could have witnessed the appearance of John Tzurillas before the judge, the former must have been more than ten years old (in view of the words he remembered hearing on this occasion, which would suggest that he was no longer a child). Grumel finally accepts the year 1010, proposed by S. Pêtridès, whose arguments are unfortunately vitiated by his identifying Euthymius of Peribleptos with Euthymius Zigabenus. (*V.Grumel*, Les invectives..., l.c.184.) We are of the opinion that this *terminus ad quem* (for the date of birth of Euthymius) will have to be pushed even further into the past in view of the fact that when Romanus Argyrus came to the diocese of Acmonia as a judge (i.e. 1025 at the latest), the young Euthymius was not merely a passive observer at the trial of John Tzurillas, but himself (together with his mother) was engaged in a lawsuit against someone.

10 ... ἐν τοῖς τῶν Θρᾳκῶν μέρεσιν ἔν τε τῇ τοποθεσίᾳ τῆς Σμύρνης καὶ ἐν ἄλλοις πολλοῖς τόποις ... (ed. *Ficker*, 67,10 - 11). Ficker correctly points out that the theme mentioned here is not the European theme τῆς Θρᾴκης, but that of τῶν Θρᾳκησίων in Asia Minor (*Ficker*, Phundagiagiten, 249, Note 2). Cf. *Hélène Glykatzi-Ahrweiler*, L'histoire et la géographie de la région de Smyrne entre les deux occupations turques (1081-1317) particulièrement au XIIIe siècle, in: Travaux et Mémoires 1 (1965) 1 - 204.

11 This large village was probably called Βάτος. This is the most likely explanation of why Tzurillas and his companion Racheas are called Βατηνοί in the epistle of Euthymius (ed. *Ficker*, 37,9; 84,4). This etymological explanation of the word Βατηνοί is suggested by Ficker himself (op. cit., 194), although he promptly rejects it on the grounds that the village in which John Tzurillas actually lived was called Χίλιοι καπνοί. This is a misunderstanding, however; the passage in question reads:"Εστι γὰρ τὸ χωρίον, ὡς οἶμαι, Χιλίων Καπνῶν, καὶ ὡς παρ' ὑμῶν ἀκριβῶς ἔμαθον, ὅτι πλείους τῶν δέκα Χριστιανοὶ οὐ κατελήφθησαν ... ...(ed. *Ficker*, 67,18 - 68,1). D.Angelov points out that Euthymius would not use the genitive here if it was really a place name. He correctly interprets the words χωρίον ... χιλίων καπνῶν as a village "of a thousand fires", i.e. peasant households (in connection with the καπνικόν tax). (*D.Angelov*, Bogomilstvoto v Bălgarija, 385 ff., n. 59). This is the clear meaning to be deduced from the whole context.

12 Euth. Acm., ed. *Ficker*, 21 ff.,37. The disciple of Euthymius here (μαθητής) refers to a young monk, a novice (ὑποτακτικός), in the spiritual charge of an older, experienced monk. See *M.Jugie*, Phoundagiagites et Bogomiles, l.c., 259.

13 Ibid., 7,15 - 16; 8,22 - 23.

14 Ibid., 66,7 - 9,19 - 21; 69,12 - 14.

15 Ibid., 66,16 - 21.

16 Ibid., 22.

17 Ibid., 33,13 - 19; cf. 32,1 - 4.

[18] Ibid., 34,11 - 16; 38,17 - 22:ὅτι πᾶσα ἡ ἐξουσία τῶν ἐπιγείων εἴτε κόλασις καὶ ὁ παράδεισος εἰς ἐξουσίαν εἰσὶν τοῦ ἄρχοντος τοῦ κόσμου τούτου, ἤτοι τοῦ διαβόλου, καὶ ὅτι τοὺς φίλους αὐτοῦ βάλλει εἰς τὸν παράδεισον, τοὺς δὲ ἐχθροὺς εἰς τὴν κόλασιν, καὶ ὅτι οὐδὲν κοινὸν ἔχει μετὰ τοῦ θεοῦ, ἀλλ' ὅτι ὁ θεὸς ἄρχει τῶν ἐπουρανίων, καὶ ὁ τοῦ κόσμου ἄρχων τῶν ἐπιγείων.

[19] Ibid., 33,25 - 34,11.

[20] Ibid., 35,18 - 37,1.

[21] Ibid., 38,15 - 16.

[22] Ibid., 30.

[23] Ibid., 24,9 - 25,6.

[24] See Note 8; cf. chap. VIII, Note 2.

[25] Euth. Acm. (Vatican text) ed. PG 131, col. 56BC.

[26] Euth. Acm., ed. *Ficker*, 37,21 - 22, PG 131, col. 56C. See the 13th article of the abjuration formula, *Tollius* 122 (= PG 131, col. 45D). Cf. *L. v. Thallóczy*, Wissenschaftl. Mitteilungen aus Bosnien u. d. Hercegovina III (1895) 367, referred to below as Thallóczy.

[27] Euth. Acm., ed. *Ficker*, 24,21 - 22.

[28] Ibid., 25,4 - 6.

[29] Ibid., 24,6 - 9. We do not know the text of the σατανικὴ ἐπῳδή, which Euthymius says the heretics called the "Revelation of St.Peter" (ἀποκάλυψις τοῦ ἁγίου Πέτρου); he claims, however, that their ἐπῳδή came not from St.Peter the Apostle, but from the heretic Peter Gnapheus, who earned the name of Lycopetrus. (Euthymius's information on Lycopetrus was included in the first article of the abjuration formula (ed. *Tollius*), where the Bogomils are in fact called Lycopetrians. This Peter Gnapheus was a historical person, the Patriarch of Antioch (468-470), who added the famous words ὁ σταυρωθεὶς δι' ἡμᾶς to the liturgical hymn Trisagion. This addition aroused a storm of protest in Constantinople, and Peter was accused of heresy; fanatical supporters of Chalcedon even went so far as to forge letters of excommunication against him (See La Narratio de rebus Armeniae, ed. *G.Garitte*, Louvain 1952, 113 ff.). The Armenian Church adopted this addition to the Trisagion hymn at the Council of Dvin (553) (ibid., 130). For the name Lycopetrus and the legend attached to it, as given by Euthymius of Acmonia (ed. *Ficker*, 51-56), see *Ficker*, op. cit., 215 ff.; cf. *Gouillard*, L'hérésie, 317 f.. See also *Garitte*, op. cit., 270. On the individual hidden under the figure of Peter's disciple Sergius (whose story was interpolated into the text of Euthymius (ed. *Ficker*, 57-58) from Κατὰ 'Αρμενίων - XXIIIrd chapter in the Panoplia dogmatica of Euthymius Zigabenus (PG 130, col. 1188, 1189), see *Ficker*, op. cit., 219 ff.; *Gouillard*, loc. cit. According to Gouillard, Euthymius of Acmonia "a désastreusement combiné un Sergios (Tychikos) paulicien avec un Pierre de nuance analogue, dont il ne savait rien, en s'inspirant d'un couple Pierre-Sergios déniché dans une liste de monophysites" (*Gouillard*, L'hérésie, 318; cf. Synodikon, 234). The explanation put forward by *Ficker*, however, is more convincing; he suggests (Phundagiagiten, 214) that the sect worshipped the apostles Peter and Paul, and that Euthymius of Acmonia was trying to convince his readers that in reality they were worshipping Peter Gnapheus and Paul of Samosata. This is one more example of this writer's method of turning facts upside down.

[30] See our study, Listy filol. 90 (1967) vol. 2, 120, n. 17.

[31] See *O.Casel*, Die Mönchsweihe, Jahrb f. Liturgiewiss. 5 (1925) 1-47; *P.Oppenheim*, Mönchsweihe und Taufritus. Ein Kommentar zur Auslegung bei Dionysius dem

Areopagiten. Miscell. liturg. in hon. L.C.Mohlberg I (1948) 259-282.

[32] There is a reference to this in the eleventh anathematizing article of this anti-Bogomil synodicon (*Tollius*, 120 = PG 131, col. 45B). H.-C.Puech had already drawn attention to this important evidence (see *Puech - Vaillant*, Le traité, 255). The eleventh article explicitly names the ἐπίκλησις, which Puech rightly identifies with the ἐπῳδὴ σατανική (*Puech - Vaillant*, op. cit., 255, n. 6).

[33] References to the Byzantine texts are given by Puech (*Puech - Vaillant*, op. cit., 150, n. 3; 151, n. 1).

[34] Euth. Acm., ed. *Ficker*, 77.

[35] Idib. 28,9 - 20 (baptism and the eucharist); 74,22 - 23 (the eucharist). See also the 10th and 11th articles of the anathema in Tollius's text (*Tollius*, 120; cf. *Thallóczy*, 366) on baptism, and the 12th article (*Tollius*, 122, cf. Thallóczy, 366) on the eucharist.

[36] Euth. Acm. (Vatican text), PG 131, col. 56B.

[37] Ibid., col. 56AB; ed. *Ficker* 27 f. The accusation in the Vatican text that heretical priests were guilty of μίξεις αἰσχράς in the proximity of the altar itself is found nowhere else.

[38] Euth. Acm., ed. *Ficker*, 26,17 - 27,12. The μιαρὰς πράξεις hinted at in the Vatican text (PG 131, col. 56A) perhaps refers to similar cases.

[39] See my study: Certains aspects du bogomilisme byzantin, ByzSlav XXVIII(1967) 40 ff.

[40] This habit of Byzantine writers confused modern scholars for a long time. It remained for *G.Ficker* (Phundagiagiten, 235-243) and H.-C.Puech (*Puech - Vaillant*, Le traité, 292-304) to lay solid foundations for the study of the relationship of the Bogomil heresy to the Messalians.

[41] Michael Psellus writes about these "Euchitae" at the beginning of his demonological treatise Περὶ ἐνεργείας δαιμόνων διάλογος ed. *J.F.Boissonade*, Nuremberg 1838, 1-10 (= PG 122, col. 824-833A). Schnitzer had already realized that these were Bogomils (*Schnitzer*, Die Euchiten im XIten Jahrhundert, in: Studien für Geistlichkeit Würtembergs XI, 1 (Stuttgart 1839) 183; I was not able to consult this work). H.-C.Puech also stresses the fact (*Puech - Vaillant*, Le traité, 326.) In the archaizing spirit of Byzantine writers Psellus is careful not to use any new names, giving the sect the name of Εὐχῖται, well known from earlier heresiological literature. (For similar cases of tendency to archaism in the heresiological writings of Byzantium see *J.Gouillard*, L'hérésie, 303). Theodore of Andida (twelfth or eleventh century) identifies the sect of "Phundaites" with the Euchites or the Messalians in his liturgical commentary quoted above.

[42] Euth. Acm., ed. *Ficker*, 3,12 - 13; 65,6 - 10.

[43] ... καὶ περιτρέχουσιν πᾶσαν τὴν τῶν 'Ρωμαίων ἐπικράτειαν καὶ εἰς ὅσους ὁ ἥλιος ἐφορᾷ Χριστιανοὺς ... Ibid., 63,25 - 26. They preach their heresies in the capital and in the provinces: Περινοστοῦσι γὰρ οἱ τοιοῦτοι ἀσεβεῖς, διδάσκοντες πανταχοῦ ἐν τῇ πόλει, καὶ ἐν τοῖς ἔξω. (Vatican text, PG 131, col. 52 C). Cf. also ed. *Ficker*, 3,11 - 13; 4,18: ... τὴν τῶν ἀθέων Φουνδαγιαγιτῶν ἀσέβειαν ... τὴν νυνὶ κρατοῦσαν εἰς πᾶσαν τὴν τῶν Χριστιανῶν ἐπικράτειαν, ...

[44] Ibid., 64,3 - 7.

Chapter VI

THE BOGOMILS AND PAULICIANS IN CONFLICT WITH THE BYZANTINE STATE POWER

Soon after the fall of Preslav (972), the Bulgarian Empire revived for more than four decades (976-1018), but the centre of political and religious life moved to Macedonia, to the region of Ohrid and Prespa. Byzantine sources tell us - albeit not in full - the exciting story of Tsar Samuel's epic struggle against Byzantium[1] until his death (1014), and the ruins still bear witness today to the greatness of Samuel's empire, consciously reviving the traditions of state and church of Symeon's time. Three charters of the Emperor Basil II have come down to us, dating from the first years after his conquest of the Bulgarian Empire; they grant to the Archbishop of Ohrid ecclesiastical jurisdiction over the whole territory of the former Bulgarian patriarchate, list the individual dioceses and confirm the continuity of their rights since the time of Peter and of Samuel.[2] This gives us some idea of the ecclesiastical organization of Bulgaria at the end of the First Empire, but we have no reliable evidence about the heretical movement in the Bulgarian-Macedonian region under the Ohrid Tsars.

Later, when the successors to Basil II (d. 1025) had forsaken his wise and moderate policy towards the conquered Bulgarians, and Greek archbishops had been established in the see of Ohrid, one of the epistles of the archbishop Theophylact devotes at least a passing reference to the appearance of heresy in the region under his ecclesiastical rule.[3] Even in Theophylact's Life of St.Clement we find the remark that heresy had spread among his flock after the saint's death.[4] Apart from these hints, however, the sources are silent for a long time about the fate of the Balkan Bogomils and Paulicians after the whole of the peninsula came once more under the rule of Byzantium (1018). It is well known that this rule was soon subjected to violent upheavals. The appointment of Greek archbishops to the Bulgarian see was by far not the only cause of discontent, brewing up in what had been Samuel's empire. After the death of Basil II the bureaucracy of Constantinople acquired a decisive voice in affairs of state, and bore down relentlessly on the conquered Slav regions with innumerable fiscal demands. Under the Ohrid and still earlier under the Preslav Tsars the Bulgarians had paid their dues in kind; now money was inexorably demanded from them.[5] Twice the widespread discontent burst into open rebellion (1040 and 1072), and in each case the rising was clearly inspired by the traditions of the independent Bulgarian Empire.[6] It can justifiably be assumed that the Bogomil sect which under Tsar Peter called for civil disobedience with regard to all in power, now turned the brunt of its criticism against the foreign oppressors. We are forced to rely on conjecture, however; nor have we any record of the heretics taking part in these anti-Byzantine rebellions.[7] Nevertheless at the end of the eleventh century the Balkan Paulicians appeared on the scene again at a moment which was highly critical for the Byzantine Empire. The civilian bureaucracy ruling in Constantinople had managed to undermine the military power of the Empire, causing the loss of the whole of its Asia Minor territories to the Seljuq Turks for a time. Not until a new dynasty, that of the Comneni, came to the throne in 1081, did things improve. They represented the military aristocracy, and succeeded in arresting the catastrophe and creating the necessary conditions for the Empire to become strong again. At the time when the first of the line, Alexius Comnenus I (1081 - 1118), came to the throne, there could as yet be no thought of regaining dominion in Asia Minor, because all forces had to be concentrated in the west to drive off the Normans. The leader of the Sicilian Normans, Robert Guiscard, attacked the eastern Adriatic coast and laid siege by land and sea to the city of Dyrrhachion (Durazzo), a key point on the road to Constantinople. Not even the help of Venice, which asked a high price for assistance in the form of far-reaching concessions, could prevent the Normans from gaining control of Durazzo for a time.[8] It was on this occasion that the Paulicians of Plovdiv played a fateful role; even in their new settlements they had remained a most unreliable element. At the critical moment their troops deserted the battlefield at Durazzo. As soon as victory over the Normans left his hands free, the Emperor Alexius took his revenge. The Paulician leaders were taken and imprisoned by a trick, their possessions confiscated, and their families treated very cruelly.[9] These stern measures did not completely solve the problem, however. The daughter of the Emperor Alexius, the learned Princess Anna, describes in her history the situation in the Paulician colony round Plovdiv, which had got very much out of hand.[10] Since the days of the Emperor John Tzimisces the numbers of the Paulicians there had grown considerably, and they lived very much as they liked, taking little notice of the orders of imperial officials. The heterodox element predominated in the varied population of Plovdiv itself and of the surrounding country; besides the Paulicians there were both followers of the Armenian church and Syrian Monophysites (Jacobites). In this context Anna Comnena also mentions the Bogomils, who formed a separate community distinct from the Paulicians.

The persecution of their individual leaders did not dismay the Paulicians of Plovdiv - indeed, it roused them to open rebellion. To a certain degree the history of the Paulician revolt of the ninth century was thus repeated in the Balkans. In this exposed region of the Byzantine Empire, too, the Paulicians were in a strong position due to their military organization and the opportunities for alliance with the enemies of Byzantium. Around the middle of the eighties they were making raids on the surrounding regions from the stronghold of Belyatovo near Plovdiv. The most dangerous thing about them, for the Byzantine state, was their alliance with the Pechenegs, Turco-Tartar nomads who were a constant threat to the Empire from the middle of the eleventh century. Traulus, a Paulician leader who married the daughter of one of the Pecheneg chieftains, instigated dangerous raids on Byzantine Thrace by the nomad hordes.[11] The army sent against them was cut to pieces, and the commander of the Byzantine armies in the Balkans, Gregory Pakurianos, a Georgian, himself fell in the battle. The struggle against the Pechenegs lasted for several years, until they were terribly defeated in 1091 by Byzantium allied with the Cumans. Force of arms alone, however, could not overcome the Paulicians of Plovdiv. It was not until late in the year 1114, when he lay in Plovdiv awaiting a Cuman attack, that the Emperor proceeded to deal with the heretics.[12] Anna Comnena describes his tireless disputations with the spiritual leaders of the Paulician sect, who skilfully refuted his arguments for days on end; two of them thereafter ended their days in prison. Dealing with the mass of the Paulicians, however, the Emperor successfully backed up his pressure with material inducements which were meant to ensure that their rebellious mood would change for good. The more important among them were given rich presents and made commanders in the army, while the ordinary people, fitted for agricultural labour, were settled on the further bank of the river Marica (Maritsa). A great new settlement grew up there, an urban centre which was given the name of Neocastrum. The fields, vineyards, houses and other property all became their own and imperial chrysobulls secured it to their heirs.

In the absence of reliable historical records we cannot guess how strong the Paulician sect was in and around Plovdiv after this, but remnants were still cropping up in Bulgaria centuries later.

Anna does not tell us whether her father also had dealings with the Bogomils when he was in Plovdiv, but gives a detailed description in her "Alexiad" of the prosecution of a number of Bogomils in the capital itself, some time at the beginning of the twelfth century.[13] The daughter of Alexius dramatically exaggerates her father's role in the affair, as if he had been the first to bring to the light of day this evil which had been cowering in the darkness. She nevertheless gives us a faithful picture of the heretics as we have known them, appearing everywhere in their monk's habit. The sect gained many adherents in Constantinople itself during the reign of Alexius, being well-received even in noble houses.[14] At last evidence came to light, pointing to the person of the principal teacher of the Bogomils, one Basil, who called his twelve chief disciples "apostles". There were also a number of women with him, fanatically devoted to his teaching. The manner in which Alexius Comnenus proved the heretic's guilt was typically Byzantine. With skilful deceit he managed to allay the old man's suspicions and persuaded him to set out in detail the complete doctrines of the Bogomil sect. The Emperor Alexius and his brother the Sebastocrator Isaac, listened to Basil while a scribe hidden behind a curtain took down all he said. This evidence was of course quite sufficient for the Permanent Synod, the session of which was attended by members of the Senate and military dignitaries as well, to condemn him as a heretic. According to the account in the "Alexiad" the Emperor used yet another ruse in order to detect the real heretics among those who were taken into custody along with Basil. They were all condemned as it were to death, and allowed - believing death was imminent - to choose between a pyre with a cross or one without; thus they revealed their orthodox or heretical beliefs. Basil himself, however, was really offered the choice between life and death. As when a group of heretics from Monteforte was burned at the stake in Milan in 1028, so now in the hippodrome of Constantinople a cross was raised; the heresiarch, perhaps dismayed at the sight of the flames, could thus repent and turn to the symbol of the true faith. According to Anna Comnena Basil did indeed waver when he came near to the great pyre whose flames thundered up into the air and cast their sparks as high as the obelisk in the hippodrome. The Bogomils madly believed that a miracle would take them safely through the flames. Nonetheless Basil's cloak, cast on to the flames as a test, caught fire at once, and straightaway the fiery element devoured the body of the Bogomil teacher in a moment. Basil's disciples were never released from prison. His doctrines were carefully noted down, however, by the learned monk Euthymius Zigabenus, who was entrusted by the Emperor Alexius with the task of compiling a work setting out all the heresies.[15]

1 *Ostrogorsky*, Geschichte^3, 249 ff.

2 The authenticity of these letters has recently been questioned by some Yugoslav scholars. See *S.Antoljak*, Samuilovata država, Skopje 1969, 72 ff.

3 'Επιστ. ΜΑ', PG 126, col. 444D - 445A.

4 See *N.Tunickij*, Grečeskoje prostrannoje žitije sv. Klimenta, Sergijev Posad 1918, 138,17 - 19.

5 See *G.G.Litavrin*, Bolgarija i Vizantija v XI - XII vv., Moscow 1960, 309-351, 363-374.

6 Ibid., 376-410; *Ostrogorsky*, Geschichte^3, 268 ff., 285 ff.

7 The chronicles record an attempt at revolt at the beginning of the reign of Nicephorus Botaneiates, instigated by Lekas. The continuation of Scylitzes states that he was a Plovdiv Paulician (Scylitzes, ed. Bonn, II, 741), but Michael Attaliates, ed. Bonn, 302, did not refer to his origin. Both authors (the first of whom relied on the second) tell us that Lekas was allied with the Pechenegs, to whom he was related. The continuator of Scylitzes also wrote that Lekas killed the bishop of Sardica, who had called on the city to remain loyal to the Emperor. Both sources mention a certain Dobromir in this context; he had apparently attempted to lead a rising in Mesembria. In the end both rebels took fright at the consequences of their deeds and submitted to the Emperor, who rewarded them with gifts and high office. See *Zlatarski*, Istorija II, 495.

8 *Ostrogorsky*, Geschichte^3, 294 ff.

9 Alexiad V, t.II, 14, 45, See *F.Chalandon*, Essai sur le règne d'Alexis Ier Comnène, Paris 1900, 81; *Zlatarski*, Istorija II, 172; *Litavrin*, op. cit. 417 ff.

10 Alexiad XIV, t.III, 178 ff.

11 Alexiad VI, t.II, 49. See *Chalandon*, op. cit., 105 ff; *Zlatarski*, Istorija II, 180 ff.; *Obolensky*, Bogomils, 191 f.; *Litavrin*, op. cit., 418 ff. According to Anna Comnena Traulus, a servant of the Emperor, was a Paulician who had received baptism; he was moved to rebel by the imprisonment of his four sisters during the persecution of the families of Paulician leaders. Traulus allied himself with the Pechenegs settled in the theme of Paristrion (Paradunavon). Tatos was also allied with the Pechenegs at this time (he was known as Chalis), as were Sestlav and Satza; all three had set up independent princedoms in this theme (see *E.Condurachi, I. Barnea, P.Diaconu*, Nouvelles recherches sur le ,limes' byzantin du Bas-Danube aux Xe-XIe siècles, in: The Proceedings of the XIIIth International Congress of Byzantine Studies (Oxford 5-10th September 1966), OUP 1967, 193.

12 Alexiad XIV, t.III, 181 ff. See *Obolensky*, Bogomils, 193 ff; *Litavrin*, op. cit., 423 ff; *Angelov*, Bogomilstvoto v Bâlgarija, 400 ff.

13 Alexiad XV, t.III, 218-228. On the date of this event see H.-C.Puech in: *Puech - Vaillant*, Le traité, 288, Note 1; *Obolensky*, Bogomils, 275 f.; *Denise U. Papachryssanthou*, La date de la mort du sébastocrator Isaac Comnène, frère d'Alexis Ier et de quelques événements contemporains, in: REB XXI (1963) 250-255: The *terminus ad quem* is the date set by the author herself for the death of Alexius's brother Isaac, who took part in the questioning of Basil. Isaac the sebastocrator died in 1102 or in November 1104 at the latest. - Anna Comnena left a space in her text, for the insertion of the correct date of the condemnation of Basil (p.218, 28). This is not the only time she did this (see *Georgina Buckler*, Anna Comnena, Oxford 1929, 251; cf. *Obolensky*, op. cit., 275).

14 Alexiad XV, t.III, 224,6 - 7.

[15] Panoplia dogmatica, PG 130, col. 20-1360. The XXVIIth chapter deals with the Bogomils (col. 1289C - 1332D. (referred to further as PG 130). A variant version preserved separately in Cod. Graec. 3 univ. Rheno-Traiectinae, fol. 80a - 96b, was printed by *Ficker*, Die Phundagiagiten, 89 - 111 (quoted as Euth. Zig., ed. *Ficker*).

Chapter VII

THE BOGOMIL MYTH AS PRESENTED BY EUTHYMIUS ZIGABENUS

It took Basil fifteen years to learn all the secret lore of the sect, and then he taught it for fifty-two years[1] before he fell into the hands of his persecutors. This at least is what one version of the work by Zigabenus says, while the other version says no more of the duration of the Bogomil leader's work than that it lasted "more than forty years".[2] We need not perhaps take these figures at their face value, but at all events the Bogomil doctrines as preached by Basil certainly went far back into the eleventh century.

Indeed the demonological tract by Michael Psellus already attests certain features typical of that form of the Bogomil cosmological myth presented to us in the record given by Zigabenus of what Basil revealed. Psellus wrote his work about 1050 or 1060[3] and it thus dates from approximately the same period as the anti-heretical invective of Euthymius of Acmonia. What the latter gives us, however, is decidedly a more primitive version, and apparently a more archaic one, whereas Zigabenus has put on record a more highly developed stage of the myth.[4]

Euthymius Zigabenus gives us, in the first place, a much clearer idea of the beginning of the cosmic story.[5] The good God created myriads of angels, among whom the most important was Samael or Satanael.[6] Both these names, the first taken from Jewish demonology, the second artificially formed in a Greek environment, came into the Bogomil myth from the apocryphal tradition. The idea of Satanael as the first of the angels also appeared in apocryphal literature[7] but in the minds of the Bogomils described by Zigabenus the idea is heightened. Satanael appears in the same form and the same robes as God the Father, and sits at His right hand. His fall was due to his arrogant desire to be equal with God.[8] In accordance with the widely known apocryphal tradition the Byzantine Bogomils leaned in this on the well-known verse in the prophecy of Isaiah (Isaiah 14, 14): "I will ascend above the heights of the clouds; I will be like the most High".

We know that some of the Bogomils whom Cosmas met called the evil creator of the world a fallen angel, while others likened him to the "unjust steward" in the Gospel of St.Luke. In the Byzantine myth the two ideas merged; we read that Samael was made steward over all "the powers of Heaven", in which office he tried to win some of the angels in his charge to join in his rebellion against God, and to that intent lightened the "burden" of their liturgical duties. The motif has clearly been taken from the parable in St.Luke's Gospel, but the sequence of events is different. While in the parable the steward did not reduce what his lord's debtors owed until after he had been accused by his master, Satanael acted in this way before his fall.[9] The conspiracy became known to God, who cast out Samael and the angels he had corrupted. Samael and his company did not remain in the void, but fell to earth; the earth was an inhospitable place, its face covered with the waters described in the Book of Genesis. It was essential for his self-preservation that Samael should use his creative powers, which like his divine appearance still remained to him. He therefore created the "second heaven", i.e. the visible sky, ordered the waters to their rightful place and arranged the earth so that he and the other fallen angels could inhabit it.

In this version of the Bogomil myth, too, Satanael's success begins to waver when the time comes to bring Man to life. In this version, however, the evil creator does not possess the soul stolen from God. He is reduced to breathing life into the body himself, but all he can manage to bring to life is the serpent. His ingenuity is of no avail, and all he can do is come to terms with God, that He may give the first men souls.[10]

The beliefs concerning the creation of man as described by the earlier and the later Euthymius have their analogy in Gnostic myths, but nevertheless the conception of the origin of the human soul is essentially different in Bogomilism and in Gnosticism. The soul is no longer a *part* of God Himself, but only His *creature*. Mediaeval dualist doctrine therefore no longer contained the conception of God's self-liberation. The ultimate aim of the cosmic drama was the restoration of the state of the universe before the fall of the angels disturbed it. The origin of the human soul was therefore always bound up in one way or another with the fallen heavenly powers. The agreement between Satanael and God described in the myth the Byzantine Bogomils believed is only one of the forms which this basic motif could take. Its substance was that the descendants of the first human beings, to whom God had given souls, would take the place in Heaven left vacant by the fallen angels.

As was only to be expected, Samael promptly broke his promise, and began to lay snares for the first human beings, out of envy. His motive recalls the rabbinical legends about how Samael tempted the first human beings, although the Christian tradition, too, attributed envy of man to Satan. There was also another motif which undoubtedly came from the Jewish legends of Samael, and may have come into the Bogomil myth through the medium of an apocryphal book; this was the birth of Cain and his sister Calomena from the union of Samael and Eve.[11] It was only after this act, which was particularly abhorrent to the ascetic sect, that Satanael lost his divine appearance and apparel, and his face became dark and uncomely. Thus in the Bogomil myth the loss of the divine glow, which in the apocryphal literature came immediately after the fall, was pushed forward to a later stage of the cosmic drama.[12]

Satanael now lost even his powers of creation (δημιουργικὸν χάρισμα), but the good God left him dominion over the world he had created. The Prince of this world, however, still repaid Him ill. With the help of Cain's "diabolic" seed he began to tyrannize over and annihilate the descendants of Adam and Eve, so that only a few human souls were left who could take the place of the fallen angels. In the end God realized that He had been deceived, and filled with compassion for the suffering souls of humanity decided to take energetic measures. It was only now that the Logos – the Son of God – was called into being; so far there had been no suggestion of His existence. This also explains the culmination of Bogomil ideas on Satanael's position in Heaven; it was only because Christ was not there in the early stages that Satanael could have divine attributes and sit on the right hand of God. Hence it was but a step to the secret doctrine which made Satanael the first-born son of God.[13] The second son, the Logos - Jesus Christ - did not appear until the middle of the sixth millennium from the creation of the world. The widely-held belief that Christ was born "per auditum" was bound up with the conception of the Logos (the Word); the Bogomils, too, believed that Christ was not really born of Mary but that he entered the world through her ear and took upon himself only the appearance of human form. His sufferings and resurrection were also no more than apparent. Finally Christ laid aside His human form, overcame the Evil One and imprisoned him in fetters, in Tartarus. It was at this point that Satanael lost the characteristic *-el (il)* of angelic names and became known from then on as Satan.[14] This change, too, is connected with the fall of Satanael in the apocryphal myths, but in Bogomil mythology it was the result of his ultimate defeat. After this victory Christ took Satanael's place at the right hand of God, and in the secret doctrine He also took his rank as the elder son. Thereupon he merged with the person of the Heavenly Father, whence he had emerged. The only *raison d'être* for this Son of God was to bring salvation to man, culminating in the victory over Satan, and therefore God appeared in three Persons only for this period of thirty-three years. It is clear from Zigabenus's account that Bogomil views on the Trinity varied considerably. In addition there are two distinct Christological conceptions combined in a somewhat forced manner.[15] The idea of the Logos emanating from God and returning whence it came is mingled with the idea of Christ as the Archangel Michael, victorious over Satan. The term Archangel was explained by the superiority of Christ's divinity over the nature of all the angels. This is clearly an echo of the opening chapter of the New Testament Epistle to the Hebrews. The Bogomils also used the term "Angel of great counsel" (Isaiah 9,6).[16] It is readily understandable that the Bogomils identified Christ with the Archangel Michael, whose role in Jewish-Christian traditions is that of commander *(archistrategos)* of all the heavenly forces. The dualist myth thus found a point of contact with the legends of the struggle between Michael and Satan. An apocryphal tradition widespread in eastern Christianity showed victorious Michael taking the place of the fallen Satan in Heaven.[17] The similarity between this tradition and the Bogomil myth is very clear, and it is highly probable that the sect simply took over the legendary story and attached it to the figure of their own Christ-Michael.

Besides these differences in Christology the Byzantine Bogomils were not very consistent in their view of Satan's present fate. Zigabenus has it that the sect believed that in His infinite goodness (against which both the Son and the Holy Ghost were powerless) God had not taken the demons' powers away from them, and had left them dominion over the world to the end of time. Their home was in the churches, where the icons were placed in their honour. Since the evil spirits still had great powers to harm human beings, it was prudent to bow down to them and not to avoid taking part in the sacrificial rites. (In other words: the ruling Church was still too powerful, and it was not wise to invite persecution.)[18] The churches were divided among the demons according to their rank and power. As the commander-in-chief of all the evil spirits, Satan first had his headquarters in the Temple of Jerusalem, later moving to occupy the church of St.Sophia in Constantinople.[19] This certainly conflicts with the idea of Satan fettered and imprisoned in the underworld.

It would appear from Zigabenus that the Bogomils' ideas about the Second Coming and the Day of Judgment, and about the final fate of Satan and the other demons, did not differ much from current ideas, and that it was only in the doctrine of the resurrection of the dead that the sect profoundly dissented from the orthodox belief. There is no systematic exposition of Bogomil eschatology in the work of Zigabenus, however.

In spite of contradictions and confused passages the basic theme stands out clearly; the whole myth is constructed round the story of the fall of Satan. Roughly sketched in what Euthymius of Acmonia writes, the myth is by now highly developed. The whole mythical story bears specific features of biblical stories and apocryphal legends, and we feel far away, now, from the abstract world of the Gnostic eons. The fact that the prince of this world has been identified with the fallen angel means that the idea of ignorance automatically disappears; a typical feature of the powers that rule the world in all the Gnostic doctrines, it would make no sense in the Bogomil myth. How could Satanael be ignorant of the existence of the good God, when he had been in His service before his fall?[20]

The explanation of the origin of evil by the falling away from God of the hitherto good angel certainly represents a significant measure of adaptation of the dualist myth to orthodox Christian doctrine. Nevertheless the author of all evil continued to fulfil the function of the Creator of the Old Testament and stood uncompromising and defiant towards the God of the invisible heavenly world. Thus the basic dualist conception originating in Gnosticism is preserved and dominates the whole of the myth. The original approach of the sect can be seen clearly in the way it was built up, motifs from the Bible or apocryphal sources being boldly incorporated into this new context. This originality can be seen in the free treatment of the parable of the unjust steward, taken from St.Luke, as well as in the "degradation" of Satanael, which is spread over three stages. As we have already seen, the Bogomils even took over current demonological ideas, giving them a new significance in their anti-orthodox teachings.

Euthymius Zigabenus also gives us plenty of insight into the way the Bogomils manipulated biblical texts. In his day the Bogomil canon was already broader than in the times of Cosmas, the Psalms and the Old Testament prophecies having been added to the four Gospels and the Apostolikon.[21] Here, too, there is a measure of adaptation to the practice of orthodox Christianity, but it should not be forgotten that the whole text of these seven books of the Bible[22] was interpreted by the Bogomils in a highly original manner. The Byzantine writer gives us a long series of examples from the Gospel according to St.Matthew to demonstrate this very curious biblical exegesis.[23]

We see how individual sentences or phrases were arbitrarily taken out of their context and in isolated cases even the wording of the text was distorted to fit the meaning desired by the interpreter of the passage. In this way quotations could be found to support individual elements of the dualist myth in all the seven books of the Bogomil canon and even from the Book of Genesis which they rejected as such.[24]

Many quotations from the Bible were interpreted as evidence of the guilt of the Orthodox Church, its priests, monks, hierarchs and Church Fathers. On the other hand the Scriptures were used to "prove" that the Bogomils were the only society of true Christians.[25] We already found this tendency among the Paulicians, although the material provides only one instance. It is remarkable, however, that both the Paulicians and the Bogomils give the same interpretation of the very same passage in the Gospel according to St.Matthew.[26]

We do not know whether the Bogomil commentary on St.Matthew's Gospel and the other biblical texts was handed down by word of mouth only at this time, or whether it had already been recorded. Nor is there any reference in Zigabenus to any other original literature in use among the sect. The synodical *tomos* already quoted, which undoubtedly derived from the condemnation of Basil, only mentions briefly the introduction of "other Scriptures" (i.e. besides the Bible). It is not impossible that this is an unclear reference to the "Vision of Isaiah", a well-known apocryphal book whose popularity among the Bogomils is attested by the preceding article in this *tomos*.[27]

No other Byzantine author has given us such a detailed account of Bogomil doctrines as Euthymius Zigabenus, allowing us to see them from all aspects and indeed revealing some of the hidden depths - the things which the sect kept carefully hidden from the uninitiated. Only gradually did the Bogomil teachers hand on to their new disciples this precious knowledge, like the pearls which the Gospel says should not be cast before swine.[28] Zigabenus acquired all this invaluable information from the notes taken down during the questioning of Basil, on which he based his account. With a certain amount of editing he later included his treatise on the Bogomils in the major work he was charged with compiling as a polemical arsenal against a number of heretical doctrines. The work, the *Panoplia dogmatica* of Euthymius Zigabenus,[29] is of the nature of an anthology and is very largely dependent on the works of earlier authors. The chapter about the Bogomils is the only section which is based on entirely original material and which provided the author's contemporaries with up-to-date information about this dangerous heresy, which was already flourishing in the provinces and the capital of Byzantium in the eleventh century.

For us Zigabenus, together with the earlier work of Euthymius of Acmonia, has opened up an aspect of eleventh century Byzantine civilization which has hitherto been hidden. The usual picture of the period figures such outstanding minds as Michael Psellus, John Italus, John Mauropous and – at the very beginning of the century – Symeon the New Theologian.[30] Their works were addressed only to the small intellectual elite. The anti-Bogomil writings, however, reflect the mental and spiritual world of a far broader stratum of the people, to whom this profound learning was never accessible. We find ourselves in an atmosphere where the imagination is fired by apocryphal legends and folk demonology, and where idiosyncratic interpretations of biblical texts are the result of the meditation of simple minds, untrained in the study of the Church Fathers. No artificial philosophical system could come out of such a milieu, which could only form a naive myth. Secular learning was firmly discountenanced here, and those equipped with it could not join the ranks of the heretical community.[31]

All this was in accord with the mental level and the ideas of the anonymous masses of monks, from whom the leading personalities of the Bogomil sect were indistinguishable to the onlooker. The Byzantine sources give no special designation for this leading element among the Bogomils, alongside whom there were the new adherents, who were only gradually initiated into the doctrines of the sect. Proselytism was a cautious matter, in any case. The Bogomil teachers based their preaching on the most general tenets of the Christian faith and the fundamental commandments of the Christian life, so that at first they created the impression of absolute orthodoxy. Not until they had gained the confidence of their listeners did they begin to reveal, bit by bit, their heterodox opinions.[32] There is an interesting remark in the epistle of Euthymius of Acmonia which suggests that even before the administration of spiritual baptism the adepts of the sect could pray together with the initiates.[33] We know, and Zigabenus confirms this, that the Bogomils took quite literally Christ's words about prayer, and recognised only the Lord's Prayer.[34] Indeed the sect, according to Zigabenus, regarded this prayer as the true bread of the eucharist, to which Christ was referring in His words at the Last Supper.[35]

Zigabenus helps us to corroborate the remarkable accuracy of what Euthymius of Acmonia said about the Bogomil rite of spiritual baptism. It would only seem that the Peribleptos monk let himself be so carried away by the urgency of his invective that he was a little summary in his account, and neglected to mention that the rite had two parts; Zigabenus distinguishes clearly between them. During the first stage only the book of the Gospels was placed on the candidate's head, while in the final phase, which followed after a certain interval of time, there followed the laying on of hands - the apostolic gesture which symbolized the gift of the Holy Spirit. All members of the sect, both men and women, took part in this ritual.[36]

To be "baptized with the Holy Spirit" meant that the candidate was accepted among the "elect", the "just", the "heirs of God".[37] These souls were no longer in the power of demonic forces; the demons fled from them, while they remained with other people even after death.[38] The Holy Spirit had taken up His abode in their souls and they became the bearers of the Word, the Logos, which was constantly being born through their teaching. In that sense each of the elect was the "Mother of God", a designation which belonged to them alone and was wrongly given to Mary, the supposed mother of Christ.[39] Hence, too, the fantasy of the Bogomils that at death they did not really die, but their souls left this world as if in slumber, and surrounded by a host of angels and apostles entered the kingdom of their Heavenly Father.[40] This was undoubtedly the curious Bogomil adaptation of the eastern Christian conception of the death of the Virgin Mary (κοίμησις, uspenije).

In Zigabenus we already come across the idea of the material body as a garment (later Latin sources use the word *tunica*) which the soul discards at death in order to put on the immortal "divine" garment in Heaven. There is also a hint of the conception of an immortal body of the same substance as the immaterial body of Christ. The souls which have been saved assume this immortal body in Heaven, while their earthly bodies fall to dust and ashes, and are never resurrected.[41] The rejection of the human body as the creation of the Devil and the prison of the soul while it is on earth, among the Bogomils took the form of an extremely ascetic way of life. Euthymius of Acmonia tells us that this asceticism, together with intense and earnest prayer, formed an essential part of the catechumens' preparations for spiritual baptism.[42] According to Zigabenus these preparatory practices were even more rigorous before the second part of the rite of initiation.[43] He leaves us in no doubt, however, that even after initiation a rigorous asceticism was demanded of members of the sect. They rejected married life, as well as the eating of meat and the drinking of wine, and even other animal products such as cheese and eggs were forbidden.[44] Other foods were limited by fasting, for which three days of the week were set aside.[45] The Bogomils regarded their stern way of life as one of the principal marks distinguishing them from the "lawgivers and Pharisees" of the Orthodox Church.[46] In contrast to the church they regarded themselves as the only Christian community, whose teachings were truer and where a purer life could be lived.[47] Western Europe was soon to know these heretics under the Greek name of Cathars - "the pure".

[1] Euth. Zig., ed. *Ficker*, 111,2. This information was also accepted by Zonaras XVIII, Vol. III, 743,12.

[2] PG 130, col. 1332C.

[3] See H.-C.Puech, in: *Puech - Vaillant*, Le traité 139, n. 5. The literature on the demonology of Michael Psellus is given by Beck, Kirche u. theol. Lit., 539, Note 2. Cf. our study, Satan als Erstgeborener Gottes (Ein Beitrag zur Analyse des bogomilischen Mythus, in: Byzantinobulgarica III (1969) 23 ff.

[4] See ByzSlav XXVIII (1967) 44 ff.

[5] There is a systematic exposition of the Bogomil myth in Ficker's edition 91, 24 - 94,22; Panoplia dogm. XXVII, PG 130, col. 1292 ff.

[6] See *Turdeanu*, Apocryphes bogomiles, 39 ff. There are also two Coptic texts in which the name Satanael is attested: the encomium in honour of the Archangel Michael, ed. *W.Budge*, in: Miscellaneous Coptic Texts in the Dialect of Upper Egypt (1915), 904 (English translation); and the encomium of John Chrysostom in honour of the Archangel Raphael, ibid., 1036.

[7] According to the apocryphal Gospel of Bartholomew, of all the angels God first created Satanael, then Michael, and then the rest in turn. See *Hennecke - Schneemelcher*, 368. Some Coptic texts refer to Satan as the first creature (archiplasma) to be formed. See *A. van Lantschoot*, Un texte palimpseste de Vat. copte 65, in: Le Muséon LX (1947) 266. See also *A.M.Kropp*, Ausgewählte koptische Zaubertexte II, Bruxelles 1931, 190 ff. In the Apocalypse of Sedrach Satan is called πρῶτος τῶν ἀγγέλων. See *M.R.James*, Apocrypha anecdota (Texts and Studies, ed. *J.Armitage Robinson* II, 3) 131, 20 ff. In a Greek legend of the creation of the world God says to Satanael: ἰδοῦ πρῶτον ἄγγελον σὲ ἐποίησα ... We then read of him: Ὁ δὲ πρῶτος τῶν ἀγγέλων ταχθείς, νῦν δὲ πρῶτος τῶν δαιμόνων ... *Ivanov*, Bog. knigi, 313, 315. In the Slavonic legend of the Sea of Tiberias (in the Russian versions) Satanael is in the highest position, above all the heavenly powers (ibid., 291, 300).

[8] For the various explanations of the fall of Satan given in patristic and apocryphal traditions see Byzantinobulgarica III, 26 f., n. 10.

[9] See ByzSlav XXVIII (1967) 46, 47, n. 33.

[10] The version given by Euthymius of Acmonia, according to which the "archon" of this world has to soil the soul stolen from God, with a mess of "unclean" animals in order to bind it to the material body, recalls the teachings of Basilides' disciple Isidore about προσφυὴς ψυχή (see *J.H.Waszink*, Basilides, in: Reallexikon f. Antike u. Christentum I, Stuttgart 1950, col.1220; cf. *Doresse*, Secret Books, 23, 72, 215 ff.). The version given by Zigabenus that man is the joint work of God and the Devil has its parallels in many Gnostic doctrines; see *K.Rudolph*, Die Mandäer I, Göttingen 1960, 151, n. 4 and 153 ff., n. 2. We find one variant of this myth in the Marcionite school, in the teachings of Marcus, the disciple of Marcion (see *Harnack*, Marcion, 202, n. 3). We do not know whether similar mythical ideas were preserved in some branch of Paulicianism unknown to us, from where they could have been handed on to the Byzantine Bogomils. In the present state of knowledge on the subject it is impossible to prove any continuity in this matter.

[11] See Byzantinobulgarica III (1969) 27, Note 12.

[12] See H.-C.Puech, in: *Puech - Vaillant*, Le traité, 201, n. 4.

[13] This motif is attested for the first time in Michael Psellus, at the beginning of his demonological treatise (ed. Boissonade, 3-5, = PG 122, col. 824 f.). For the whole of this question see our study in Byzantinobulgarica III.

[14] See the Greek version of the "Martyrdom of St. Paul and St. Julian", quoted

by Turdeanu, Apocryphes bogomiles, 41; and also the apocryphal Gospel of Bartholomew (Bartholomew's Questions), see *Hennecke - Schneemelcher* I, 361; and also the Greek legend of the creation of the world, ed. *Ivanov*, Bog. knigi, 315 ff. The motif of the loss of the syllable - il is frequent in the Slavonic apocryphal tradition (see the texts quoted in Byzantinobulgarica III (1969) 32, n. 35).

[15] See Byzantinobulgarica III (1969) 32, n. 29. The Bogomil etymological arguments round the names Ἰησοῦς and Χριστός (Euth. Zig., ed. *Ficker*, 93,29-30 = PG 130, col. 1301B) have a remarkable parallel in the Gospel of Bartholomew quoted above (see *Vasiliev*, Anecdota, 21; *Hennecke - Schneemelcher* I, 371). For the views on the Trinity attested by Zigabenus, see Puech: *Puech - Vaillant*, Le traité,179 f.

[16] *W.Lueken*, Michael, Göttingen 1898, gives cases of this name for Christ attested in patristic literature. There is also epigraphical testimony (CIG Add. IV, 95a = *Kaibel*, Epigr. Graec., No 726).

[17] See Byzantinobulgarica III (1969) 32, n. 35 (references to the Slavonic and Coptic texts).

[18] For the Bogomils' own re-evaluation of common demonological ideas see ibid., 24 f., n. 8.

[19] Euth. Zig., ed. *Ficker*, 96, 19-27 = PG 130, col. 1313CD.

[20] See our study in: Listy filol. 90 (1967) Vol. 2, 122. In the same article (116) the author gives his view of the attempt by *H.Söderberg* to trace, by careful but not ultimately convincing comparison, models for individual motifs of the Bogomil-Cathar myth in those of the Gnostics and Manichaeans. (*H.Söderberg*, La religion des Cathares, Uppsala 1949).

[21] Euth. Zig., ed. *Ficker*, 98, 10-20 = PG 130, col. 1292BC.

[22] These are "seven pillars" holding up the house which "Wisdom hath builded" (Proverbs 9,1). By "house" they meant their own sect (ibid.)

[23] Euth. Zig., ed. *Ficker*, 102,18 - 110,8 = PG 130, col. 1321A - 1332D.

[24] There is an interesting example of their highly original exegesis of Christ's words about the Law of Moses. The Gospel of St. Matthew reads: Ἰωτά φασιν ἓν ἢ μία κεραία οὐ μὴ παρέλθῃ ἀπὸ τοῦ νόμου, ἕως ἂν πάντα γένηται (Matth. 5,18). The Byzantine Bogomils took ἰῶτα to mean the Ten Commandments of the Old Testament, apparently because *iota* is used in Greek to indicate the figure ten. They attributed the same meaning to the word κεραία in the text quoted here, with this ingenious argument: τὴν γὰρ κεραίαν εἴ τις ὀρθώσει πλαγίως κειμένην, ἰῶτα ποιήσει. They thus recognized that the Ten Commandments of the Law would remain valid, but with this significant limitation: they would remain valid *for the Jews*, who would keep the Ten Commandments until the end of the world. On the other hand the Bogomils gave a forced meaning to the words spoken by Christ just previously (Matth. 5,17), to fit them to their own myth. According to the Bogomils Christ came into the world not to "fulfil" the Law (πληρῶσαι) but to "fill" the place left empty in Heaven after the fall of the angels. (See Euth. Zig., ed. *Ficker*, 106,1-10 = PG 130, col. 1325B).

The mythical idea that Cain was born of union between Eve and *the Devil* was adduced by the sect from I. John 3,12: Κάϊν ἐκ τοῦ πονηροῦ ἦν. The Bogomils, according to Zigabenus, were bold enough to adduce the creation of man by the combined efforts of Satanael and the Heavenly Father from Genesis 1,26: ποιήσωμεν ἄνθρωπον κατ' εἰκόνα ἡμετέραν καὶ καθ' ὁμοίωσιν. Here the plural "Let us make" fitted their purposes ideally, as did the first person plural possessive: before the fall, Satanael had the same likeness of God the Father, hence the words *"our* image, after *our* likeness". (Euth. Zig., ed. *Ficker*, 94,18 - 22. In fact the Bogomils used the first verse of the Book of Genesis (Gen. 1,1; see ed. *Ficker*, 92,9 - 10 = PG 130, col- 1296D).

Examples of the Bogomil distortion of the wording of the Gospels: 1) ... ὀδόντα ἀντὶ ὀδόντος (Matth. 5,38) was understood by the Bogomils as τὰς δύο ὁδούς, τὴν πλατεῖαν καὶ τὴν στενήν (ed. *Ficker*, 107,10-13 = PG 130, col. 1328A). 2)......... ἐνεβριμήσατο αὐτοῖς ὁ 'Ιησοῦς λέγων ..(Matth.9,30) was read by them as ἐνεβρωμάτισεν αὐτοῖς ὁ 'Ιησοῦς λόγον (ed. *Ficker*, 110,4-8 = PG 130, col. 1332A. Cf. *Angelov*, Bogomilstvoto v Bălgarija, 219.). 3) ὑποδήματα (Matth. 3,11) was interpreted as ὑποδείγματα τῶν σημείων (ed. *Ficker*, 104,31-32 = PG 130, col. 1324C).

25 Euth.Zig., ed. *Ficker*, 102,20-23 = PG 130, col. 1321A.

26 This was Matthew 7,22-23. Cf. Peter of Sicily, 'Ιστορία χρειώδης, ed.*D.Papachryssanthou*, *J.Gouillard*, in: Travaux et Mémoires 4 (1970) 53,30-32; 55,30-33; Euth.Zig., ed. *Ficker*, 108,15-26 = PG 130, col. 1328D-1329A. *Gouillard*, L'hérésie, 317, sees this as indicating that the Bogomils took their commentary to St.Matthew from the Paulicians. The coincidence could simply have arisen from the tendency both sects shared, however. - We have been unable to discover the origin of the story of Rachel, based on Matthew 2,18. It is an apocryphal motif which was of course interpreted by the Bogomils in the spirit of their own teachings. See Euth.Zig., ed. *Ficker*, 103,24 - 104,3 = PG 130, col. 1321D - 1324A.

27 Art. 4,5. *Tollius*, 116 (= PG 131, col. 44A). *Thallóczy*, 364.

28 Euth. Zig., ed. *Ficker*, 94,29-32; 107,27 - 108,7 = PG 130, col. 1328BC.

29 For Zigabenus and his Panoplia see *M.Jugie*, La vie et les oeuvres d'Euthyme Zigabène, in: Échos d'Orient 15 (1912) 215-225. Further literature is given by *Beck*, Kirche u. theol. Lit., 614, n. 2. Cf. *Gouillard*, L'hérésie, 301.

30 See *Joan M.Hussey*, Ascetics and Humanists in eleventh century Byzantium. Friends of Dr. Williams's Library 1960.

31 See the comment by Zigabenus on the Bogomil interpretation of Matthew 8,19-20 (ed. *Ficker*, 109,1-9 = PG 130, col. 1329B). Cf. ByzSlav XXVIII (1967) 46, n. 25.

32 Euth. Acm., ed. *Ficker*, 30; Euth. Zig., ibid., 91,14-17.

33 Euth. Acm., ed. *Ficker*, 85,10-15. Cf. Chapter XVII, n. 71.

34 On prayer among the Bogomils see H.-C.Puech, in: *Puech - Vaillant*, Le traité, 245 ff. That they recognised only the Lord's Prayer, rejecting all other prayers and hymns used by the Church, which in their eyes were mere βαττολογίαι, is attested not only by Euthymius of Acmonia but also by Theodore of Andida (PG 140, col. 461B). See also the 7th article (*Tollius*, 118, cf. Thallóczy, 365) of the synodicon against the Bogomils; here we find a single instance stating that the sect did not accept the *doxology*. The Bogomils prayed μετὰ τῆς εἰς γῆν κατακλίσεως ibid.; Euth. Zig., ed. *Ficker*, 100,24 = PG 130, col. 1316A: μετὰ γονυκλισίας). Unlike Euth. Acm., giving the number of prayer sessions as seven for the day and seven for the night (PG 131, col. 56BC), Zigabenus says that the Bogomils prayed seven times during the day and five times during the night, each time with several repetitions of the Lord's Prayer (ed. *Ficker*, 100, 21-23 = PG 130, col. 1313D). They were convinced that they were building their house upon the "rock" of the Lord's Prayer while the Orthodox Church was building on the "sand" of all kinds of prayers it had made up itself (see the Bogomil interpretation of Matthew 7,24 and 26; Euth. Zig., ed. *Ficker*, 108,27-32 = PG 130, col. 1329A). In Matth. 6,6 the word ταμιεῖον was interpreted as νοῦς (Euth. Zig., ed. *Ficker*, 107,14-18 = PG 130, col. 1328AB; cf. H.-C.Puech in: *Puech - Vaillant*, Le traité, 247, n. 1). Byzantine texts confirm

Cosmas's observation that the heretics did not cross themselves when they prayed (see *Tollius*, 118, anathema 7), and give the same reason for the profound aversion the Bogomils felt concerning the cross itself (see H.-C.Puech in: *Puech - Vaillant*, Le traité, 234 ff). *Tollius'* text (120, anathema 11) says that σταυρὸν φούλκαν ἀποκαλοῦσι. *Thallóczy's* text (366) gives what is undoubtedly the correct reading: φοῦρκαν.

35 For the Bogomils the true eucharist was the Lord's Prayer and His New Testament (referring to Luke 22,20). See Euth. Zig., ed. *Ficker*, 97,6-13 = PG 130, col. 1313B. Cf. H.-C.Puech, in: *Puech - Vaillant*, Le traité, 227.

36 See Euth. Zig., ed. *Ficker*, 100,30 - 101,7 = PG 130, col. 1312C. H.-C.Puech (in *Puech - Vaillant*, Le traité) believes that the liturgical formula sung at the *end* of the ceremony and described in the Panoplia as εὐχαριστήριος ὕμνος was in fact identical with the σατανικὴ ἐπῳδή referred to by Euthymius of Acmonia. In our opinion, this argument is invalidated by the identity of this σατανικὴ ἐπῳδή with the ἐπίκλησις - calling upon the Holy Ghost (see Chapter V above, n. 32). Zigabenus refers to the epiclesis in connection with the *first* stage of the Bogomil initiation rite: καὶ τὸ παρ' αὐτοῖς ἅγιον Πνεῦμα ἐπικαλούμενοι ... (Euth. Zig., ed. *Ficker*, 100,33-34).

37 ... τοὺς ἐκλεκτοὺς καὶ δικαίους καὶ κληρονόμους τοῦ θεοῦ ... Euth. Zig., ed *Ficker*, 102,24-25 = PG 130, col. 1321A). It is also attested that they referred to themselves as Χριστοπολῖται (*Tollius*, anathema 12; cf. *Thallóczy*, 367).

38 Euth. Zig., ed. *Ficker*, 100,5-10 = PG 130, col. 1309C.

39 Ed. *Ficker*, 100,11-14 = PG 130, col. 1317B.

40 Ed. *Ficker*, 100,15-20 = PG 130, col. 1317C.

41 Ibid.

42 See Chapter V, n. 25.

43 Euth. Zig., ed. *Ficker*, 100,34-36 = PG 130, col. 1312C.

44 ... ἀπεχομένους κρεωφαγίας καὶ γάμου καὶ τῶν τοιούτων. Euth.Zig., ed. *Ficker*, 106,18. The relevant passage in the Panoplia (PG 130 col. 1325C) also mentions the prohibition of cheese and eggs as food. On the subject of marriage see the commentary to Matthew 5,32 and 22,30 (ed. *Ficker*, 106,25 - 107,2 = PG 130, col. 1325D). Cf. also *Tollius*, 117 - 118, anathema 6; *Thallóczy*, 365. An outline of Bogomil asceticism is given by H.-C.Puech, in: *Puech - Vaillant*, Le traité, 260 - 272.

45 Euth. Zig., ed. *Ficker*, 101,30-34 = PG 130, col. 1320C.

46 Ed. *Ficker*, 106,13-14 = PG 130 , col. 1325BC.

47 διδάσκειν ἀληθέστερα καὶ μετιέναι πολιτείαν ἀκριβεστέραν τε καὶ καθαρωτέραν .. Ed. *Ficker*, 106,16-18.

Chapter VIII

BYZANTINE BOGOMILISM AFTER ALEXIUS COMNENUS

Byzantium, even after the reign of Alexius I, continued to see in the Bogomil heresy a very real danger. Every year the churches echoed again and again with *anathema* against the sect, which was spreading "in every town, every country district and province".[1] A special synodicon against the Bogomils seems to have been read on the anniversary of the condemnation of the heretics in Constantinople.[2] In one of the dioceses of the Athenian Metropolis, however, it was incorporated in the Byzantine Synodicon for the Sunday of Orthodoxy, read on the first Sunday in Lent.[3] The articles directed against individual heretical doctrines of the Bogomils were also used in drawing up a formula for the reacceptance of repentant Bogomils into the Church.[4] This latter text also includes a number of articles of a Messalian character which have in fact nothing to do with the teachings of the Bogomils.[5] Indeed, even in the original thirteen articles the latter were identified with the *Messalians*, and names familiar in early literature against the Messalians crop up again. On the other hand, the father of the "Messalian-Bogomil" heresy is specified as the Patriarch of Antioch, Peter Gnapheus (468-470), whom Euthymius of Acmonia by some curious logic had already implicated in the affairs of his Phundagiagitae.[6] In the amplified text, serving as the abjuration formula, we even find names familiar to us in the polemics against the Manichees and the Paulicians.[7] It is only the real teachers of the Bogomils who are missing; the name of Basil is not mentioned, not to speak of the Bulgarian priest Bogomil. Indeed Zigabenus himself, who was well aware of the Bulgarian origin of the name Βογόμιλοι ,[8] seems to have had no idea of the existence of the first Bulgarian teacher of this dualist heresy. The name of the priest Bogomil did, however, appear in one provincial version of the Byzantine *Synodicon for the Sunday of Orthodoxy.*[9]

The constant vigilance of the Byzantine Church to keep the Bogomil heresy at bay during the reigns which followed closely on that of Alexius I can be clearly traced in the records of the permanent synod of Constantinople.[10] It is here that we learn of heretical writings by a deceased monk, Constantine Chrysomallus, which were being read in some monasteries in the reign of John II Comnenus (1118-1143). Informers brought them to the attention of the Patriarch, who had them burnt on the grounds that they contained Bogomil and Messalian heretical beliefs.[11] The doctrines expounded by Chrysomallus, as reproduced in part by the synodical document, however, do not suggest that this was an accurate description.

His mystical method is entirely based on instruction and initiation (κατήχησις καὶ μύησις), conferring on the initiate the ability to perceive the presence of the Holy Spirit. There is nothing to indicate, however, that Chrysomallus shared the coarsely materialistic conception of mystical experience peculiar to the *Messalians*. On the other hand, the idea of "escape from the power of Satan" which Chrysomallus preached was certainly not derived from the dualist metaphysics of the Bogomils. The whole intellectual context is quite different, postulating the inner transformation of man in the sense of the Eastern Christian doctrine of "deification".[12]

Only those specially chosen were able to pass on, by the laying on of hands, the gift of mystical perception of the grace of the Holy Spirit, changing human nature from its very foundations and lifting it beyond reach of sin. This is contrasted with the church rite of baptism, administerad to tiny children who cannot be instructed in advance and cannot therefore comprehend the essential transformation which is the real meaning of baptism. It follows that orthodox baptism cannot free the soul from its helplessness and subjugation to sin, and cannot liberate it from the power of evil. For the same reason it is no help for a Christian to lead the most virtuous life, observing the regulations of the Church strictly; nor, indeed, is it any good to him to be a dignitary of the Church, however highly placed. All these are of no significance without instruction and the inner revival of the soul, acquired through the initiation rite of Chrysomallus.

His contemporaries had no hesitation in accusing Chrysomallus of the Bogomil heresy, for the Bogomils, too, declared church baptism ineffective; the new ceremony they insisted on also included inspiration by the Holy Spirit, through the laying on of hands. It is nevertheless quite clear that in the mystical doctrine of Chrysomallus the rite of initiation had a very different significance from that of the Bogomils. In itself the laying on of hands was not a feature of Bogomilism alone, and Chrysomallus may have been inspired purely by the Acts of the Apostles.

There is nothing in the report of his heresy to suggest that there was an organized sect round Chrysomallus. Indeed, the reference to the chosen individuals who are capable of instructing and of administering initiation is more in the line of tradition of "spiritual fathers", who enjoyed great authority in Byzantine monastic life.[13]

This synodical act was dated May 1140. Three years later, at the beginning of the reign of Manuel I Comnenus (1143 - 1180), two bishops in Asia Minor were accused of Bogomilism; they were Clement of Sasima (or Sosandra) and Leontius of Balbissa.[14] In fact, however, the points which the accused themselves admitted, and on which their condemnation rested, are in themselves no proof of heresy. The main stumbling block seems to have been the second baptism administered by Leontius to a child previously baptized by an unworthy priest. The simple fact of *rebaptism*, however, seemed to the Byzantine ecclesiastical authorities to signal the danger of Bogomil heresy.[15]

That autumn of 1143 a monk by the name of Niphon was also brought before the Patriarchal Synod, suspected of being a heretic.[16] At the beginning of the following year the Synod decided that he remain in solitary confinement in the Peribleptos monastery, which we have heard of before, where he had been interned before his trial. The only clear objection to his orthodoxy was that he "cursed the God of the Jews". Here, too, it is doubtful whether this could be considered proof of Bogomilism; in any case there was an unexpected turn to the case. The new Patriarch, Cosmas II Atticus (1146-1147) set Niphon free and maintained friendly relations with him. The Patriarch's enemies welcomed this as an opportunity to come out against him and ultimately secured his deposition.[17]

Under Cosmas's predecessor, however, the Patriarchal Synod had condemned several Bogomil heretics to the stake in Constantinople.[18] No synodical document recording this cruel sentence has been preserved, but many years later Theodore Balsamon recalled the event in his commentary to tit.IX, Chapter 25, of the Nomocanon of XIV tit.[19] The scholion referring to the burning of these heretics is appended to the chapter on magic practices, which was subject to the same penalty under Roman law. It should be noted that it was not difficult to liken heresy to this crime, which is perhaps why the late third century Imperial constitution prescribed death at the stake for Manichaean teachers.[20]

The Byzantine lawyer accuses the Synod of irregular procedure, since there was no provision for physical punishment of heretics in canon law; this was the province of the secular arm. It is characteristic, however, that he did not object on principle to the death sentence for heresy. The principle that the law should be interpreted "in the more human fashion" (ἐπὶ τὸ φιλανθρωπότερον), which was accepted elsewhere, as in cases of magic practices and similar crimes, was in his view out of place here, especially since the Bogomils themselves chose this death as martyrdom for their faith. As this shows, the sect was already facing baptism by fire in Byzantium, as so often later, in the West. It remained a rare phenomenon for Bogomils to die at the stake in Byzantium, however, confined to Constantinople. Elsewhere Balsamon himself reproaches the Byzantine Church for its drastic treatment of individual cases of heresy in the capital, while allowing whole towns and villages in the provinces to lie at the mercy of the Bogomil plague.[21]

A later Slavonic source, dealing with the reign of Manuel, does in fact refer to dualist heresy in one provincial region of the Empire, in Macedonia. The existence of heresy in these parts was already mentioned in an anonymous Latin chronicle of the First Crusade. We read there that in the region of Pelagonia, round Bitolj and Prilep, the Crusaders stormed and burned a fortified town *(castrum)* which was inhabited by heretics.[22] The Bulgarian Life of Hilarion[23] takes us to the region of Măglena, a town in Macedonia. The bishop of the diocese, Hilarion, was ordered by the Emperor Manuel to purge his flock of the Bogomil heresy. This may have been an imperial edict of broader impact. The Life of St.Hilarion also reproduces the bishop's polemics against the Armenians and the Paulicians,[24] describing their conversion to the true faith. In a Bulgarian prologue of the 13th – 14th centuries St. Hilarion is praised as "a fighter against the Bogomils".[25]

The most interesting of all the documents testifying to the Byzantine Bogomils in the reign of Manuel[26] is a *Latin* polemic against them, written in Constantinople.[27] It was not mere chance that produced this work at that time. The increased contacts between Byzantium and the western world which resulted from the Crusades were strongly reflected in changes in the way of life at the imperial court. The western orientation of Byzantine foreign policy brought men from the West to Manuel's court, among them two Italians from Pisa. Leo Tuscus was an interpreter in the Emperor's chancellery, while his brother Hugo Etherianus was Manuel's adviser on western theological questions. The latter's long treatise on the procession of the Holy Spirit aroused great interest among the theologians of Byzantium. His polemic against the Byzantine Bogomils, however, is today of unique value as a historical document – not on account of its conventional contents, but (as A.Dondaine wrote) "by the very fact of its existence". The title itself is striking: ***Adversus Patherenos***. The Italian author gave to the Byzantine sect the name by which it was known in his own country and elsewhere in the West.[28] And indeed, in the reign of Manuel, when Hugo was writing his polemic, the dualist heresy had crossed over the borders of the Byzantine state and spread over parts of Western Europe with a speed which amazes us today.

1 ... ἐν πάσῃ πόλει καὶ χώρᾳ καὶ ἐπαρχίᾳ *Tollius*, 106.

2 *J.Gouillard* (Synodikon, 232 f.) has shown conclusively that Tollius's text (see Chapter V, Note 8), where the thirteen articles of the anathema are preceded by introductory instructions and followed by diptychs and a concluding exhortation, is an independent liturgical synodicon which was read on the anniversary of the condemnation of the Constantinople Bogomils under Alexius I. He points out the testimony, hitherto ignored, of the homily by Pseudo-Xiphilinos which quotes in the preamble some of the articles of that synodical document, and refers to the death of the Bogomil leader at the stake (PG 120, col. 1289A - 1292A).

3 This is a provincial version of the Synodicon for the Sunday of Orthodoxy, preserved in five copies and used in one of the bishoprics under the Athenian Metropolitan. See *S.Eustratiadis*, Τὸ συνοδικὸν τῆς ἐκκλησίας τῆς Ἑλλάδος, in: ΕΕΒΣ 13 (1937) 8-16. The text was included in *Gouillard's* edition of the Synodicon for the Sunday of Orthodoxy (Synodikon, ll. 277 - 387; cf. commentary, 232 f.). Compared with Tollius' text the title and the diptych are missing here. One MS (Bruxellensis II 4836, fol. 76r - 78v) gives only the thirteen articles of the anathema, the introductory and concluding passages are missing. See *J.Davreux*, in: Byzantion 10 (1935) 96; cf. Gouillard, Synodikon, 232).

4 The text has survived in two MSS: Cod. Vindob. Theol. Graec. 306 and Cod. Taurinensis Bibl. Nat. 185 (B. IV. 22). The anathema articles were published by Thallóczy according to the Vienna MS. The introductory passage, prescribing the rules for re-admission of Bogomils to the Church, was edited by *G.Ficker*, Phundagiagiten, 172 - 175; on p. 175 he corrects some inaccuracies in Thallóczy's reading.

5 See *Ficker*, Phundagiagiten, 229 ff. Neither these anathema articles, nor the synodal record of the teachings of Eleutherius of Paphlagonia, nor the extracts preserved from the works of Chrysomallus, seem to us convincing proof of a renaissance of Messalianism, as *J.Gouillard* assumed (L'hérésie, 319 ff.). For Chrysomallus see below.

6 The invective of Euthymius of Acmonia against the Bogomils left its mark on the introductory part of the form for re-admission of repentant followers of the Bogomil heresy (see n. 4 above). Here the treatment to be accorded to the former heretic depends in each case on whether the fateful ἐπῳδή had or had not been read over him; if so, the new adept of the sect was in the power of devil-worship for ever. Euthymius of Peribleptos used similar words to describe his young disciple: οὔτε τὴν ἐπῳδὴν ἐπάνω αὐτοῦ ἀνέγνωσαν, οὔτε τὸν διάβολον αἰσθητῶς προσεκύνησεν (Euth. Acm., ed. *Ficker*, 85,12-13). The way the abjuration formula derives from Euthymius's work is evident.

7 The model used was the text of the chronicle of George Monachus (see *Ficker*, Phundagiagiten, 226).

8 PG 130, col. 1289C.

9 The text is given by *Gouillard*, Synodikon, 59 (apparatus); commentary pp. 228 f. See below, chapter XVI. The same recension of the Synodicon for the Sunday of Orthodoxy (Gouillard's Cc) mentions the Bogomils in Panormos (text printed in the apparatus, p. 61). There are some indications that Palermo, Sicily is meant, but the localization remains uncertain in view of the frequent occurrence of the toponym Panormos (see commentary pp. 229 f.).

10 "Synodos endemusa". For this institution see *P.J.Hajjar*, Le synode permanent dans l'Eglise byzantine des origines au XIe siècle (Orient. Christ. Anal., 164) Rome 1962.

[11] *Grumel,* Regestes, N. 1007 (references to all editions of the text quoted).

[12] See ByzSlav XXVIII (1967) 41 ff, n. 12.

[13] See *Gouillard,* L'hérésie, 320.

[14] *Grumel,* Regestes, NN. 1011, 1012, 1014.

[15] See ByzSlav XXVIII (1967) 48, n. 38.

[16] *Grumel,* Regestes, NN. 1013, 1015.

[17] *Cinnamus,* ed. Bonn, 63-66; Nicetas Choniates, lib. II, ed. Bonn, 107. See *Obolensky,* Bogomils. There are also allusions to Bogomilism in an epistle of the Ephesian Metropolitan Georgius Tornices dating from 1153-1155. See Georges et Démétrios Tornikès, Lettres et discours, ed. *J. Darrouzès,* Paris 1970, 211,4, 215,5.

[18] *Grumel,* Regestes, N. 1020.

[19] See the scholion to tit. IX, Chapter 25 of the Nomocanon of XIV tit., PG 104, col. 1112B.

[20] See *Maisonneuve,* 35.

[21] Scholion to tit. X, Chapter 8 of the Nomocanon of XIV tit., PG 104, col. 1148D.

[22] Anonymi Gesta Francorum et aliorum Hierosolymitanorum, ed. *B.A.Lees,* Oxford 1924, 8 (= Histoire anonyme de la Première Croisade, éd. et trad. par *L.Bréhier,* Paris 1924, 22 ff). See *Obolensky,* Bogomils, 163 n. 2; *Angelov,* Bogomilstvoto v Bǎlgarija, 374.

[23] It was written by the Patriarch of Trnovo, Euthymius; edited by *E.Kalužniacki,* Die Werke des Patriarchen Euthymius, Vienna 1901, 27 - 58. This hagiographical text was incorporated in an abbreviated form in a Serbian chronicle dating from 1650, the Chronograph of Vrchobreznica (Vrkhobreznitsa), preserved in a MS in the collection of the National Museum, Prague (IX G 6, f. 225 - 228v). This shorter version was published by *A.I.Jacimirskij,* Mel'kije teksty i zametki po starinnoj i russkoj literaturam. LI, in:Izv. otd. rus. jaz. i slov. XIII (1908) kn. 2, 150 - 160. *A.Solovjev* (Svedočanstva, 12 ff.) believed this shorter version to be the original Life of St.Hilarion, written some time at the beginning of the thirteenth century by the higumen Peter, who took charge of the monastery founded by the holy man after the latter's death. This view was based on Jacimirskij's general notes, who had come to the conclusion, after a very superficial comparison of the two versions of the Life of St.Hilarion, that "the simple narrative of the Chronograph appears in Euthymius with somewhat rhetorical ornament in the manner of the later Byzantine school of hagiography" (*Jacimirskij,* l.c., 150). In reality, the passage in the Vrchobreznica Chronograph presents a compressed extract from the narrative of Patriarch Euthymius and from the polemical arguments which the Bulgarian writer puts into the mouth of Hilarion. In the Chronograph, to make the presentation of the text clearer, the tenets of the heretics are first given in brief form; in the Prague MS they are written in red letters. The way the text was summarized was not always very happy, amd this is particularly true of the passage dealing with the Bogomils. In the version by Euthymius of Trnovo the reference to Manuel's orders to stamp out the Bogomil heresy is followed logically by a paragraph on Hilarion's campaign against the Bogomils (see the text in *Kalužniacki,* op. cit., 53). In the Serbian Chronograph, on the other hand,there is a reference to the Emperor's orders but no further mention of the Bogomils. The text goes on somewhat illogically to an account of Hilarion's successes in the conversion of "Armenians" and Paulicians (see the text in *Jacimirskij,* l.c., 156).

[24] The arguments against the "Armenians" and Paulicians were taken from Euth.

Zig., Panoplia dogm., tit. XXIII and XXIV.

[25] See *Ivanov*, Bog. knigi, 35.

[26] Two other passing references to Bogomil heresies in the provinces of Byzantium can be quoted: a letter from Theodore Prodromos to Michael Italicus, Archbishop of Philippopolis (1147 - 1166) refers to (unspecified) heretics in the vicinity of Plovdiv; see R.Browning, Unpublished correspondence between Michael Italicus, Archbishop of Philippopolis, and Theodore Prodromos, in: Byzantinobulgarica I (1962) 279 - 297; see text, p. 287, 24 - 25. A reference to "Euchites" (i.e. Bogomils) in the encomium of Michael Choniates on Nicetas, Metropolitan of Chonai, concerns Asia Minor. Bogomilism is treated as a real and continuing danger in this document. See Μιχαὴλ τοῦ 'Ακομινάτου, Τὰ σωζόμενα I, Athens 1879, 53. Attention was drawn to this passage by *A.P.Każdan*, Novyje materialy o bogomilach (?) v Vizantii XII v., in: Byzantinobulgarica II (1966) 275 - 277.

[27] See *A.Dondaine*, Hugues Ethérien et Léon Toscan, in: Archives d'Hist. doctrinale et littéraire du Moyen Age XIX (1952) 67 - 134.

[28] In the study quoted here (pp.111 f.) *A.Dondaine* tries to prove that the name Patareni (παταρηνοί) came from the name of the Byzantine town of Patara on the Lycian coast. *D.Angelov* supported this hypothesis with a reference to the existence of unspecified heretics in Lycia taken from a hagiographical source of the eleventh century. (De Sancto Lazaro monacho in monte Gelsio. ASS Novembris III, Bruxellis 1910, col. 543. See *Angelov*, Bogomilstvoto v Bălgarija, 386). There is, however, no evidence to connect the Bogomil sect with Patara in Lycia. Dondaine himself was aware that the hypothesis did not rest on very firm ground, and proposed a new etymology in a later work, deriving the name from the word πατέριτσα which meant a T-shaped crook, the mediaeval symbol of the pastor (*Dondaine*, Durand de Huesca, 275 ff.) *I.Dujčev* derives the presumed Greek form *Pater-inoi* from the frequent repetition of the Paternoster (ByzSlav XIX (1958) 318 f.). We shall be returning to this in the twelfth chapter.

Additional note:

For Constantine Chrysomallus see *J.Gouillard*, Constantin Chrysomallos sous la masque de Syméon le Nouveau Théologien, in: Travaux et Mémoires 5 (1973) 313 - 327. It is remarkable to see that some of the Orations of Symeon the New Theologian contain either the literal quotation or a more fully developed version of the tenets of Constantine Chrysomallus as recorded in the protocol of the 1140 Synod.

Chapter IX

THE WESTERN IDEAL OF THE *VITA APOSTOLICA* AND THE HERETICAL APOSTLES FROM THE EAST

There is evidence that the heresy from the East overran the countries of western Europe with the force of an avalanche from about the middle of the twelfth century. It is of course not out of the question that the dualist faith of the Bogomils found its way there much earlier, in isolated cases, but there is not a single reliable proof that it did so. Yet as early as the turn of the millennium foci of heresy kept appearing at different points of the western world, and did not escape the notice of contemporary writers.[1] Gerhard, Bishop of Csanád in Hungary (1037-1046), himself of Italian birth, was aware of the existence of heretics in northern Italy. This learned bishop also knew of the heresies which always existed in Byzantium, but he wrote not a word to suggest that there was any connection between them and the sectarian phenomena in the West.

Contemporary western writers certainly do not confirm the Hungarian bishop's idea of "fortunate" Gaul, innocent of heretical teachings.[2] In fact preachers of unorthodox ideas began to appear from the early eleventh century in various parts of what is now France, but then was far from being a united country with a single government. The reference to "Manichees" in Aquitaine, in the chronicle of Ademar de Chabannes, refers to the year 1018; Duke Guillaume called an ecclesiastical council to deal with them. These heretics of monastic appearence were strictly abstemious as regards food, and rejected baptism, the cross, and all that belongs to "the right faith".[3] We cannot fail to recall the eastern Bogomils, but the very brief chronicle report tells us nothing about the origin of these "Manichees", nor does it allow us to form reliable judgment of the nature of their teachings. This same western chronicler gives a laconic reference to the appearance of "Manichees" in Toulouse.[4]

Where the sources give a fuller description of the teachings of the different heretical groups, we see that however resolutely they turned their backs on the reality of this world, their faith was not based on dualist metaphysics.[5] The handful of learned clerics burnt at the stake in Orleans in 1022 believed in one God, "the creator of all things". Equally clearly monotheist were the beliefs of the heretics of Monteforte (near Turin) who met their death in the flames on the soil of Milan in 1028. These two groups, so far apart, are brought close together not only by the tragic fate they shared, but also by a similar spiritual interpretation of the Christian faith.[6] For the heretics of Monteforte the birth of Christ as the son of Mary was the symbol of the birth of the soul from the Holy Scriptures. The clerics of Orleans, too, understood the Son of God in a purely spiritual way, and did not believe in His real birth, death and resurrection. In the teachings of the Monteforte heretics the Holy Ghost symbolized understanding of Holy Writ and was the hidden force ruling everything in the world. The Law of God, as the heretics of Orleans understood it, was written within man himself *(in interiori homine)*. Their faith remained inaccessible, therefore, to those able to understand only earthly things and keeping the letter of the law as written on parchment. Free enthusiasm - the direct inspiration of man by the Holy Spirit - took away all meaning from the sacraments of the Church and robbed the hierarchy of its role as intermediary. The heretics of Monteforte believed that neither priest nor bishop, nor even the Pope of Rome, had power to forgive sins. The Holy Ghost visited their brethren every day, scattered throughout the world, and He was the only bishop *(pontifex)* they recognised. In Orleans new adepts of the heretical faith were accepted into the community with a symbolical laying on of hands; cleansed of sin, they were filled with the gift of the Holy Spirit, revealing to them the profundities of Holy Scripture. The soul, reborn through this divine food, was blessed with angelic visions ... The kingdom of people inspired by such a faith was indeed not of this world. The strictly ascetic life of the Monteforte heretics led them towards a martyr's death, and if the opportunity did not present itself they deliberately arranged it. Their spokesman, brought to face the Archbishop of Milan, in his ecstatic expectation of martyrdom made no secret of the beliefs he shared with the rest. Few of the Monteforte group took advantage of the chance to save their lives, offered at the last moment; most of them leaped bravely into the flames. Six years earlier in the cathedral of Orleans, face to face with the king, the queen and the ecclesiastical council, and surrounded by a hostile crowd outside the building, the heretics gazed steadfastly at the fate the world had prepared for them. Forced to confess their beliefs, they then refused to argue further with their judges. Their souls saw God in Heaven, waiting to elevate them to celestial beatitude.

Three years after this martyrdom in Orleans, other heretics brought before the synod in Arras were not forced to end their lives in the flames, but to sign a confession of the Catholic faith. Here, as in Orleans, the records of the synodical proceedings have preserved their authentic testimony. The teachings of this group were not concerned with the spiritual sphere,[7] but the daily practice of the Christian life in accordance with the commandments of Christ and His apostles. In their own interpretation of it this meant they should turn their backs on the world, live abstemiously on the work of their hands, harm no-one and live in charity with those who shared their resolve. Their "Pelagian" conception of salvation was based entirely on Christian "righteousness" in this sense, and the sacraments of the Church therefore lost all significance. The efficacy of baptism was entirely annulled by the moral defects of the priest performing the ceremony and by the fact that the baptized Christian was bound to sin during his lifetime. Moreover, baptism lost all meaning since it was administered to infants who were ignorant of the Christian faith and had as yet no will of their own. These and similar arguments against the validity of baptism as administered by the Church were to recur again and again among future generations. In this matter the heretics of Arras were presaging future beliefs, and no continuity can be proved between them and later groups holding the same views. It is not without interest, however, that in the stormy upheavals of ecclesiastical reform half a century later the question of the validity of sacraments administered by an unworthy priest provoked a tragic conflict in Cambrai, in the immediate vicinity of Arras.

A letter written by the bishop of Arras-Cambrai (1025) reveals that the heretics came to his diocese from Châlons-sur-Marne; this is not the only text to mention this locality in connection with heresy. A reference in the History by Raoul Glaber concerns the year 1000;[8] even later, Roger II Bishop of Châlons asked Wazo, Bishop of Liège (1042 - 1048), for advice on how to deal with the "Manichees" who had appeared in his diocese.[9]

The confessions of the heretics before the synod in Arras in 1025 clearly point to an Italian origin, at least as far as their "teacher" Gundulphus is concerned.[10] There is more evidence in contemporary sources that there were important centres of heresy in the north of the Apennine peninsula, whose influence spread far indeed.[11] The sectarians were firmly convinced that their fellows lived everywhere in the world as then known.[12]. Undoubtedly there were links between some groups, but their preaching did not affect large areas. Further development was necessary before the West was ready for really vast heretical movements.

From the middle of the eleventh century reports of active heretical groups in the West are not found; while the ecclesiastical synods were still pronouncing excommunications of heretics and their protectors, and a papal epistle of c.1060 still issued a warning against the danger of heresy,[13] nevertheless the main concern of ecclesiastical circles was for wider matters.

From the middle of the eleventh century supporters of the movement for reform were decidedly in the majority in the Roman Curia; this movement had been active in the Church from the tenth century onwards, but the struggle against simony and "Nicolaism" now turned against the very foundations of the existing system. It was no longer only a question of discarding the evil practice of selling ecclesiastical benefices, nor of raising the level of the clergy by deliberate selection and preventing the alienation of church property by insisting on a celibate clergy. There was now a strong attack on all lay influence on ecclesiastical appointments. This meant that the existing order, established for centuries albeit contrary to the ancient traditions of the Church, was now to be superseded. It was essential to overcome the resistance of the ruling feudal class, in which family ties and mutual interests united the secular aristocracy and the ecclesiastical hierarchy. In Italy the protagonists of reform had a powerful ally in the city people, as yet without any rights in public life. Feelings ran highest in Milan, where as early as 1046 the city people organized in defence of their own interests during the disturbances over the succession to the archbishopric. Ten years later two priests, Landulphus and Arialdus, with their demagogy directed the movement of revolt against the simonist clergy which became known as the *Pataria*.[14] The original meaning of this name was already obscure to its contemporaries, and research has not solved the problem. The oath taken by the followers of the Milan Pataria heralded the idea of a Community, and the atmosphere was the same in other Italian towns where the Milanese Pataria found eager imitators. In the given circumstances the form the movement assumed was in full agreement with the interests of Rome, but it offered a dangerous precedent. Members of the clergy who owed their benefices to simony or who lived with wives or concubines were subject to persecution, their property confiscated, and no tithes paid to them. The sacraments they administered were considered invalid and treated with contempt. The people did not attend Divine Service, but indulged in passionate discussion; lay criticism of the Church was given free rein. Rome itself poured oil on the flames by calling upon the congregation not to attend services held by priests who did not practise celibacy. The legates of the Papal Curia gave their full support to the movement, which had the favour of Pope Alexander II and later of Gregory VII, both leading figures working for reform of the Church. Outside Italy, as well, the campaign against simonist clergy was supported by the Papal Curia. Gregory VII decreed that those who took part in the burning of one Ramihrdus in 1077,[15] in Cambrai in France, should be excommunicated; his only crime had been to refuse the Host from the hands of a simonist priest. Ramihrdus had followers who still took the same stand even after his death. The Pataria movement reached its peak in Milan in the early seventies. Another dispute over the archbishopric of Milan then led to sharp conflict between the power of the Western Empire and the bold theocratic conception of Gregory VII, aiming at putting in the hands of the Papacy supreme rule over the whole Christian world.

Victorious, the Church established itself as a special hierarchical society in which there was no longer any place for lay participation in action. Canon law was elaborated, setting the clergy aside as a superior caste; the concentration of all church administration in the hands of the Papal Curia involved the organization of a costly bureaucracy. The interest of the Church was much taken up at this time with questions of inalienability of ecclesiastical possessions, immunities and various juridical matters over which there was constant friction with the secular power. The Church demanded the restoration of confiscated property and revenues, and everywhere provoked violent opposition by demanding the restoration of tithes, taken over by the secular authorities. Material wealth had an undesirable effect on the way of life of the higher hierarchy, in sharp contrast to the miserable social standing of the lower clergy.[16]

Disillusionment at this state of affairs found strong expression in Arnold de Brescia (d.1155);[17] this disciple of Abelard demanded with rationalist logic that the Church should return to apostolic poverty and finally did not flinch from the conclusion that the contemporary Church was not the Church of God and its leaders not bishops. Arnold's birthplace was one of the cities affected by the Pataria movement in the previous century, and he took up this heritage. His passionate criticism of the Church found its most powerful outlet in Rome, in the stormy times of the city community's fight to throw off the domination of the Holy See. Arnold de Brescia inevitably came into conflict with the Church and his end was tragic indeed. Yet even within the ecclesiastical hierarchy no less sharp condemnation of the contemporary Church could be heard in many places. Bernard of Clairvaux himself, the tireless persecutor of Arnold de Brescia, could not help at the end of his life thinking nostalgically of the original state of the Church "long ago", when the apostles did not let down their nets to catch silver and gold, but the souls of men.[18]

The Church was indeed far removed from the ideal of its early days, but the longing to imitate the "apostolic life" did not die down. From the end of the eleventh century onwards individuals or groups were constantly appearing, leaving everything around them in order to follow the commandments of the Gospel and the living example of the apostles. For these men the *vita apostolica* meant a life of complete poverty, and following Christ as His apostles did.[19] Long before the time of Peter Valdès and St.Francis of Assisi, this ideal found its devoted followers among the clergy, individuals devoted to church reform and seeking a more perfect way of Christian devotion, at first secluded from the world.

At this time there was a large area of dense forest between Maine and Bretagne, which had not been systematically cleared as forests had been elsewhere when the great monasteries were founded. Nevertheless, at many points hermitages now appeared, to be followed in time by the first settlements.[20]

At the end of the eleventh century the Breton priest Robert d'Arbrissel left the centre of learning in Angers and went deep into the forest of Craon, where he subjected himself to the most severe ascetic discipline. He was already famous as a preacher, however, and younger members of the clergy who had already come under his thrall in Angers soon came in large numbers to seek him out in his solitude, and listen to his eloquence. Robert organized his followers in a monastic community according to the rule of St.Augustine, but soon left them and together with a few companions set out to wander through western France, preaching as he went.[21] His mission took him as far south as to Toulouse. His appearance and manner were clearly modelled on the apostles; barefóot, clad in a habit of rough homespun, he wore his beard and hair flowing long. Like Christ and His apostles, this wandering preacher and his companions often had not where to lay their heads. Among Robert's closest companions were Bernard and Vitalis and the spiritual development of Giraldus de Salis, the Périgord preacher, was also profoundly influenced by him. The sermons hitherto delivered in the quiet of the woods now resounded through the marketplaces. In those days the spoken word had far greater power to spellbind than it has today, and all the more so here since it was accompanied by visible acts.[22] Robert d'Arbrissel and his disciples prepared the way for a new conception of Christian poverty, unknown in monastic circles hitherto: not only to imitate the life of the poor, but to share it with them and amongst them.[23] Robert refused no-one, however abject, and cared for them all with devotion. The numbers thronging after him grew daily as he went on his pilgrimage, and women predominated; coming from all social classes, they deserted their homes to follow the preacher in an ecstasy of enthusiasm. Wherever he came, Robert's influence undoubtedly overshadowed that of the little respected parochial clergy. The churches were empty while crowds listened to the Breton preaching in the open air, with his disciples. They were sharp in their condemnation of the spiritual pastors of the people,[24] and spared no-one, even calling for the faithful to break off relations with unworthy priests.[25] In his native Brittany Robert d'Arbrissel had already made many enemies when by virtue of his office by the side of the Bishop of Rennes he had uncompromisingly carried out the principles of church reform.[26] His present activity could not fail to arouse opposition among the hierarchy, who feared that the preaching of this Breton hermit might undermine the authority of the Church entirely. Criticism was also directed against the crowds of men and women following Robert, because their vagrant life fitted under no monastic rule. There is no doubt that such a varied and uncontrollable crowd must often have been a nuisance to its surroundings.[27] Robert himself soon realized the need for a more stable form of life for his followers, whom he called only by the name of *pauperes Christi*.[28] While Bernard and Vitalis set up the monasteries of Tiron and Savigny for the men, Robert and a handful of his best disciples took care of the vast crowd of women. The first primitive buildings were put up to house the women, at a deserted spot called Fontevrault, with sanctuaries for Divine Service. Besides women suffering from leprosy there was a separate section for those who had been living as prostitutes. The great numbers of these women, arising from the loose life of many of the clergy, presented a serious moral and social problem of the time. Robert tried to reform these women by having them lead a monastic life together, while Vitalis de Savigny preferred to persuade them to marry and settle down to a normal life.[29] Even after he had founded Fontevrault, Robert continued his tireless preaching throughout the French countryside. A year after his death (1117) Norbert of Xanten took to the same way of life. This noble-born German canon was suddenly moved to drop a promising career and spent two eyars in meditation and solitude, practising the strictest asceticism. The disfavour with which he was regarded by the ecclesiastical synod of Fritzlar cemented his resolution (1118). He gave away all he had and went to France - the home of wandering preachers. In St.Gilles in Languedoc he found Pope Gelasius II who gave him explicit permission to preach. Norbert then made for eastern and north-eastern France, and following Christ's instructions to His apostles (Matthew 10,10) took "nor scrip for (his) journey, neither two coats, neither shoes". As a "stranger and pilgrim upon the earth" he gave to the poor and the lepers, all that was brought to him as a gift. His contemporaries regarded with amazement this manner of living "on the earth but demanding nothing from it".[30] Norbert's eloquence as a preacher, accompanied - in the spirit of the times - by the performance of miracles, had an immense effect everywhere, and his authority was truly great. Nevertheless the following year (1119) the new Pope, Callixtus II, did not approve of his mission.[31] Finally Norbert found a deserted spot near Laon, Prémontré, where he established a community of clergy to live by the rule of St.Augustine with special emphasis on poverty, symbolized by the white habit of these "white canons".

Like the Cistercians, Norbert and the other wandering preachers were moved by the need to escape to solitude when choosing sites for their new monasteries. Thus isolated from the world, the monks did not fulfil the mission of working among the people, so desperately needed in view of the utter inability of the parish clergy to do so.[32] At this time the Church was still uncompromising in insisting that only ordained priests in the apostolic succession could have the care of souls and the right to preach. Neither monks nor laymen could take on this work, except in rare cases where special permission was given, and with the consent of the local bishop concerned.[33] The longing to be able to take an active part in religious life was nevertheless growing stronger in all levels of society. It was one of the signs of the rising standard of economic and cultural life as urban life developed rapidly, and at a time when the social position of women was also undergoing a noticeable change. In the first half of the twelfth century women were playing a very large part in the religious movement. On the other hand, the existing monastic orders and those which were now being founded gave them little opportunity for active work. At first the Praemonstratensians founded mainly double convents (with a branch for women and one for men), but they soon closed their doors to women. Soon afterwards the Cistercians, too, refused the entry of women. It became clear, too, that the purely feminine foundations like those of the Fontevrault congregation were not very viable without the strength of a men's order to lean on.[34]

The time was still far distant when leading circles in the Church itself would realise the need to turn this movement to its own account; meanwhile it developed along its own lines, outside the Church and inevitably in opposition to it.

The life of the *monk Henry*[35] followed different lines from that of Robert d'Arbrissel, Giraldus de Salis or Norbert of Xanten. No hagiographical life of the monk, later persecuted by the Church, was written down, nor do we know anything of his origins. He first appeared in Lausanne, coming to Le Mans in 1116; here he obtained the bishop's permission to preach, just as Giraldus de Salis was officially allowed to do so by the Bishop of Poitiers.[36] Although the chronicle of the see of Le Mans gives a deliberately distorted picture of the "false hermit Henricus", it is clear that he had much in common with other wandering preachers of the day. As described in the Chronicle, his appearance bore a striking resemblance to that of Robert of Arbrissel, and his ascetic mode of life also appears to have been similar. Henry's approach to the problem of reforming women of evil repute was similar to that of Vitalis de Savigny. His attacks on the clergy, however, at least according to the writer of the Chronicle, not only alienated the people from their priests but made them openly hostile, so that when the bishop of Le Mans returned from Rome the people would have nothing to do with him, recognizing only Henry as their bishop.[37]

There are whole periods in the life of Henry for which we have no source of information. We know nothing about his life after he was driven from Le Mans, until in 1135 he appeared before the synod of Pisa,[38] where he was forced to recant his "heresy". It seems that he was then to enter the monastery of Cîteaux, but this did not in fact happen, and Henry went on preaching as he moved from place to place.[39] References to Poitiers and Bordeaux indicate the direction of his wanderings. Finally in Languedoc he found a powerful patron in the person of Alphonse Jourdain, Count of Toulouse, to whom Bernard of Clairvaux himself wrote a warning letter on the subject.[40] His account of the consequences of Henry's preaching in the south is strongly reminiscent of the much earlier complaints made by Bishop Marbod of Rennes against another wandering preacher, Robert of Arbrissel. The picture is remarkably similar: churches without worshippers, the people without priests, priests without the respect of the people.[41] Bernard, accompanying the papal legate on his mission (1145), was undoubtedly very successful in the south. In Albi the legate was met with ridicule, but shortly afterwards the church could not hold the crowds anxious to hear the eloquence of the great Cistercian Abbot. Nevertheless Bernard came up against stiff opposition from the numerous noble families of Languedoc who had listened with sympathy to Henry's verbal attacks on the clergy. Ultimately, however, Bernard managed to get a statute issued in Toulouse, against heretics and all who protected them. Henry was out of the city by that time, seeking safety in flight.[42] Later he was captured and handed over to a certain bishop, but nothing further is known of his fate.[43]

Henry the wandering preacher was in the end described as a heretic, and his case is a good illustration of the changes in meaning the term underwent as the Church developed. At the time when the campaign for ecclesiastical reform was in full swing, in the eleventh century, leading circles in Rome themselves instigated opposition to the clergy, encumbered by the evils of simony and "Nicolaism". Ultimately, however, the Gregorian reforms achieved their aim of freeing ecclesiastical institutions from dependence upon secular authority, and of establishing a hierarchical ordo. It was no longer permissible to doubt the validity of sacraments administered by ordained priests. Anyone who continued to declare that the validity of the sacraments depended on the purity of life of the priest administering them could be accused of heresy.[44] Yet it was no longer possible to silence the criticism, once it had been aroused. Henry the monk was not the only voice heard to declare that the clergy's sinful way of life had robbed them of the power to "bind and to loose" *(ligandi et solvendi)*.[45] At the time when Henry's teachings were profoundly disturbing the southern regions of France, far to the north, in Cologne on the Rhine, a group of "heretics" were attacking the dignity of the Papacy, accusing the successors of St.Peter of meddling in the things of this world. The Popes and the higher clergy subordinated to them had lost the ability to hand on consecration, yet unconsecrated priests could not administer the sacraments; the bread on the altar was thus not changed into the body of Christ.[46] Henry had also taught that the *corpus Christi per indignum ministrum non conficitur*, but he explicitly said that Mass could be offered wherever a person worthy of the office could be found.[47] It is obvious that the object of attack in both cases was not the sacrament of the Eucharist but the ecclesiastical hierarchy whose manner of living was such as to render them for ever unworthy of administering the sacraments.[48]

One of the disciples of Henry the monk noted down his words in writing, but all we know of his preaching today comes from the polemic of one Guillaume, a monk.[49] Henry derived his mission from Christ himself, who said: "Go ye therefore, and teach all nations".[50] He gave voice to an idea which was to be heard again, and with great force, from the Waldensians: "It is better to obey God than man".[51] Henry justified his teachings by the New Testament alone, although he did not spurn the works of the Church Fathers.[52] He quoted St.Mark's Gospel (Mark 16,16) to prove that infant baptism was valueless, since the children could not consciously believe. Christ's famous "Suffer the little children to come unto me" showed clearly that these innocent souls could be saved without baptism - a view which was of course contrary to the doctrine of original sin.[53] Everyone is either saved or damned after death, so that no sacrifices for the soul of the departed can change his fate.[54] Agreement between the two parties was sufficient basis for marriage, without any church ceremony.[55] Henry had no use for priestly vestments and the outward symbols of pastoral office, nor for church buildings.[56] His views already show the tendency towards consistent "Biblicism" which later marked one of the branches of the Waldensian sect. This involved rejection of the ecclesiastical institutions and customs which could not be attested in the Scriptures.[57] Thus the Gospels do not command the use of chrism and oil in baptism, and the same objection was raised against the practice of confession which indeed runs counter to the words of St.James the Apostle, "confess your faults one to another".[58]

It is a question whether Henry reached these conclusions through his own interpretation of the Biblical texts or whether he had come under the influence of earlier teachers. There was a strong anti-church movement in existence in the south of what is now France, even before Henry's time. In 1119 the Council of Toulouse threatened with excommunication all heretics who rejected the ecclesiastical hierarchy or the sacraments.[59] It was about the same time that Peter de Bruis was active in the south; twenty years later Peter Venerabilis wrote a great polemical work against his teachings, and in the Preface added a little later reported the violent death of this heretic.[60] The work of the Abbot of Cluny is now our only source of information on the nature of the "Petrobrusian" heresy, which spread from the Alps on both banks of the Rhône and further west into Gascony.[61] Peter de Bruis was not moved to follow the "apostolic life", but to rebel against the ruling Church. The movement took on very destructive forms, Peter's followers attacking and destroying churches and using physical force against monks and priests, forcing the former to marry. Their contempt for all consecrated customs of the church culminated in such ostentatious blasphemy as roasting meat over a bonfire of wooden crosses on Good Friday.[62] There is no sign that the secular authorities in the South took any steps to deal with this destructive wave of heresy, but Peter himself was lynched by an angry crowd in St.Gilles when he publicly burned a cross.[63]

The intense hatred of Peter de Bruis for the Church may perhaps be explained by his personal experience, if it is true that he had been deprived of his priesthood by the Church.[64] The origin of his teachings remains obscure, however; the only parts of the Bible he recognised were the Gospels, and the only part of Christian worship he allowed was prayer and adult baptism.[65] Certainly the separate elements of this heretical doctrine had appeared at least a century earlier in the West; there is a surprising analogy in the Italian group attested in the diocese of Châlons and later in Arras (1025).[66] Like Peter de Bruis later, these early heretics rejected infant baptism and the sacrament of the eucharist, and regarded church buildings as valueless. Peter ordered his followers to destroy churches, for man can pray anywhere, in a stable as well as before an altar.[67] The contempt for the cross noted among these Arras heretics was attested almost at the same time among the "Manichees" of Aquitaine, of whom we know so little. The Petrobrusians, however, hated the cross as the cruel instrument of the martyrdom of Christ, and their line of argument was very close to that heard by Presbyter Cosmas from the Bulgarian heretics. There is no need to go to the East to find the origin of this single item of the Petrobrusian heresy, however, for in other respects it was quite distinct from the dualist faith of the Bogomils.[68]

There is a considerable lapse of time between Peter de Bruis and the sectarian groups of the first half of the eleventh century, and continuity of development cannot be proved. There can be no doubt, however, that the Petrobrusian movement, which was still a vital force, influenced Henry the monk, who was preaching in the region of southern France in the early forties of the twelfth century. Peter Venerabilis had before him a work putting forward the doctrines Henry preached, and he declared that the monk had changed some points in the teachings of Peter de Bruis.[69] Comparison of the two shows that Henry did not go as far as his predecessor in rejecting the Eucharist and the cross.[70] Fundamentally, indeed, Henry did not go beyond the general religious movement of the time, which sprang from the atmosphere of ecclesiastical reform even if it sometimes strayed into heretical positions. It was a trend which had never yet taken on organized form.

Sometimes we see only small groups of local significance, as in a village near Soissons, where men and women lived in the manner of the apostles.[71] On the other hand, Henry the monk wandered about the countryside preaching, followed by his closest disciples, and the movement he called into being depended entirely on his personality.[72]

From the middle of the century, however, a really organized sect rapidly gained ground in the West, with a very effective way of achieving the western ideal of the *vita apostolica*. Christ lived without worldly goods and commanded his followers to live in poverty, and so the members of this sect had neither houses nor fields nor other property. They, too, were *pauperes Christi* who cared not for the things of this world. They lived as sheep among wolves, fleeing from town to town before persecution, barefoot, tireless in work and in prayer.[73] Rejecting the hierarchy and the worship of the Church, the sect agreed with earlier "heretical" groups, one of which is attested for Cologne alongside the new sect.[74] Yet the fundamental differences and even conflict between the ideas of the two trends could not fail to strike their contemporaries.[75]

It was nothing unusual in these times to reject infant baptism, and the arguments brought forward by this new sect were similar to those of other heretics in contemporary Western Europe.[76] It did not remain secret for long, however, that the sect was against the sacrament of baptism altogether, and administered a new baptism to their initiates, according to their own rite. This was known to consist of the laying on of hands, and in addition the future member of the sect had a holy book placed on his head. Eckbert, who later became Abbot of Schönau, writing in the early sixties of the twelfth century, has left us a vivid picture of this heretical rite. It often took place in dark cellars or in weavers' shops, where the members of the sect could gather in safety. All around, on the walls, lights were blazing as a symbol of the words of the Gospel, that this was baptism by the Holy Spirit and by fire.[77] This initiation rite placed the adept among the "elect", who were called Cathars, "the pure". They lived very ascetical lives, but the Church was well aware that their asceticism did not spring from the mere longing for a more perfect form of the Christian life, but expressed a fundamental rejection of certain things which the sect believed were sinful and impure. The Cathar sect had this negative attitude to married life as well as to the eating of meat. According to Eckbert of Schönau they would not eat meat even in dire need. Earlier we are told by Everwin of Steinfeld that they felt the same repugnance towards other foods of animal provenance, that is to say, towards milk products. Both German writers give the same Cathar argument, that all these things essentially come *de concubitu, de coitu.*[78] Nor was this idea new in the West: it was already found among the group of Soissons heretics (1114).[79] Eckbert of Schönau, however, knew of a deeper reason for this principle, one which the Cathars kept hidden from others. This was the belief that all animal bodies were created not by God, but by the Devil.[80] This fundamental negation of the flesh is clearly seen, too, in the views held by the sect on the nature of human beings and the human nature of Christ. They believed that the souls of men were really angels, imprisoned in mortal bodies after their fall. The Saviour had only apparently taken on Himself a human body.[81] These metaphysical ideas were only handed on to a select circle of initiates.[82] Nevertheless these secret teachings did come to light, revealed by people who had belonged to the sect for a time. Eckbert had the opportunity to acquire a thorough knowledge of the Cathar doctrines; he had come up against them when he was a canon in Bonn,[83] and later, when he was already Abbot of Schönau, he was called upon to combat this dangerous sect in Mainz.[84] This was after he had written his polemical work against the Cathars, addressed to the Archbishop of Cologne (1163).[85] It was very natural that the learned Benedictine should classify the obviously dualist doctrine of this contemporary sect as Manichaean, and indeed he attributed several specifically Manichaean ideas to the Cathars.[86] Nevertheless the actual features he describes are quite sufficient to designate the same sect described so long before in the Eastern Christian world by Cosmas the Presbyter, by Euthymius of Acmonia and Euthymius Zigabenus. In the West we see clearly for the first time how the sect was organized - a matter to which the Eastern sources paid very little attention. The true members of the sect, for whom the name *perfecti* became accepted in the West, are already clearly distinguished from the believers *(credentes)* who had not yet received spiritual baptism.[87] There were women among both the "perfect" and the "believers"; from the outset they played a very significant role in the movement.[88] The earliest sources in the West already refer to Cathar bishops, presbyters and deacons. At the same time, however, we sometimes find a single leading figure accompanied by twelve "teachers" *(magistri)*, obviously modelled on the twelve apostles.[89]

Some decades earlier, Basil had had his twelve "apostles", in Constantinople. According to the evidence of Euthymius of Acmonia, the Bogomil teachers followed the example of the Apostles and drew lots to decide where each was to work. Their missionary work reached into every corner of the Byzantine Empire and "to all Christians under the sun".[90] These words of the Byzantine monk seem to presage the sect's victorious advance.

The apostolic life and the consciousness of their apostolic mission, which the sect brought with it to the West, could not fail to evoke a powerful response there. Similarly the sect's attacks on the ecclesiastical hierarchy found ready support in the atmosphere of stormy criticism of conditions in the contemporary church. This criticism, already of long duration, had undoubtedly been instigated by the eleventh century movement for ecclesiastical reform. Paradoxically, however, as we shall see below, this mood agreed remarkably well with the material interests of certain circles, running counter to the goal of the Gregorian reforms. In this respect, too, the ground was already prepared for the Cathars' anti-church propaganda in the West.[91]

Heretical missions from the East came to Western Europe at a time when the crusades were providing much intensified opportunities for contact between the latter and the Byzantine Empire. We do not know how the sect reached the West, but from the outset it is depicted in western sources with distinctive features which cannot be confused. The first reports came from Cologne on the Rhine, when a letter from the Provost Everwin of Steinfeld, to Bernard of Clairvaux attests the existence of the sect about 1143-1144.[92] This new heresy progressed rapidly through the Rhine basin and in Flanders, following the route taken by merchants from the Low Countries on their way to the markets of Champagne. Perhaps this provenance in Flanders explains why the French used the name "weavers" *(textores, texerants)* for these heretics.[93] Immediately after Everwin's letter to Bernard of Clairvaux, Pope Lucius II was also informed sbout the new sect. The clergy of Liège sent quite concrete information about the Cathar heresy to Rome; the centre is named for the first time as Mont-Wimer (Mont-Aimé) in Champagne.[94] In 1157 the Council of Rheims took energetic measures against the "Manichaean" sect now spreading everywhere.[95] A number of Cathars of Flemish origin, both men and women, were discovered a few years later in a suburb of Cologne.[96] The sources give the names of three of the leading figures, knowledgeable in Latin and well-read in the Scriptures. Shortly before some of these Flemish Cathars met their death in the flames in Cologne, a company of German-speaking sectarians landed on the coast of England (about 1162); they were all simple folk, looking like peasants. They were soon put in prison, and after they had been publicly convicted of heresy in Oxford were branded and turned loose to freeze and starve.[97] This attempt to proselytize on English ground failed tragically, but on the Continent the sect grew so fast that it could not be rooted out.A letter from the monk Heribert attests the presence of Cathar "apostles" far south of the Loire,in Périgord.[98] The Council of Tours (1163) was forced to admit that the plague of heresy which had appeared long before in the region of Toulouse had now affected Gascony and other neighbouring provinces.[99] This would suggest that the Cathars had taken root in the County of Toulouse as early as in the forties. This means that the success of the mission led by Bernard of Clairvaux had been ephemeral, and that soon after his departure a new wave of heresy swept the region. As we shall see below, the Cathar sect had taken root on the other side of the Alps since the middle of the twelfth century, for it was already widespread in the northern and central regions of the Apennine peninsula before the year 1167. It all happened in a very short time. Contemporaries reported in a panic that those who believed these heretical doctrines had "multiplied throughout all lands"[100] and were as numerous "as the sands of the sea".[101] Nothing could quench the missionary ardour of the heretical apostles, ready to travel far and wide "over land and sea" to gain a single adherent.[102] The victorious progress of the sect was not without its sacrifices, but a martyr's death was accepted and even joyfully welcomed.[103] More than one chronicler recorded for the year 1163 the martyrdom of a Flemish girl who flung herself voluntarily into the flames.[104] A strange power carried these people over the boundaries of death, drawing on their conviction that they bore the message of the true Christian faith handed on in secret since the days of the Apostles, and preserved "in Greece (i.e. Byzantium) and other countries". Everwin of Steinfeld remarked on this in his letter of 1143-1144 to Bernard of Clairvaux.[105]

The sect brought the name *Cathar* with it from the East; the original meaning of the Greek name was soon forgotten in the West and in its German form *Ketzer* the word was a general term for "heretic". Eckbert of Schönau was well acquainted with the original sense of the word Cathari, first attested in his work; the name expressed the pride of the members of the sect in their "pure" way of life, as Euthymius Zigabenus already commented.[106]

The eastern provenance of the sect is also clearly indicated by the designation "Bulgarian heresy" *(Bulgarorum haeresis)* which emerged at the beginning of the thirteenth century.[107] Even earlier we find the curious name *Popelicans, Popelicani,*[108] appearing in various forms including that of *Publicani,*[109] clearly a popular etymology based on associations with the text of the Gospels. The name "Popelicans" seems to have been derived from an abbreviated form of the name of the town Plovdiv - Philippopolis, often given in western sources as *Phinepople* or *Phinepôpel.*[110] It is possible that another abbreviation of Philippopolis can be discerned in the name *Piphles (Piphli, Piphili)* used in Flanders and attested for the first time in the 1st canon of the Council of Rheims in 1157.[111] Plovdiv was famous as a centre of dualist doctrines, and it is therefore not impossible that it was from here that the first apostles of the new faith came to Western Europe, and that this was the reason why the name of this city stuck to them - a name well known from the time of the first crusade.

The sect looked constantly to the East, and the new Cathar churches in the West maintained lively contact with their mother churches "across the sea". It was from here that they drew their strange myths of the origin of Evil, the creation of the world, redemption; and it was from here that they were ordained bishops, in an unbroken line of apostolic succession.

[1] References to the sources for the history of the various sectarian groups in the West in the first half of the eleventh century are given by *Theloe,* 5 - 29; cf. also *Grundmann,* Bewegungen, 476 - 483. The best analysis was presented by *Ilarino*

da Milano, Le eresie popolari del secolo XI nell Europa occidentale, in: Studi Gregoriani 2, Rome 1947, 43 - 89. This study was written before the publication by *A.Dondaine* of the texts he discovered, which irrefutably prove the eastern origin of western Catharism. Taking certain heretical phenomena in this early period as representing the initial phase of Catharism, Ilarino did not consider the possibility of Bogomil influence. Dondaine himself, in his study of the question written in 1950, viewed with great reserve the possible penetration into western Europe of eastern dualist ideas, before the middle of the twelfth century. He referred in this connection to the earlier unsubstantiated attempts to interpret the source material in this sense, and in particular corrected the dates hitherto accepted for the life of Raoul Ardent, who gave a characteristic description of Cathar heresy in the Agen region, in one of his homilies (Homiliae, pars II, XIX, PL 155, col. 2010 - 2013). His "Speculum universale de virtutibus" was written between 1179 and 1200, so that the homilies could not have been written before the second half of the twelfth century. In so far as Bogomil influence penetrated to western Europe in the eleventh century in isolated cases, it was still too weak to be felt (*Dondaine*, Hiérarchie II, 268 ff.). Somewhat later the same author came to a different conclusion, however. Early instances of heresy in the West, in the eleventh century, now took on for him the appearence of a single stream of influence from the East, coming to the surface at different points in the western world. (See *A.Dondaine*, L'origine de l'hérésie médiévale, in: Riv. di Storia della Chiesa in Italia 6 (1952) 47 - 78. Cf. *E.Werner*, Bogomilstvoto i rannosrednovekovnite eresi v latinskija Zapad, in: Ist. Pregled XIII (1957) kn. 6,16-31. *Puech*, Catharisme médiéval, 77 f., on the other hand rightly defends Dondaine's original point of view. *Borst*, 72, n.2, tends more to the conclusions put forward by Dondaine in 1952, but sees the basis of the early expressions of heresy not in dogmatic speculation inspired by Bogomil dualism, but in the desire to live the true Christian life, springing from native tendencies.

[2] "... Italia non consuevit haereses nutrire, ad praesens in quibusdam partibus haeresium fomentis abundare auditur. Gallia vero felix, quae his munda perhibetur. Graecia infelix, sine quibus nunquam vivere voluit. Verona, urbium Italiae nobilissima, his gravida redditur. Illustris Ravenna, et beata Venetia, quae nunquam inimicos Dei passae sunt ferre. Infelices, et miserrimi, qui Dei, et Ecclesiae suae adversarios questu lucri attollunt". Deliberatio Gerardi Moresenae ecclesiae ep(iscop)i supra hymnum trium puerorum ad Isimgrimum liberalem, in: S.Gerardus episcopus Chanadiensis, Scripta et Acta, ed. *J.Graf Battyány*, Karlsburg - Ungarn 1790, 99. On Gerhard see *J.A.Endres*, Studien zur Geschichte der Frühscholastik, in: Philos. Jahrb. d. Görres Gesellsch. 26 (1913) 349-359. Further literature is given by *Borst*, 79, n.20. Whereas in the lines quoted above Gerhard is speaking of heresy in general, earlier (98 f.) he makes the character of the heresy which appeared in his time a little more precise: "Hoc tempore omnes maledixerunt apud nos concitati zelo non solum divinis ritibus, et Ecclesiae, et Sacerdotibus, quin etiam ipsi Dei filio Jesu Christo Domino nostro. Omnibus inaudita saeculis haeresis repetere eleemosynas pro animabus defunctorum christianorum more expensas." From the words *apud nos* it could be assumed that the writer is referring to conditions in Hungary. See *Ilarino*, art. cit., 51. There are no grounds for assuming that Gerhard is referring to the Bogomils, and thence to assume a connection between the Italian heretics referred to and those of "Greece" (= Byzantium), as Borst does (78). *Puech*, Catharisme médiéval, 81 f., accepted Borst's interpretation: "s.Gerhard, Évêque de Czanád... rapporte des hérétiques de Vérone, Venise et Ravenne, qu'ils avaient en

Grèce beaucoup de congénères, de coreligionnaires". The new Atlas of Church history shows these three Italian towns (Verona, Venice and Ravenna) and the see of Csanád in Hungary with the note "Orte oder Gebiete mit einer-möglicherweise bogumilischen-Häresie". See Atlas zur Kirchengeschichte, ed. *H.Jedin* - *K.S.Latourette* - *J.Martin*, Freiburg i. Br. 1970, Map 57A. Both Puech and the author of the map seem to have been somewhat misled by Borst's not very clear formulation ("in Venedig und Ravenna hat man sie nicht geduldet") as far as Venice and Ravenna are concerned. In reality, the quotation from Gerhard's text says quite clearly that these places never suffered from the presence of heretics. E.v.Ivánka connects yet another passage in the work of Gerhard of Csanád with the Bogomils; it concerns the Archangel Uriel, known to the Greeks and called upon particularly by the heretics (ed. *Battyány*, 233). See *E.v.Ivánka*, Gerardus Moresanus, der Erzengel Uriel und die Bogomilen, in: Orient. Christ. Period. XXI (1955) 143-146. Since however we have no evidence whatever that the figure of the Archangel Uriel known from a number of apocryphal texts ever found its way into the Bogomil myth, there is no reason to see in Gerhard's comment "eine der ersten Erwähnungen des Bogomilismus im lateinischen Schrifttum".

[3] ...et quicquid sanae doctrinae est. Ademari historiarum liber III, MGH SS IV, 138 (ad 1018).

[4] Ibid., 143.

[5] *Dondaine*, L'origine ... 62 f., finds practically all the basic features of Bogomilism present in the teachings of western sectarian groups in the first half of the eleventh century. The fact that details correspond does not prove anything, however, for the same elements may have a different significance in a different system of ideas.

[6] The most reliable source for Orleans is Gesta synodi Aurelianensis, in: *Bouquet*, Recueil X, 536 ff.; for Monteforte, Landulfi historia Mediolanensis, lib. II, cap. 27, MGH SS VIII, 65 f. Ilarino (art. cit., 72) rightly sensed the essentially spiritual character of the doctrines of both these groups; his attempt to derive this spirituality from dualist metaphysics, however, is not convincing (85 f.). The fact that the sources are silent on this point cannot be taken to mean that the heretics deliberately hid the truth. This was decidedly not a valid explanation in the case of the Monteforte group, whose longing for martyrdom led them to confess the tenets of their faith quite openly. It should also be noted in this connection that the practice of martyrdom inflicted on one another attested for the Monteforte heretics, had not the same significance as the *endura* of later Catharism (see below, Chapter XVIII) with which Ilarino da Milano identified it (86).

[7] *Ilarino* (85 ff.) makes a distinction between the teachings of the Arras heretics and those of the groups in Orleans and Monteforte, classifying the latter groups as *corrente a evoluzione catara*.

[8] Glabri Rodulphi Hist. lib. II, in: *Bouquet*, Recueil X, 23. There is no confirmation in any other source of this story of the simple peasant Leutard preaching heresy. In view of the general character of Rudolf Glaber's work this evidence should be treated with caution.

[9] Anselmi Gesta episc. Leod., MGH SS VII, 226. These "Manichees" rejected marriage and not only refused to eat meat, but believed it was wrong to kill any living thing. In this they agreed with the group of heretics condemned in Goslar, in Germany, in the year 1051. The same source mentions this latter group (228). Further references can be found in *Theloe*, 20 f., n. 71. In his reply Bishop Wazo accused the Châlons "Manichees" of inconsistency, in that they extended the Old Testament

"Thou shalt not kill" to all living things, but did not refrain from eating foods of vegetable origin (Anselmi Gesta, l.c., 227). This makes it quite clear, of course, that these heretics could not have been real Manichees.

[10] Ex synodo Atrebatensi, 541. Cf. the preface to these Acta of Arras (ibid., 540).

[11] See Glabri Rodulphi Hist. lib. III, in: *Bouquet*, Recueil X, 35, on the year 1022, in connection with the outbreak of heresy in Orleans. Cf. *Ilarino*, art.cit., 61.

[12] Ibid., 88.

[13] For examples, see ibid., 77.

[14] On the Pataria see *Volpe*, Movimenti, 1 ff.; *S.M.Brown*, Movimenti politico-religiosi a Milano ai tempi della Pataria, in: Arch. stor. lomb. LVIII (1931); *Dupré-Theseider*, Introduzione, 77 ff.; *Borst*, 82, n. 4.

[15] References in *Theloe*, 27 f.

[16] See *Volpe*, Movimenti, 38 ff.; *Dupré-Theseider*, Introduzione, 102 ff.

[17] See *A.Frugoni*, Arnaldo da Brescia nelle fonti del secolo XII (= Ist. stor. ital. per il medioevo. Studi Storici, 8-9), Rome 1954. References to earlier literature are in *Borst* 88, n.22,23.

[18] See *Dupré-Theseider*, Introduzione, 124.

[19] *J.v.Walter*, Die ersten Wanderprediger Frankreichs, Studien zur Geschichte des Mönchtums I, II. Leipzig 1903, 1906; *Grundmann*, Bewegungen; id., Eresie; *E. Werner*, Pauperes Christi, Leipzig 1956.

[20] On the spread of hermitages in western Europe and its importance for the settlement of uninhabited areas, see *Werner*, op. cit., 25 ff. (with plentiful references to the relevant literature).

[21] It is not certain that Robert's mission as a preacher was laid upon him by Pope Urban II himself. The statement to that effect in Baldrich's Life of Robert d'Arbrissel, not confirmed in any other source, is regarded with justified doubt. See *Werner*, Pauperes Christi, 45; *Selge*, Die ersten Waldenser I, 246, n. 47.

[22] "Nec erat sermo illius sterilis, nec otiosus, quem commendabat laudabilis et verborum et operum comitatus. Quod praedicabat, complebat operibus..." Vita B. Roberti de Arbrissello, auctore Baldrico episcopo Dolensi, PL 162, col. 1051A.

[23] See *M.Mollat*, Le problème de la pauvreté au XIIe siècle, in: Vaudois Languedociens et Pauvres catholiques (= Cahiers de Fanjeaux, fasc. 2), 1967, 42.

[24] See the letter of Marbod, Bishop of Rennes, ed. *Walter*, op. cit. I, 187.

[25] *J.de Petigny*, Robert d'Arbrissel et Geoffroi de Vendôme. Bibl. de l'École des Chartes 5 (1854) 12 f.

[26] Ibid., 3.

[27] Ibid., 6.

[28] This term, which had been used since the tenth century to denote monks, acquired a special connotation in the twelfth century. See *Werner*, Pauperes Christi, 19.

[29] See *Walter*, op. cit. II, 85 f.

[30] "...in terra degere et nil de terra quaerere." Vita S.Norberti archiepiscopi Magdeburgensis, MGH SS XII, 675.

[31] See *A.Hauck*, Kirchengeschichte Deutschlands IV, Leipzig 1913, 372 f.; *Grundmann*, Bewegungen, 44; *Werner*, Pauperes Christi, 50.

[32] See *Grundmann*, Eresie, 386 f.

[33] *Grundmann*, Bewegungen, 63 f.

[34] See *Grundmann*, Bewegungen, 46 ff., 203 ff.; id., Eresie, 400.

[35] For references to the sources see *Theloe*, 63, n. 237. To this list we must add the polemic of the monk Guillaume, first published (according to the MS of the Bibliothèque Nationale, lat. 3371) by R.Manselli. See *R.Manselli*, Il monaco Enrico e la sua eresia, in: Bull. dell'Ist. stor. ital. per il medio evo e Archivio Muratoriano 65 (1953) 1 - 63 (text 44 - 62). Cf. *Dondaine*, Durand de Huesca, 256 ff.

[36] Vita B.Giraldi de Salis, ASS X, 254 - 267.

[37] Ex gestis pontificum Cenomanensium, in: *Bouquet*, Recueil XII, 547 - 551.

[38] For the date of this synod see *Hefele - Leclercq* V 1, 706 f.; cf. *Theloe*, 65, n. 241.

[39] Ex gestis pontificum Cenomanensium, l.c., 554. Cf. *Theloe*, 65, n. 242.

[40] S.Bernardi epist. CCXLI, PL 182, 434 ff.

[41] Epistola ad Robertum, l.c.: "Videmus egentes presbyteros a suis desertos gregibus velut indignos, quibus se offerant, quorum se commendent orationibus, a quibus injunctionem poenitentiae accipiant, quibus solvant decimas vel primitias..." Cf. S.Bernardi epist. CCXLI, l.c., 434A : "Basilicae sine plebibus, plebes sine sacerdotibus, sacerdotes sine debita reverentia..." Here the picture is still further emphasized: "Ecclesiae synagogae reputantur; sanctuarium Dei sanctum esse negatur: Sacramenta non sacra censentur: dies festivis frustrantur solemniis. Moriuntur homines in peccatis suis... Parvulis Christianorum ... Baptismi negatur gratia."

[42] Epistola Gaufredi, monachi Claravallensis... Pl 185, col. 410B - 416A. For Henry's influence in Toulouse see ibid., col. 411D : "Paucos quidem habebat civitas illa, qui haeretico faverent: de textoribus, quos Arianos ipsi nominant, nonnullos. Ex his vero qui favebant haeresi illi plurimi erant, et maximi civitatis illius." This difficult passage has been variously interpreted. Sufficient attention has apparently not been paid to the fine distinction expressed by the words *haeretico* and *haeresi illi*. There were not many people in Toulouse who were personally devoted to Henry; they were disciples he had won over in the city. The circle of those who sympathized with his opposition to the Church, however, was much greater, and included (as Bernard's biographer explicitly tells us) people of the first families in the city. This interpretation of the given passage is fully confirmed by the following lines, distinguishing between Henry and the "Arians" who escaped with him, on the one hand, and their supporters *(fautores)* in the city, who renounced them, on the other. The supporters of Henry are called *textores* not only by the biographer of Bernard of Clairvaux but also Bernard himself: "Clerici et sacerdotes, populis ecclesiisque relictis, intonsi et barbati apud eos inter textores et textrices plerumque inventi sunt." (Sermones in Cantica 65, PL 183, col. 1092B). According to Gaufridus, these *textores* among whom the monk made converts were called *Ariani* in Toulouse. (Cf. *Grundmann*, Bewegungen, 31, n. 35). From the context it appears that this was a group of people which already existed in Toulouse before the arrival of Henry. The terms *Ariani* and *textores* led some authors to the conclusion that these were already Cathars (see *R.Manselli*, Una designazione dell'eresia catara: "Arriana Heresis", in: Bull. dell'Ist. stor. ital. per il medio evo e Archivio Muratoriano 68 (1956) 238 f.

In his analysis of this passage in Gaufredus, *E.Griffe* takes the words *haeresi illi* to refer not to the teachings of the monk Henry himself, but to the heresy of the "Arians" of Toulouse, whom he sees as Cathars (*Griffe*, Aventure, 34 ff.). It is his contention that the movement brought into being by the preaching of Henry significantly reinforced the sporadic foci of Cathar heresy he believes had existed ever since the early twelfth century, in the region east of Toulouse and reaching as far as Albi in the north (*planities Tolosae*, ibid., 174 f.). It is certainly true that the designation *Ariani* was one of those used in the West for the Cathar sect in its beginnings (see Chapter XII, n.21). However, it goes no further towards proving that these "Arians" of Toulouse, among whom were close disciples of the monk Henry, were Cathar in their beliefs. An argument against such an interpretation is the fact that from the outset the Cathar sect appeared as quite distinct both in organization and in doctrine, never merging with local groups (see below, on the situation in Cologne). The question as to who the *Ariani* mentioned here really were has still to be answered.

43 See *Theloe*, 69.

44 See *Grundmann*, Eresie, 389.

45 See the polemic by the monk Guillaume, ed. *Manselli*, l.c., 56.

46 Everw. Steinf., col. 678D f.

47 Monk Guillaume, ed. *Manselli*, l.c., 53.

48 Cf. *Grundmann*, Eresie, 392.

49 See n. 35 above.

50 Monk Guillaume, ed. *Manselli*, l.c., 44 f.

51 Ibid., 44. (See Chapter XII below).

52 Ibid., 45.

53 Ibid., 47 ff.

54 Ibid., 62.

55 Ibid., 55.

56 Ibid., 61.

57 See Chapter XIV below.

58 Monk Guillaume, ed. *Manselli*, l.c.,51, 58.

59 *Hefele - Leclercq* V 1, 570. The second Lateran Council again considered heresy in the south of France, in 1139 (see ibid., 731 f.), as did the Council of Rheims in 1148)ibid., V 1, 826). *Grundmann*, Bewegungen, 52, points out that all the decisions taken by the councils against heresy in the twelfth century referred to the south France.

60 Petrus Venerabilis, Tractatus adversus Petrobrusianos, PL 189, col. 719 - 850 (quoted further as Peter Ven.) See *J.Fearns*, Peter von Bruis und die religiöse Bewegung des 12. Jh., in: Archiv für Kulturgeschichte 48 (1966) 311 - 355. The author convincingly dates this work by Peter Ven. in 1139/1140. The *terminus ad quem* for the Preface is 1141 (314 ff.).

61 Peter Ven., col. 721A.

62 Ibid., col. 726B, 771C.

63 Ibid., col. 723A.

64 See Abelard, Introd. ad theol., lib.II, c.4, PL 178, col. 1056AB. Cf. *Fearns*, art. cit., 317.

65 Ibid., 319.

66 Ibid., 332 f.

67 Peter Ven., col. 722B.

68 In his admirable analysis of the Petrobrusian heresy Fearns gives a very good account of the fundamental differences between this sect and that of the Bogomils of the East. He nevertheless considers the specific arguments explaining the Petrobrusian hatred of the cross (Peter Ven., col. 773D - 774A) as of Bogomil origin

(324 f., 331 f.). In spite of remarkable similarities, however, the line of argument is not identical in both cases. A fundamental attitude of hatred of the cross as the instrument of Christ's passion could arise independently in various heretical groups, and give rise to similar lines of thought. Fearns also considered the possibility of Bogomil influence on the Petrobrusian attitude to the books of the Bible (318 f.). The followers of Peter de Bruis based their beliefs only on the Gospels, however, and either rejected the rest of the Bible outright, or viewed some of the books with doubt. Unlike the Bogomils they were equally sceptical of the epistles of St. Paul and the other apostles (Peter Ven., col. 737CD, 828C).

69 Peter Ven., col. 723A.

70 *Fearns*, art. cit., 329.

71 See *Theloe*, 33 f.; *Grundmann*, Bewegungen, 17; *Koch*, Frauenfrage, 41 f.

72 A special case was that of Tanchelm in the region of Antwerp, 1110 - 1115 (see *Theloe*, 31 ff.; *Borst*, 84 f.). His success, too, was facilitated by the inefficiency and low moral standards of the parochial clergy (see Vita Norberti archiepiscopi Magdeburgensis (A), cap. 16, MGH SS XII, 690 f.). Tanchelm called upon the people not to partake of the Eucharist and not to pay tithes (see the letter sent by the chapter of Utrecht to Frederick, Bishop of Cologne, *Frédéricq* I, No.11, p. 16); contemporary accounts agree that he also called himself the Son of God. He thus stands quite apart from the contemporary trends aiming at *imitatio Christi* and modelled on the example of the apostles (see *Grundmann*, Bewegungen 35 f., n. 45; *Werner*, Pauperes Christi, 194). The Breton Eon de Stella is quite outside the main trend (see *Theloe*, 39 f.; *Alphandéry*, Idées morales, 102 ff., n. 4; *Borst*, 87 f.; *Maisonneuve*, 106 f.).

73 Everw. Steinf., col. 677CD; Herib. Mon., PL 181, col. 1721.

74 Everw. Steinf., col. 678D - 679C.

75 Everwin makes a clear distinction between the Cathars as a special sect, and the other heretical group in Cologne. Bernard de Clairvaux, on the other hand, who was informed by Everwin and perhaps another source as well, made no distinction between them. He saw them as one stream of heretical thought in which however serious contradictions might be observed. (See S.Bernardus Claravallensis, Sermones in Cantica, 66, PL 183, col. 1093 - 1101).

The Cathars taught, just as did that local group in Cologne, "quod sacerdotalis ordo defecerit in Romana Ecclesia" (Ecb. Schon., col. 71A). From what Eckbert goes on to say: "opportet ut nobis demonstretis sub quo papa acciderit iste defectus", it can safely be concluded that the Cathars he knew did not as yet believe that the apostolic succession had been broken by Pope Silvester (see Borst, 215, n. 9; cf. below, Chapter XVII, 266). According to Eckbert, the Cathars repeated the usual argument that the Eucharist was not valid when administered by an unworthy priest: "Quomodo fieri potest, ut qui tam irrationabiliter vivunt, distribuant in Ecclesia corpus Domini?" (Ibid., col. 88B). He leaves no doubt, however, that they rejected the sacrament of communion as such: "omnino renuitis credere, quod ab aliquo sacerdote sive bono sive malo possit ulla consecratio fieri corpus Domini" (ibid., col. 84B). Cf. also the words of Elisabeth of Schönau: "...infelices katari ... lacerant vulnera mea effundentes corpus et sanguinem sacramentorum meorum..." *F.W.E.Roth*, Die Visionen und Briefe der hl. Elisabeth ... Brno, 1886[2], 76.

76 "De baptismo varie loquuntur. Baptismum nihil prodesse dicunt parvulis qui baptizantur, quia per se ipsos baptismum petere non possunt, quia nullam fidem possunt profiteri." (Ecb. Schon., col. 15A).

[77] "Est autem aliud quod inde communius sed occultius loquuntur, videlicet quod nullus baptismus in aqua prosit ad salutem. Inde et eos qui ad sectam illorum transeunt, suo quodam occulto modo rebaptizant, quem baptismum in Spiritu sancto et igne fieri dicunt" (col. 15A). For a description of the spiritual baptism of the Cathars see Ecb. Schon., col. 51C - 52A.

[78] Everw. Steinf., col. 678A; Ecb. Schon., col. 14B.

[79] Guibert of Nogent, De vita sua lib. III, c. 17, PL 156, col. 951C.

[80] "Et hanc quidem rationem manifestius dicunt, sed in occultis suis, quod pejus est dicunt, videlicet omnem carnem facturam diaboli esse, ideoque neque in summis necessitatibus eam ullatenus gustant." (Ecb. Schon., col. 14C - 15A).

[81] Ibid., 15D - 16B, 94C - 98B. On the Christology of the Cathars see also Elisabeth of Schönau, in: *Roth*, op. cit., 76.

[82] Ecb. Schon., col. 19C.

[83] Ibid., p. 13 - 14.

[84] See Vita Ecberti, in: *Roth*, op. cit., 352.

[85] On the date of this work see *Borst*, 7.

[86] Ecb. Schon., col. 17D. Eckbert identifies the Cathars with the Catharists, a branch of Manichaeism (col. 18A).

[87] Ecb. Schon., col. 14B, 31A. Everw. Steinf. (col. 678CD) distinguishes three ranks: *auditores*, *credentes*, *electi*. We do not find the term *auditores* in any other source, however.

[88] Everw. Steinf., col. 679D - 680A.

[89] Ibid., col. 677B: "...qui dicebatur episcopus eorum cum socio suo...": Ecb. Schon., col. 17A:"...ex electis suis habent duodecim quos appellant magistros, et tertium decimum principem ipsorum: episcopus autem septuaginta duos qui ordinantur a magistris, et presbyteros et diaconos qui ab episcopis ordinantur, et hi electi inter eos vocantur." The monk Heribert knew the name of the leading Cathar in Pêrigord: "Princeps eorum Ponnus vocatur" (Herib. Mon., col. 1722).

[90] See Chapter V, n. 43.

[91] See Chapter XII below.

[92] On the date see *Borst*, 4, n. 4.

[93] *Grundmann*, Bewegungen, 31 f., gives all the examples of the term *textores*, *texerants*. See also *Borst*, 248; *Koch*, Frauenfrage, 16 ff.; *Grundmann*, Eresie, 399. Borst rightly points out that the word has a pejorative nuance. We might add that proof of this can be found in St.Jerome: "...et sum illas didiceris, inter mulierum *textrinas* cantato, imo legendas propone *in tabernis* tuis..." S.Eusebii Hieronymi Stridonensis presbyteri Contra Vigilantium Liber unus, PL 23, col. 360C. Cf. Ecb. Schon., col. 90A: "...tu, infelix Cathare, ...omnia quae a sacerdotibus nostris in ecclesia aguntur, *in textrina* tua irrides..." and below, col. 90B: "Ab uno viro qui *de angulis* vestris exierat unam talem sapientiam vestram audivi" (my italics, ML). The word *textrinae*, associated with *tabernae* or *anguli*, here symbolizes the low and obscure milieu in which ecclesiastical polemists placed heresy.

[94] References to the heretics in Mont-Wimer (Mont-Aimê) and the diocese of Liège are given by *Borst*, 91 f., n. 11.

[95] *Mansi* XXI, 843; *Hefele - Leclercq* V 2, 913.

[96] For references to the sources, see *Theloe*, 57 ff.

[97] Guillelmus Novoburgensis, Historia rerum Anglicarum, in: *Bouquet*, Recueil XIII, 108. Cf. *Borst*, 93 f. Other evidence of the occurrence of a single case of heresy (perhaps Catharism) in England is given by Guilberti Foliot Epistola CCL (Ad

Rogerium Wigorniensem episcopum) PL 190, col. 935 - 936: "quid super his *textoribus* sentiamus, qui vestra nuper ingressi diocesim..." (my italics, ML). Cf. *Theloe,* 54.

98 Herib. Mon., PL 181, col. 1721 - 1722. Cf. *Grundmann,* Bewegungen, 18, n. 8; *Borst,* 4, n. 8.; 92.

99 *Mansi* XXI, col. 1177 - 1178; *Hefele - Leclercq* V 2, 971 f.

100 Ecb. Schon., col. 13B.

101 Guillelmus Novoburgensis, l.c., 108.

102 Ecb. Schon., col. 16C.

103 See Everw. Steinf., col. 677B.

104 She belonged to the group of heretics from the Flanders who were discovered in Cologne (see n. 96.).

105 Everw. Steinf., col. 679D.

106 Ecb. Schon.,31A; cf. the 1st canon of the Council of Rheims, 1157: "ut quasi puriores mundioresque valeant apparere." On the name "Cathar" see Borst, 240 ff. The earliest diplomatic documentary evidence of the term dates from 1164-1167 (*Frédêricq* I, No. 33, p. 35); see *Christine Thouzellier,* in: École pratique des Hautes Études, Annuaire 1971 - 1972, 359.

107 Robert d'Auxerre, Chronicon, MGH SS XXVI, 260, 21-24; 270, 28 (referring to the years 1201 and 1206). Cf. *Thouzellier,* Languedoc, 151, n. 82.

108 The name is first found (in the form Populicani) in 1162, in a letter from the French king Louis VII to Pope Alexander III (*Bouquet,* Recueil XV, 790). See also Historia Vizeliacensis Monasterii, ad 1167: *Deonarii, seu Poplicani* (*Bouquet,* Recueil XII, 343).

109 The anonymous Gesta Francorum mentions the Publicani in several passages, in the list of peoples making up the army of the "infidels" fighting against the crusaders. The significance of the term is not clear in this context, but there is nothing to justify the assumption that it refers to a religious sect. (Cf. *Runciman,* Medieval Manichee, 46, who identifies these Publicani with the Paulicians).

110 Our interpretation is supported by the context in which the word Popelicans occurs in Villehardouin's description of the events in Plovdiv in 1205. See Chapter XVI, n. 9.

111 See n. 95. The name occurs in the title: *De Piphilis.* Ecb. Schon., col. 13C, attests the use of the name in Flanders.

Chapter X

THE EMERGENCE OF THE CATHAR CHURCHES AND THEIR RELATIONS WITH THE CHRISTIAN EAST

These relationships as seen in the twelfth century are remarkably well illustrated by three documents which are quite independent of each other.[1] The record of the proceedings of the great assembly of Cathars which met in 1167 in the town of St.Félix de Caraman, near Toulouse in southern France, is a very valuable document indeed.[2] Here we find the first references to the territorial organization of the dualist churches of western Europe. There was a bishopric somewhere in the north of France, while the second was centred in the south, near the Languedoc town of Albi,[3] whence the name Albigenses, often applied to the Cathars. In northern Italy at that time there was only one bishopric, but in the Languedoc, in southern France, where the followers of the Cathars were most numerous, two more sees were to be established, in Toulouse and Carcassonne. It seems that the third bishopric to be set up at that time, and mentioned in our text, was also in southern France, in the region of Agen.[4] At the assembly in St.Félix the representatives of the churches concerned elected the new bishops, who then received the "consolamentum" and were consecrated. Both rites were performed by a dignitary of the sect, here called Papa Niquinta. *Consolamentum* is the Latin term for the Cathar spiritual baptism; derived from the verb *consolare*, to comfort, it is associated with the concept of the Holy Spirit - the Paraclete (Greek: *parakletos*). It can be deduced from later sources that not only the consolamentum, but also the ordination *(ordo episcopi)* was administered by the laying on of hands. Papa Niquinta, who appears in this document as the supreme authority, was well acquainted with the organization of the dualist sect in the eastern regions of the Christian world. It was there that the "original churches" *(primitivae ecclesiae)* were to be found, and it was their order and customs that the new churches now forming in France and Italy should take as their models. The mysterious reference to the "seven churches of Asia" takes us back to the far-off beginnings of the sect, in Asia Minor.[5] The founding of the "seven churches of Asia" with its clear reminiscence of the text of the Apocalypse, can be readily understood as part of the sect's conscious integration in the traditions of apostolic times. When the representatives of the Cathar communities were meeting in St.Félix those seven churches were already a thing of the past. Indeed we can assume that their existence was no more than ephemeral. Nevertheless as late as the second half of the thirteenth century there was still a dualist church in the town of Philadelphia in Asia Minor, quoted in the Apocalypse as the seat of one of the seven churches. It can thus readily be assumed that one of those "seven churches of Asia" founded by the Bogomils of Asia Minor according to the seven churches of the Apocalypse survived into the thirteenth century. This *ecclesia Philadelphiae in Romania* attested 1250 in Rainerio Sacchoni can perhaps be identified with the *ecclesia Romanae* (i.e. *ecclesia Romana,* or *Romaniae*) referred to in the records of the St.Félix gathering. It is mentioned there as one of the "original churches" still in existence, and other churches are given at the same time: *ecclesia Drogometiae, Melenguiae, Bulgariae, Dalmatiae.*[6] It must be borne in mind that Greek and Slavonic toponyms are often very distorted in western sources. Thus the form Drogometia and many other variants refer to the original form Dragovitia.[7] This name is connected with the name of a Slav tribe, the Dragoviči (Dragovitchi). It is noteworthy that the Metropolitan of Plovdiv was called the Exarch of Thrace and of Dragovitia.[8] This would suggest that the heretical church bearing the same name, and attested only in western sources, should also be sought for in the Plovdiv region.

Scholars have generally associated the *ecclesia Melenguiae* with the Bulgarian town of Melnik, in the Rhodope Mountains,[9] but we could also consider the region inhabited by the Slavonic tribe of the Melingi, in the Peloponnese.[10] The name *Bulgaria* was used at this time to denote roughly the territory covered by Samuel's empire, which was still under the ecclesiastical jurisdiction of the Archbishopric of Ohrid. In the narrower sense the term Bulgaria could refer to the Bulgarian theme, which covered the regions of Macedonia formerly the nucleus of Samuel's state.[11] We are not given the necessary information, however, for a more exact location of the *ecclesia Bulgariae* of our heretical document. The reference to the church of Dalmatia is in accordance with the sources of the end of the twelfth century which mention the appearance of heresy on the Dalmatian coast.[12]

All the "primitive churches", according to the words of Niquinta, had clear territorial boundaries, and maintained peaceful relations with each other. In the same spirit the assembly in St.Félix drew clear boundaries between the newly established Cathar churches - that of Toulouse and that of Carcassonne. Each see coincided with the Catholic diocese.

We thus learn surprising details about the different dualist churches in the East, but nevertheless there is much that remains un-clarified. Who was Papa Niquinta? Where did he come from, and on what was his authority based? Why did he administer the *consolamentum* and consecrate not only the newly elected bishops of the churches of Toulouse, Carcassonne and Agen, but also the bishop of Albi, one from northern France and one from Lombardy, who had already been previously ordained? The answer to these questions is provided by a slight work, of a narrative character, which has only been known to us in full for about twenty years. It was written about the beginning of the thirteenth century by a Catholic who was very well acquainted with the history of the Cathars in Lombardy. Scholars know it as *De heresi Catharorum in Lombardia*. There is also a much later work entitled *Tractatus de hereticis* (1260-1270) which deals with the early history of the Lombard Cathars, but the lapse of time has caused some distortion; the author of *De heresi* undoubtedly gives a more accurate account of some events. On the other hand, the *Tractatus de hereticis* supplies interesting additional information about the earliest beginnings of the sect in Lombardy. They were bound up with Bishop Marcus, whose name also appears in the record of the assembly in St.Félix. We learn from the *Tractatus* that he was a gravedigger, who was won over to the heretics by a notary who had come to Lombardy from France. Two of Marcus' friends were converted to the heretical faith at the same time, one a weaver and the other a carpenter. In a locality called Roccavione there was even a whole group of Cathars who had emigrated there from France. According to the account in *De heresi*, Marcus was ordained from Bulgaria, i.e. from the *ecclesia Bulgariae* mentioned in the record of the proceedings in St.Félix de Caraman. Suddenly, however, this considerable community of Italian Cathars was disturbed by a foreign bishop, from as far away as Constantinople, who brought doubts into their midst. The name Papasnicheta is easily recognized as that of the chief figure in the St.Félix assembly of 1167. The Latin record of that gathering, of course, has distorted the Greek Nicetas into Niquinta. The word *Papas*, as we know, was the usual designation for monks and priests in Byzantium and as we have seen, the Bogomil leader John Tzurillas was so called. Nicetas, who had been ordained bishop by the "Dragovitsian" church, began to shake their confidence in the validity of the *ordo Bulgariae*. The upshot was that Marcus was ordained anew, thus taking his place in the episcopal succession of Drugonthia (i.e. Dragovitsa). Soon after his death, however, new doubts arose, this time questioning the validity of the ordination administered by Nicetas. Once again it was a foreigner from "overseas" who voiced his suspicions, one Petracius; his accusation went a generation further back, to the Dragovitsian bishop Simon (Symeon?) who had ordained Nicetas. This led to the first split, which lasted for several years. Finally some perspicacious members turned to a transalpine bishop for advice. It looked as though the whole of the Italian Cathar community would accept his decision, when unexpected complications put an end to this attempt at reunion. Individual schismatic groups then began to send their emissaries "overseas" *(ultra mare),* to be ordained again by one of the mother churches in the Balkans. Besides Bulgaria and Drugonthia (i.e. Dragovitsa) we find for the first time "Sclavonia". In itself this is not an unambiguous term, for it may refer to any of the Slavonic regions of the Balkan peninsula. In the twelfth century however, the region most likely to come to mind is Dalmatia, where, as we know, one Cathar church was already attested.

The Italian Cathars never managed to unite their church organization again. Yet another journey *ultra montes* failed to help - the decision of a French bishop no longer carried enough weight to help restore a single authority which would be recognised without protest by all the Cathars in northern and central Italy. The original single church was now replaced by six different churches, each submitting to one of three different *ordines: ordo Bulgariae, Drugonthiae, Sclavoniae.* At the end of *De heresi* the author lists the members of the hierarchy of four of these churches: each of the bishops had two assistants, known as the *filius maior* and the *filius minor.*[13] In the course of time they became his successors. There can be no doubt that this hierarchical system was brought from the East, and later western sources confirm this.

Throughout the history of these internal conflicts among the Lombardy Cathars, described by the author of *De heresi,* purely personal motives keep coming to the surface. Nevertheless the real cause is to be found in the constant doubts about the validity of the consolamentum and the consecration of the Bishops ordained by one or other of the mother churches. This in itself derived from the very nature of the mediaeval dualist faith, a full knowledge of which can only be gained by study of the western sources.

[1] Before the discovery of *De heresi catharorum* and *Tractatus de hereticis,* scholars considered the possibility of a continuous line of development leading up to mediaeval Catharism from remnants of Manichaeanism or Gnosticism which could have survived on the soil of western Europe for centuries. The Priscillian sect was thought to be an important link in the chain. (For the latest study see *B.Vollmann,* Studien zum Priszillianismus. Die Forschung, die Quellen, der fünfzehnte Brief Papst Leos des Grossen. St.Ottilien 1965.) *Lucie Varga* (Les Cathares sont-ils des Néo-

manichéens ou des Néognostiques? in: Rev. de l'Hist. des Relig. CXX (1939) 175 - 193) sees Catharism as Neo-gnosticism. *P.Alphandéry* on the other hand cautiously suggests the possibility of certain gnostic elements surviving in the doctrines of small and little known mediaeval sects (Le gnosticisme dans les sectes médiévales latines, in:Rev. d'hist. et de philos. relig. VII (1927) 395 - 411).

In his study dated 1950 (see Chapter IX above, n. 1), *A.Dondaine* rightly concludes that it was not until the dualist teachings of the Christian East penetrated to the West around the middle of the twelfth century that Catharism came into being. His only misconception is in the stress he lays on the *radical form* of Cathar dualism as though it was the appearance of this particular form that marked a new epoch in the history of western heresy.

2 The text has survived in the version printed by *G.Besse* in his Histoire des ducs, marquis et comtes de Narbonne, Paris 1660, 483 - 486, from a manuscript he acquired from a priest of the Toulouse chapter. A.Dondaine, who reprinted this document, convincingly refutes the attacks on its authenticity, made primarily by *L.de Lacger* (see *A.Dondaine,* Les actes du concile albigeois de Saint-Félix de Caraman, in: Studi e Testi 125 (Miscell. G.Mercati 5), Città del Vaticano 1946, 324 - 355). Cf. also *Borst,* 11 ff., n. 26. The doubts later raised by Y.Dossat can do little to shake this. (See *Y.Dossat,* Remarques sur un prétendu évêque cathare du Val d'Aran en 1167, in: Bull. philol. et hist. (jusqu'à 1715) du Comité des travaux hist. et scient., Années 1955 et 1956, Paris 1957, 339 - 347; and the same author's latest article, A propos du concile cathare de Saint-Félix: Les Milingues, in: Cathares en Languedoc, 201 ff.). *E.Griffe* (Le catharisme dans le diocèse de Carcassonne et le Lauragais au XIIe siècle, in: Cathares en Languedoc, 220 ff.; cf. id., Aventure, 67 ff.) defends the trustworthiness of Guillaume Besse against Dossat's scepticism; the former is well known for his painstaking efforts to reproduce documents exactly. Dossat's analysis of the records of the St.Félix assembly has produced a valuable emendation of the text, however, where the *ecclesia Aranensis* is mentioned (Acta St.Félix 326,13,18,28). In his view the reading should be *Agenensis*. (On this point see *Thouzellier,* Languedoc, 14, n. 7.) See also: *F.Šanjek,* Le rassemblement hérétique de Saint-Félix-de-Caraman (1167) et les Églises cathares au XIIe siècle, in: RHE LXVII (1972) No 3-4, 767 - 799.

3 Bishop Sicard Cellerier, who was present in St.Félix de Caraman (see Acta St. Félix, 326,11-12), was not from Albi itself but from Lombers, a fortified locality (castrum) twelve kilometres to the south of Albi. This had been the seat of Sicard's predecessor Olivier, who held a public disputation there in 1165, with the Catholic bishop of Lodève. It can justifiably be assumed that there had been a strong Cathar group in Lombers for some years prior to this date. (See *Borst,* 232; *M.Bécamel,* Le catharisme dans le diocèse d'Albi, in: Cathares en Languedoc, 237).

According to Anselm of Alexandria (Tract.her., 308,14-21) the people of Languedoc *(provinciales)* acquired their heretical ideas from the Cathars of northern France *(francigene),* who had already established a bishopric. The wave of Catharism seems indeed to have moved from north to south.

4 See n. 2.

5 On this point see our study, La question de l'origine... (quoted in Chapter IV above, n. 48), 270.

6 Acta St.Félix, 326,29 - 327,35.

7 For the different forms of this name found in western sources, see *A.Dondaine,* Liber, 63.

[8] On the question of the existence of the Dragoviči tribe (Δραγουβῖται) in southern Macedonia, of a bishopric of the same name in the region, of the designation of the Plovdiv Metropolitan referred to, and geographical names in the region of Plovdiv derived from the root *drag-*, see: *K.Jireček*, Die Heerstrasse von Belgrad nach Constantinopel und die Balkanpässe, Prague 1887, 72; *F.Dvorník*, Les Slaves, Byzance et Rome au IXe siècle, Paris 1926, 235 ff.; *I.Dujčev*, Dragvista - Dragovitia, in: REB XXII (1964) 215 - 221; *P.Gautier*, Clément d'Ohrid, évêque de Dragvista, ibid., 199 - 214. Cf. *Genoveva Cankova-Petkova*, Njakoi momenti ot razselvaneto na slavjanskite plemena ot iztočnija djal na južnite slavjani i ustanovjavaneto im na Balkanskija poluostrov, in: Slavjanska filologija 14 (Sofia 1973) 37 ff.

[9] For the mediaeval stronghold of Melnik see *I.Dujčev*, Melnik au Moyen Age, in: Byzantion XXXVIII (1968) 28 - 41.

[10] To our knowledge, *J.Duvernoy* and *Y.Dossat* are the only writers to associate the reference to the ecclesia Melenguiae in the record of the assembly in St.Félix not with the Bulgarian town of Melnik but with the Slavonic tribe of the Milingoi, which survived until well into the Middle Ages in the Peloponnese (*Duvernoy* I, 28; *Dossat*, in: Cathares en Languedoc, 210 ff.). For this tribe see *D.Zakythinos*, Despotat de Morée I, Paris 1932, 32; *A.Bon*, Le Péloponnèse byzantin jusqu'en 1204, Paris 1951, 48, 63, 71, 73, n. 1, 80, 102, 163. The name Μιλιγκοί occurs in the Morée chronicle in the form Μελιγκοί Etymologically it derives from the same root as the name of the town Melnik (see *Dujčev*, art. cit.). In view of the frequent distortions of Greek and Slavonic names in western sources, it is not impossible that the latinized form *Melenguia*, in spite of the change of vowel in the second syllable, refers to the region of the *Melingoi*. Historically, too, it cannot be decided *a priori* that the Bogomil doctrines could not have taken root among the Slavs in the Peloponnese. The fact that the sources are silent on this point is not decisive, for neither do they provide evidence that the sect existed in the Rhodopes, around Melnik.

[11] For Kekaumenos' use of the word Bulgaria in a wider sense see *P.Lemerle*, Prolégomènes à une édition critique et commentée des "Conseils et récits" de Kékauménos (Acad. Royale de Belgique, Cl. des lettres et des sc. mor. et polit., Mémoires, t. LIV, fasc. 1) Bruxelles 1960, 21 f., n. 4.; Vizantijski izvori za istoriju naroda Jugoslavije III, Belgrade 1966, 200, n. 16. The historians of the Crusades used the geographical term for a very wide territory. See *Jireček*, Heerstrasse, 86; *Zlatarski*, Istorija II, 14.

[12] See Chapter XV below.

[13] *D.Mandić* derives these names from I.Timothy 3,4, which speaks of the bishop's "sons" (*Mandić*, Bogomilska crkva, 270). This is pure hypothesis. This institution of bishop's assistants may have been what Everwin of Steinfeld was referring to (Everw. Steinf., col. 677B) in the words: "episcopus eorum *cum socio suo*" (our italics).

Chapter XI

THE DUALIST DOCTRINES AS SEEN BY WESTERN AUTHORS. THE MYTH AND THE ETHICS

Naturally the doctrines of the sect were never a set of unchangeable dogmas. In the tenth century Cosmas had allowed himself the ironical comment that each of the Bulgarian Bogomils invented something for himself, and less than two centuries later the Benedictine Eckbert reproached the German Cathars for the same variety of opinions.[1] Already in the Byzantine world the teachings of the sect were constantly enriched by the results of free speculation on the texts of the New Testament, and by the creations of the folk-imagination, inspired by apocryphal writings. In the West the variety and permutations of the dualist doctrines can be traced quite distinctly, although of course it is difficult to say what has been taken over from the eastern branches of the sect and what is the later contribution of the western Cathar groups.

Nevertheless we quite definitely know something of the original Bogomil heritage brought to the West; we have to hand an authentic Bulgarian Bogomil book. The Slavonic version has been lost, and it cannot be discounted that it was perhaps based on a Greek text now lost.[2] It is certain, however, that this text was in use in the "Bulgarian" church in the twelfth century, and that it was taken over from them by the Lombard Cathars. It was brought to them from Bulgaria by one Nazarius, who was then *filius major* of the Cathar community in Concorezzo near Milan. This "secret book" *(secretum)* was preserved in the church of Concorezzo, but a copy was also passed to the Cathars of Languedoc. Ultimately it came to rest in the archives of the Inquisition in Carcassonne, after the destruction of which it survived in a copy until our own day.[3] In parts the Latin text is decidedly distorted,[4] and the same is true of the second Latin version, preserved in the Vienna MS.[5] The interesting marginal glosses in the Vienna MS point to a Bosnian and Italian ambiance.[6]

In the form of question and answer *(erotoapokriseis)* so popular with Byzantine and Slavonic authors,[7] this heretical document presents the secret knowledge as if it were passed on to His best beloved disciple John by Christ, at the Last Supper.[8] As a whole, however, this Interrogatio Iohannis is not identical with any of the known apocryphal Apocalypses of St.John.[9] It presents the cosmological myth of the dualist sect, taken to its eschatological conclusion. Naturally apocryphal literature has left its mark on the Bogomil *Interrogatio Iohannis;* motifs from the Vision of Isaiah can be traced, as from the Slavonic Book of Enoch, the legend of the Wood of the Cross, and even from one of the apocryphal Apocalypses of St.John.[10] In some cases, however, the themes concerned are of such broad popularity that it is impossible to decide more precisely where they were taken from.[11] The *Interrogatio Iohannis* gives us a version of the Bogomil myth which differs in some respects from that recorded by Euthymius Zigabenus; it assumes the pre-existence of Christ, which in itself excludes the idea that Satanael was the first-born of God. Christ sits at the right hand of God at the time when Satanael, as the heavenly steward, is moving freely through the universe giving orders to the angels in charge of the elements.[12] The primordial cosmos is given more detailed form here than that imagined by the Bogomils of Zigabenus: it comprises seven invisible heavens, the firmament, and below the firmament the four material elements in layers. Satanael penetrates the layer of air and reaches the waters covering the face of the earth, and finds that the earth is supported by two fishes.[13] He cannot penetrate further because the heat of the layer of the fiery element is too great.

This image of Satan wandering through the universe was further developed in the myth of the Lombard Cathars. The Cathars who lived in the locality of Concorezzo mentioned above were the original Italian community founded by Marcus, the first Cathar bishop in northern Italy. His successors remained loyal to the original *ordo Bulgariae,* returning to the fold after his death. The anonymous author who gave us the early history of the Lombard Cathars in his *De heresi* knew a Bishop of Concorezzo (Iohannes Iudeus) who undertook a long voyage by sea in order to be ordained bishop in Bulgaria. The voyage made by one of his successors - Nazarius is proof that the contacts between Concorezzo and the mother church in Bulgaria were maintained. Their contemporaries used the simple term *Bulgari* to describe the Concorezzo heretics. In a similar fashion the name *Sclavini* stuck to another group, whose leader Caloiannes travelled to the church overseas, in "Sclavonia", for ordination as bishop in those confused times. This Italian bishop with a Greek name had his seat in Mantua, but under his successor Orto the church moved its headquarters to a nearby village, Bagnolo.

Essentially the faith of these Italian "Sclavini" and "Bulgari" did not depart from the original Bogomil view of the origin of evil. "Lucifer", later the evil creator and prince of our world, was-in the eyes of these Italian followers of the dualist faith-first a good angel, and only became evil after his fall away from God. Yet not even this

Secret Book brought from Bulgaria by the Concorezzo Cathars could answer the everlasting and disturbing question of where, ultimately, evil came from. The evil actions of heavenly creatures (Lucifer and some of the angels) could not be explained as long as God was assumed to be the motive force behind all that happened. It was therefore necessary to look for the primary impulse to evil *beyond* His sphere of power, and in this sense imagination completed the myth: we see a mysterious evil creature, whom Lucifer (still the highest of all the angels) meets on his wanderings through the universe. This was a mysterious spirit who rose from the depths of primordial Chaos. His four faces took the form of man, a bird, a fish and an animal - clearly a pre-image of the creatures to come. For Lucifer this was a fateful meeting; he succumbed to the charms of the unknown being and was instigated by it to prepare a rebellion in Heaven, in consequence of which he fell, taking with him the angels he had seduced.

This esoteric teaching of the Lombard Cathars distinctly changes the original Bogomil idea of the world, however. It assumes the existence of an evil being which has not been created, and therefore does not owe its existence to God, although in itself it has no powers of creation. From here it is but a short step to the extreme conclusion of dualist speculation, to belief in the two eternal principles of Good and Evil. This "radical dualism" is attested as early as 1176 -1190 in Italy, by Buonaccorso, a former *magister* and *doctor* of the Cathars of Milan.[14] According to our Lombard history the focus of this extreme dualist trend was the village of Desenzano, on the shore of Lake Garda. This heretical community set itself up as an independent church during the conflicts which rent the sect in northern Italy, and made contact on its own account with the eastern Dragovitsian church. Iohannes Bellus, who received the *ordinem Drugonthiae* (i.e. *Dragovitiae*) was no longer living when *De heresi* was written. His successors, Bishop Marchisius de Soiano and his *filius major* Amizo, were then preaching the radical doctrine of "two rulers, two Gods without beginning and without end". The Devil is no longer presented as a creature once good, who had merely succumbed to an evil power. Lucifer is now the direct *creature of the forces of evil*, the son of "the God of darkness"; the latter is an active creative principle. This evil God is the creator of all visible, ephemeral things, while all that is invisible and everlasting springs from the good God. The good God lives in His "celestial home" *(patria coelestis)*, while his antagonist is omnipotent in the visible, earthly world *(in tota hac mundiali machina)*. The ruler of darkness has his (evil) angels, while on the other hand the "invisible Father" also has His own world, made up of four immaterial elements.[15]

This absolutely symmetrical image of two dominions, two worlds, has nevertheless a number of points connecting it with the original Bogomil myth in which all derives from one single principle. The idea that the evil angels were created by the evil ruler of our material world can already be glimpsed in the Bogomil secret book brought to Concorezzo by Nazarius. Nor can we fail to see that in the radical dualist myth the figure of the *fallen angel* does not disappear from view; his origin is merely shifted *away* from the realm of the good God. Even the story of Satanael (Lucifer) as the "steward of the heavenly forces" remains, although it had of course to be adapted to fit the new scheme of things. Lucifer, born *in machina mundi* had at first to use deceit to get into the kingdom of Heaven, and to make himself popular among the angels, in order to reach his high position. Ultimately he was vanquished in an apocalyptic struggle and cast back to earth, carrying some of the angels with him.

In Buonaccorso's text we can already distinguish two separate Cathar ideas on the creation of the world. Some believed that the material elements were created by God Himself, while others attributed this work to the Devil. They were all agreed, however, that it was the Devil who *separated* these four elements. This precisely defines the creative role of the Devil or the Evil God. It is only in the extreme dualist versions that the principle of evil itself appears as the *creator*, who creates *ex nihilo* the mass of chaos and after separating the four elements forms the visible world from them. In the myth propagated by the Concorezzo and Bagnolo branches of the sect, however, Lucifer is not the *creator*, but merely the *factor*.[16] Again we find the idea of co-operation between God and the Devil, as the Bogomils of Zigabenus believed, but in this later case the two worked together not when it came to the creation of man, but from the very creation of the world. God sent his angel to help separate the four elements, when Lucifer and the four-faced spirit of evil could not manage it. (In a later Italian source their inefficiency was explained by their number: there were not three of them.)[17]

As in the myth recounted by Zigabenus, here too God and the Devil follow different ends with the creation of man. It was God's intention to use man to restore the original numbers of those in Heaven, broken up by the fall of some of the angels. It was only for this reason that He allowed Lucifer to retain his original capacities and gave him time to carry out the Creation. The Devil, however, was aiming at eternal dominion over the world, and intended that man should serve him. The victim of his tyranny was an angel who had been helping him, and whom Lucifer imprisoned in a material body. The sect found the evidence of this perfidy in the Gospel story of the king's servant who was to be sold into slavery because he could not pay his lord what he owed him. According to St.Matthew the servant begged to be given time to pay, but when his lord forgave him the debt he went out and pitilessly extorted from one of his fellow servants the debt that the man owed him.[18]

The Bulgarian secret book follows only the first part of the parable, going on to describe the tragedy not of one but of *two angels*, imprisoned against their will in the bodies of Adam and Eve.[19]

Both these versions are closer to the myth of Satan *stealing* the human soul, as we found it in Euthymius of Acmonia, than to that given by Euthymius Zigabenus, according to which God Himself breathed the soul into man.

In another variant of the Cathar myth, attested in a later source, we find at least the motif of the Devil's inability to put life into the bodies of the first human beings, but the myth develops in a different direction than as told by Zigabenus. At the Devil's request God sent two angels to earth, who disregarded His orders and fell asleep on their way. This made them forget their "heavenly home" and spellbound in the bodies of Adam and Eve they began their age-long pilgrimage from one body to another through all the generations of the Old Testament. Not until after the birth of Christ did they find salvation, in the bodies of Symeon and Anna.[20]

Another version supposes some degree of personal guilt on the part of the angel sent by God to help separate the four elements.[21] Such changes helped to lessen the impression of cruelty left by the Devil's actions shown in all its drastic brutality in the *Interrogatio Iohannis* and in the account given in *De heresi*. Like the wicked servant in Matthew's parable, seizing his fellow-servant by the throat, Lucifer began to strangle the good angel and pitilessly demand that he should "pay what he owed". This meant that he should be subject to the laws of the flesh established by the Devil. The secret book brought by Nazarius already knew of the *opus carnale* as the means to enslave the human race. The idea that the forbidden fruit of the Garden of Eden was the sinful passion which Satan, in the form of a serpent, first roused in Eve, and which Adam later fell victim to, remained firmly entrenched in the mediaeval dualist myth. In the western variants, too, Cain was born of Eve's union with the Devil, while Abel was her son with Adam. There is no mention here of the cruel tyranny wielded by Cain's diabolic descendants over those of Abel, which Zigabenus recounts. Nor is this found in the *Interrogatio Iohannis*, where the words "sons of the Devil" is used to designate the whole of the human race. (This reminds us of the "children of the Devil" which Cosmas heard used by the Bulgarian heretics.)

The begetting of children was thus begun at the instigation of Satan, and he plays his role at the birth of every individual. While body is born of body, so soul is born of soul. In the secret book brought over by Nazarius the souls of men are traced back to the two angels forced to inhabit the bodies of Adam and Eve. The Lombard heretics of the Concorezzo church believed that all souls come *ex traduce*[22] from the angel sent by God and then imprisoned in Adam's body by the devil. Ultimately this way of multiplying souls will create enough of them to fill the place left vacant in Heaven by the fall of the angels. In the *Interrogatio Iohannis* we are told that the Last Judgment will be held when "the numbers of the just are fulfilled".[23]

The theory of *traducianism* did not satisfy all the Lombard Cathars. Many believed that the new souls predestined for salvation were created all at the same time, and placed by God in the newly-born human beings one by one. At the same time the wicked souls, that is to say the fallen angels, were reincarnated from one generation to the next without hope of salvation. At the end of the world there will be the same number of just and of wicked souls, the former automatically taking the place of the latter, who will be cast into eternal destruction at the Last Judgment.

This idea of inexorable justice punishing all the fallen angels without distinction was not the general belief, however. Voices were not lacking to suggest that God would be more merciful to those angels who had fallen into sin in ignorance, or because they were forced to. In this view, it was only to replace the damned angels that new souls were created.[24]

According to the *Interrogatio Iohannis* the number of the "just" will correspond to the number of crowns the fallen angels were forced to leave behind in Heaven, when they had to put off their heavenly apparel. It is certainly not by chance that we find a similar idea in the "Vision of Isaiah", the apocryphal book used by the sect both in the East and the West. There we read of "countless angels and just men", clad in glorious robes and waiting for the moment when Christ triumphs over death and they can attain the crown of eternal glory and take their appointed seat in Heaven.[25] (In this connection the apocryphal book does not mention *fallen* angels, whose heavenly role with all its attributes was to be taken over by other souls.)

The motif of the crown, the robes and the throne left behind by each of the angels cast out of Heaven crops up again in the teachings of the Cathar church on the shore of Lake Garda, which believed in the two eternal principles. The whole story of the fall of the angels is fitted into this sharply dualist view of the world. The Archangel Michael overcomes the power of Lucifer in battle, and the son of the prince of darkness is cast back into the evil world. Some versions of the dualist myth describe two battles in Heaven, or explain things by saying that Lucifer climbed up to Heaven again after he was cast out, accompanied by a company of his fallen angels, and that the battle in Heaven took place then. The battle between Michael and Satan is of course a motif taken from the Apocalypse, as is the statement that Satan pulled one third of the angels down with him as he fell. There is a reference to this in the secret book of Nazarius. In the myth of the *radical dualists*, however, this idea of one third of the angels was given a curious interpretation based on the belief in the *trichotomy* of angelic creatures. They believed that the *populus Dei* which lived in the invisible supernatural world was made up of the *spiritus*, the soul *(anima)* and the celestial body *(corpus coeleste)*. In this sense, therefore, it was the *third part* of each angel that fell with Satan to earth. The *bodies* of the angels remained lying in Heaven as if dead - the "dry bones" of which the prophet Ezekiel spoke. The *spirits* of the angels proved very resistent, but their *souls* fell to the dark earth and the Devil put them into human bodies one by one. Clad in this earthly garment *(tunica)* the souls forgot their heavenly home and their initial guilt.

In all the versions of the mediaeval dualist myth the origin and story of the human soul is bound up in one way or another with the tragedy of the fallen angels. Whether human souls are identical with the fallen angels or form only a part of them, whether they are the descendants of these angels or were created in order to replace them, their earthly existence has only one purpose — through repentance to restore the broken integrity of the world of Heaven.[26]

This goal is far off, however, for God Himself gave the Devil "seven days" of freedom, which means seven ages. From his throne above the clouds Satan rules the world unchecked.[27] Up to the coming of Christ Satan managed to delude the human race into believing that he was the only God, and his obedient servants to this end were the figures of the Old Testament, Moses, Enoch and the like. Christ was to reveal the true state of affairs to man, and that was why Satan prepared His death on the cross long before. (In the *Interrogatio Iohannis* the apocryphal motif of the wood of the cross is presented in a new way, as the wood of three trees entrusted to Moses.) Mary the mother of Christ was an angel, sent to earth; this we read in the secret book brought to Concorezzo by Nazarius, who had come across the same belief in Bulgaria. It was told to him there by the *bishop and filius major* of the "Bulgarian" church.[28] In the teachings of the Bagnolo Cathars not only Mary, but Christ himself, and St.John the Evangelist, were also angels.[29] This was of course all bound up with the belief that Christ's *incarnation was only an illusion.* In the secret book we again find the idea that Christ was born *per auditum*,[30] a conception recorded by Euthymius Zigabenus too. Among the Lombard Cathars who followed the *ordo Bulgariae*, however, there soon appeared a division of opinion on the question of the essential nature of the body of Christ. Some believed it to be a real body, with the reservation that Christ laid it aside when He returned to Heaven.[31]

The Son of God came into the world to "seek and to save that which was lost", but at the last moment Satan tried to hinder this mission by sending his own angel - John the Baptist. This conception of John Prodromos was closely connected with the rejection of the rite of baptism by water, which had been instituted by John. The *Interrogatio Iohannis* refers to the words of the Gospel, that Christ will baptize "with the Holy Ghost, and with fire", and adds that this baptism alone has the power to transmit the *forgiveness of sins.* Here, of course, this concept has its specific meaning which can only be understood in the light of the dualist myth.[32]

It cannot be too often repeated that our sect did not share the common Christian idea of original sin. This is quite clear from the Bulgarian secret book, and a later western source repeats the categorical belief of the Cathars that *peccatum originale nichil est.*[33] The first real sin was for them the *rebellion of the angels against God*, which led to the soul's imprisonment in the mortal body and its subjection to human passions. This was all the consequence of that fateful error, which happened before our world was created. The *penance* which was necessary to redeem that tragic error in the superterrestrial sphere could only be achieved in the sect of the "true Christians", the "pure" (the *Cathars*). It was attained at the moment when the administration of spiritual baptism *(consolamentum)* wrenched the soul from the power of the prince of this world and filled it with the *Holy Spirit.*

The followers of the *extreme radical dualist* trend were not thinking, in this context, of the third Person of the Trinity, but of that *spiritus* which was an essential part of every celestial being. Unlike the soul *(anima)* the spirit, as we know, resisted temptation and thus acquired the designation *spiritus firmus.* The spirit then descended to our regions and searched for *its* own soul. Hearing the spirit's voice, the soul responds, and at that moment remembers its guilt against Heaven. This mysterious contact between the spirit and the soul is consummated at the moment of the laying on of hands, the supreme moment of the rite of *consolamentum.* Confirmed by the spirit, which was given by God to lead and defend the soul, the soul then begins to repent both for its original sin in Heaven, and for all the sins committed during mortal life.[34] After death this soul, saved, awaits the hour when it will return to the kingdom of the Father of the Just.

But those human souls which do not find their corresponding spirit during their life on earth, do not die with the death of the body. They are condemned to incarnation in other human bodies and other living creatures, and their incarnations continue until they find the only path to salvation, in the Cathar faith. Ultimately, when all the souls have completed their "repentance" on earth, they will all together enter their heavenly home. Then they will take upon themselves once more the bodies which have lain there as if dead ever since the moment of their fall.

The resurrection of the material earthly body would certainly have been quite inacceptable to the followers of the dualist sect. Only the *heavenly body* in the existence of which the *radical dualists* believed, could come to life again. The radical branch of the Cathars understood the resurrection of Christ in this sense, too, for He was not God, but a *creatura,* essentially similar to the rest of the "heavenly host". Christ, too, was thus composed of spirit, soul and celestial body. Thus He appeared on earth, when He was sent to save the souls of the "lost sheep of the house of Israel". He brought with Him from Heaven an immaterial body, so that he partook of nothing of his "mother" Mary. She, in any case, was not a woman but an angel, and her body was just as immaterial as that of Christ. It is clear that the Mariology and the Christology of the *radical dualists* was not very different from that professed by other Cathar groups, not only in northern Italy but in Bulgaria and perhaps in "Sclavonia" as well. The idea of the trichotomy of the heavenly beings, and therefore of Mary and Christ as well, was however characteristic of the extreme dualist trend. The followers of this trend believed that Christ died on the cross in his celestial body; His death was painless, but meant that for a time His body was separated from His spirit and His soul. They were rejoined together three days later, here on earth, so that forty days after His "resurrection" Christ returned to Heaven as a complete heavenly being. Until the resurrection of Christ no human being had received the *Holy Ghost for the forgiveness of sins;* it was only after this that Christ delegated this power to His apostles.[35]

The Acts of the Apostles (19, 1-7), describing how Paul came to Ephesus, distinguished between "John's baptism" and baptism "in the name of the Lord Jesus". The text goes on to describe how Paul "laid his hands upon" his disciples, and the "Holy Ghost came on them". In the

Cathar interpretation the baptism in the name of Jesus and the laying on of hands merged into one rite. The significance of the concepts is deliberately shifted to justify the distinction of the Cathar sect from the official Church. In the Roman Catholic church baptism by water was still performed, the baptism instituted by John the Baptist at the instigation of the Devil, in order to prevent the Saviour from carrying out His mission. In contrast, the Cathars, the true *ecclesia Christi*, had preserved the baptism of Christ from the times of the apostles, and by the laying on of hands invested the faithful with the grace of the Holy Spirit and the forgiveness of sins.[36]

It should not be forgotten, however, that spiritual baptism did not make the new "perfect" member of the sect *immune from sin*. Quite the contrary - it put him under the obligation to strict asceticism for the rest of his life. To break these rules was "mortal sin" which would cast his soul back into bondage. Only the administration of the rite of *consolamentum* over again could save him from the consequences of mortal sin. The certainty of salvation was not dependent only on an ascetic life following the administration of spiritual baptism; it also depended on the moral purity of the Cathar dignitary who performed the rite. Any moral lapse on the part of a Cather bishop or other dignitary admimistering the *consolamentum* acted in reverse and had its consequences not only for the living, but even for dead members of the sect. Even the souls of Cathar martyrs could thus lose their state of heavenly bliss and fall back into the mortal world, to take their place again in the sorrowful cycle of reincarnation.

Hence the ceaseless anxiety about the salvation of the soul, attested so eloquently by later sources.[37] The account given in *De heresi* is the first to reveal the consequences which could be feared from so labile a state. We have seen how the doubts expressed at the expense of the "Dragovitsian" bishop Simon inevitably shook the confidence of believers in the *ordo Drugonthiae*, to which Nicetas had recently succeeded in uniting the whole of the Cathar community in Italy and France It was typical of this state of mind that the accusations levelled by Petracius concerned primarily the intimate relationship between Bishop Simon and a certain woman. If the accusation were founded on fact, there was mortal sin involved. It is therefore very understandable that most of the Cathars of northern Italy were torn by doubts about the validity of the *consolamentum* and the episcopal consecration administered by Nicetas, who had himself been ordained by Simon, now, perhaps, suspect. There were similar reasons behind the failure of the attempt to settle the dispute roused by Petracius and his followers; all seemed to be going well, and both sides had agreed that Garattus should be the next bishop to accept ordination in the *ordo Bulgariae;* the date of his departure for Bulgaria had been set, the money collected and reliable companions chosen for the journey. An embarassing revelation at the last moment undermined the moral authority of the bishop-designate, and Garattus never regained general recognition among his fellow-believers in Italy.[38]

We know the result: the Cathar community in Northern and Central Italy divided into six churches; yet even this was not permanent, and our Lombard author describes the first steps towards reunion. Garattus became reconciled with Caloiannes, the only one of the Italian bishops he mentioned in his prayers, and to whom he gave the appropriate honours.[39]

[1] Ecb. Schon., col. 96C.

[2] On this question see *Turdeanu,* Apocryphes bogomiles II, 208 ff.; *Borst* 8, Note 11; *H.-C.Puech,* in: *Hennecke - Schneemelcher* I, 234. It has so far proved impossible to identify the presumed Greek original on which this text was based.

[3] The Carcassonne text has survived in a single copy made in the seventeenth century for Doat, and in a printed edition by *J.Benoist,* in: Histoire des Albigeois et des Vaudois, Paris 1691, Vol. I, 283 - 286. A later edition was published by *J. Thilo,* Codex apocryphus Novi Testamenti, Lipsiae 1832, 884 - 896.

In a MS volume probably copied in the Inquisition archives in Carcassonne, *A. Dondaine* found a text more or less identical with that preserved for us by Doat and Benoist, and therefore also probably derived from the lost Carcassonne MS. (Ms.109, Bibl. Publique de Dôle, pp. 44r - 46r. See *Dondaine,* Manuel, 125). The origin of the secret heretical book is revealed in a note at the end of the Carcassonne text: "hoc est secretum haereticorum de Concorrezio portatum de Bulgaria (a) Nazario suo episcopo plenum erroribus." See *R.Reitzenstein,* Zur Vorgeschichte der christlichen Taufe, Leipzig - Berlin 1929, 309. This sentence had already been added to the manuscript of the secret book which the author of the Tractatus de hereticis saw in the second half of the thirteenth century, and must therefore undoubtedly be attributed to one of the thirteenth century Italian inquisitors. In Anselm's additions to this treatise the note is given thus: "Hoc est secretum hereticorum de Concorezo portatum de Bulgaria plenum erroribus" (Tract. her., p. 319,29-30), this addition is numbered rubric 13. On the reading of the preceding sentence, which is not clear, see the commentary by *Dondaine,* Hiérarchie II, 249.) The treatise itself says of Nazarius: "Item Nazarius tenet quoddam scriptum quod secretum vocat" (Tract. her., 311,6-7). The Italian inquisitor Rainerio Sacchoni himself, when still a member of the Cathar sect (i.e. 1228 - 1245, see Chapter XVII below), heard Nazarius describe Mary as an angel (see p. 242 below). Nazarius then said "(dixit) quod habuit hunc errorem ab episcopo et filio Maiore Ecclesiae Bulgariae jam fere elapsis annis LX" (Rain. Sacch., 76). These sixty years are usually calculated from the year 1250 when Rainerio was writing his Summa which gives the year 1190 as the date of Nazarius' journey to Bulgaria. It cannot be excluded, however, that Nazarius himself was reminiscencing, when Rainerio heard him talk of his journey some time between 1228 and 1245, *Obolensky* (Bogomils, 242 f.) seems to take this view. If this were so, it would only be possible to give a very approximate date to the journey of Nazarius, within the limits of 1168 - 1185.

[4] In the Addition to the Tract. her. mentioned above (rubric 13) we read that the secret book of the Concorezzo heretics was written in *falsis latinis* (319,30-31). From Dondaine's edition and commentary it is not unequivocally clear whether this accusation of bad Latin comes from Anselm of Alexandria himself, or whether it was part of the sentence he quotes as coming at the end of the secret book (Hoc est secretum...).

[5] *Ivanov*, Bog. knigi, 73 - 87, printed the two versions (Carcassonne and Vienna) as parallel texts, as did *R.Reitzenstein* after him (op. cit., 297 - 309). The references in this study are to Ivanov's edition. For the Vienna MS 1137 see *Reitzenstein*, op. cit., 295 f.

[6] *Reitzenstein*, op. cit., 311; cf. 312.

[7] See *V.Močul'skij*, Sledy narodnoj Biblii v slavjanskoj i v drevnerusskoj pismennosti, Odessa 1893.

[8] This is explicit in the sub-title of the text in the Vienna MS: Joannis et Apostoli et Evangelistae in coena sancta regni coelorum de ordinatione mundi et de Principe et de Adam.

[9] For the latter see *Hennecke - Schneemelcher* II, 535.

[10] 'Αποκάλυψις τοῦ ἁγίου 'Ιωάννου τοῦ θεολόγου, ed. *C.Tischendorf*, Apocalypses Apocryphae, Lipsiae 1866, 70 - 94. The borrowing concerns the fiery abyss into which sinners will be cast at the Last Judgment. According to the Concorezzo secret book the abyss is so deep that a stone heavy enough for a thirty-year-old man to lift and throw into it would not reach the bottom for three years. See Interrog. Ioh., 86, Carcass. version (in the Vienna MS the text breaks off at this point, the end being lacking). In Tischendorf's Greek Apocalypse of St.John (87 f.) the fiery abyss is prepared for τὰ πνεύματα τὰ ἀκάθαρτα μετὰ τοῦ ἀντικειμένου and the time the stone would take to fall is 20 years; in some MSS it is τρία ἔτη (ibid., 88, apparatus). The Slavonic version also speaks of three years (*Ivanov*, Bog. knigi, 71). In various forms this motif also crops up in Slavonic versions of the argument between Christ and the Devil (ibid., 253, cf. n. 1).

[11] Turdeanu's study (Apocryphes bogomiles) is of fundamental importance for the distinction it draws between the Interrog. Ioh. on the one hand, an apocryphal work created by the Bogomils and presenting their dualist myth; and the older apocryphal literature on the other hand, which had come into being without Bogomil influence or participation, and from which the sect only took various elements to build up their own myth. Turdeanu accepts only the Vision of Isaiah as an exceptional case; here he finds signs that it has been altered to fit the Bogomil doctrines. This remains a doubtful hypothesis, although Turdeanu rightly points out that Vaillant's objection (concerning the exclusion of the names of the "just" of the Old Testament) is based on a corrupt text and is therefore not conclusive. (See *A.Vaillant*, Un apocryphe pseudobogomile: la Vision d'Isaïe, RESl 42 (1963) 120; and the opposing view: *E.Turdeanu*, La vision d'Isaïe, tradition orthodoxe et tradition hérétique, in: Κυρίλλῳ καὶ Μεθοδίῳ τόμος ἑόρτιος II, Thessaloniki 1968, 308 ff.). Vaillant sets out to show that both the Slavonic and the Latin version of the Vision of

Isaiah derived from a common Greek text. This lost Byzantine version was rewritten from the original apocryphal story known in its entirety today only in an Ethiopian translation. The Byzantine Bogomils were already using this rewritten version, as can be seen from their curious conception of the Trinity which (according to the anti-Bogomil synodical *tomos*, *Tollius*, 116, art. 4), was taken from the Vision of Isaiah. This motif occurs in both the Slavonic and the Latin version, and was therefore present in their common Greek model (*Vaillant*, art. cit., 112 f., 119). Turdeanu also acknowledges now that both the Slavonic and the Latin texts of the "Vision" came from a Greek version no longer in existence. Unlike Vaillant, who assumes that the Latin translation came into being in an Italian Catholic environment, Turdeanu puts forward the justifiable hypothesis that the Greek text of the Apocryphon came from the Byzantine Bogomils to the Italian Cathars, who translated it into Latin (*Turdeanu*, La vision, 317 ff.).

[12] See Byzantinobulgarica III, 34.

[13] On the structure of the cosmos in horizontal layers see *Linda Sadnik*, Der Balkan und die Hochkulturen des Vorderen Orients, in: Saeculum V (1954) 34 - 40, with references to the author's earlier studies. Certain parallels to the idea of two fish holding up the earth can be found in the apocryphal text (also of the erotoapokriseis type) "Beseda trech svjatitelej" ("Razumnik"). See *Ivanov*, Bog. knigi, 260; cf. also one of the Russian versions of the Legend of the Sea of Tiberias, ibid., 296. (Cf. *N.Minissi*, La tradizione apocrifa e le origini del bogomilismo, in: Ricerche slav. III (1954) 110. See also *A.Afanas'jev*, Poetičeskije vozzrenija slavjan na prirodu II, Moscow 1868, 161 f.)

[14] For Buonaccorso see: *Ilarino da Milano*, La "Manifestatio heresis catarorum quam fecit Bonacursus", secondo il. cod. Ottob. lat. 136 della Biblioteca Vaticana, in: Aevum XII (1938) 281 - 333. *R.Manselli*, Per la storia dell'eresia nel secolo XII, in: Bull. dell'Ist. stor. ital. per il med. evo e Archivio Muratoriano 67 (1955) 189 - 264. Cf. *Thouzellier*, Languedoc, 45, Note 122; 107. The figure of Buonaccorso remains mysterious. The "Manifestatio heresis catarorum quam fecit Bonacursus qui olim fuit magister eorum Mediolani coram populo" must be accepted as his authentic work (text in *Migne*, PL 204, col. 775 - 777; *Manselli*, art. cit. 207 - 211). He was not, however, the author of the anti-heretical compilation which follows the "Manifestatio" in Migne's edition, and the first ten chapters of which are devoted to the Cathar heresy. The relationship of the "Manifestatio" to the "Confessio" of Iacobus and Micasmus remains unclear. *Manselli*, who published the text on the basis of the Paris MS, Bibl. Nat. lat. 14927, fol. 13a (art. cit., 206 - 210), believes that it is the authentic confession of Buonaccorso while the Manifestatio is a later version. The question cannot be definitely decided until all the relevant MSS are known. A.Dondaine printed the testimony of two people, Mirisona and Jacobus, concerning Patarene errors; in the Paris MS this text follows that of the Confessio (*Dondaine*, Durand de Huesca, 272 - 273). He also referred to the existence of another MS containing the "Confessio Iacobi et Mirisona" (ibid., 264 ff.).

Note: *Manselli* (art. cit., 212 ff.) independent of the texts discussed above, published a hitherto unknown Cathar abjuration of the twelfth century.

[15] De heresi 309,1-5; Moneta, 3. For the two basic variants of the dualist myth see our studies, ByzSlav XXVIII (1967) 51 ff. and Listy filol. 90 (1967) 122 f. K.F.Radčenko is certainly not right when he claims that already in the Interrog.

Ioh. the two versions - the radical and the more moderate - conflict (*K.F. Radčenko*, Zamečanija otnositel'no otdel'nych mest knigi Ioanna Bogoslova po spisku izdannomu Döllingerom. SPb 1904, 65). Nevertheless the motif of Satan's creation of the wicked angels (Interrog. Ioh., 77) certainly points towards a new form of the myth in which the evil as well as the good God rules over his own angels and his own world.

16 Bonaccursus, Manifestatio, art. 1. The Devil as *creator* and *factor*: Moneta 109 (cf. ibid., 124; *creatio* and *factura*); Jacobus a Capellis, *Doell.* II, 274: "diabolum vero non creatorem, sed factorem dicunt."

17 Salvo Burci, *Doell.* II, 60; *Ilarino*, 314.

18 Matthew 18,23-30. De heresi, 310,26-28.

19 Interrog. Ioh., 78.

20 Georg. Disputatio, 1720; Peter the Martyr, 326.

21 Rain. Sacch., 76.

22 Interrog. Ioh. 80: "...et nascitur spiritus de spiritu et caro de carne." Similarly De heresi, 310,29-31; Rein. Sacch., 76: "...omnes animae sunt ex traduce ab illo angelo."

23 Interrog. Ioh., 84.

24 De heresi, 310,32 - 311,11.

25 The Slavonic and Latin texts, ed. *Ivanov*, Bog. knigi, 142, 143 (cap. XI).

26 Interrog. Ioh., 76, C: "traxit cum cauda tertiam partem angelorum Dei;" V: "et habuit VII caudas trahentes tertiam partem angelorum dei." On the trichotomy of the angels: De heresi, 309,16-20; Peter Sarn. I, 13; Peter the Martyr, 326; Salvo Burci, *Doell.* II, 58, *Ilarino* 313; Moneta, 3 ff., 105 ff.; Brevis Summula, 317. For the concept of the *tunica*: *Duvernoy* II, 34 ff. Cf. *Borst*, 150, n. 23, with references to parallels to these Cathar ideas.

27 Interrog. Ioh., 80 ff. The idea was widespread among the Cathars of the West and in Byzantium "quod diabolus faciat pluere et ningere, et tonitrua et tempestates et concitaciones ventorum." (Tract. her., 313,19-21) Cf.: ... καὶ οἰκονόμον αὐτὸν ἀποκαλοῦσι τῆς τε βροχῆς, τῆς χαλάζης ... 'Επιστ. Κοσμᾶ 371,30 and hence: Sinodik of Boril, Palauz., No 43.

The idea that evil angels were responsible for natural and other catastrophes already occurs in Clement of Alexandria. See DTC 4, 347. Cf. also the articles of the Council of Braga, against the Priscillian heresy (*Lucie Varga*, in: Rev. de l'Hist. des Relig. CXX (1939) 189).

28 See n. 3 here.

29 De heresi, 311,26-27.

30 Interrog. Ioh., 82.

31 De heresi, 311,30-33.

32 Interrog. Ioh., 82.

33 Brevis Summula, 132.

34 Salvo Burci, *Doell.* II, 58; *Ilarino*, 313; Moneta, 4.

35 Moneta, 5.

36 Ibid., 280 ff.

37 Rain. Sacch., 69 f.

38 De heresi, 307.

39 Ibid., 308,32-34.

Chapter XII

CATHARISM AND THE WESTERN WORLD

This anonymous Lombard author cannot help giving us a picture of the sect closed in as it were upon itself, suffering from frequent upheavals flowing from the eastern epicentre and aggravated by local factors.

To see the development of Catharism in Italy from another angle, and to understand its place in the tumultuous evolution of society in the patchwork of urban communities of the northern and central regions of the Apennine peninsula, we must turn to the testimony of other contemporaries.

In 1173 the sect provided the occasion for an interdict to be laid on the city of Florence. The brief report of this event in the "Annals of Florence" is also the first time the name *Paterini* is attested. We know that it was about the same time that the Italian Hugo Etherianus used the word in the treatise against the Byzantine Bogomils which he wrote in Constantinople. The term is also used later in Byzantine sources. Various forms of the word were used in the West (Paterini, Patrini, Patareni etc.) as one of the most usual names for the sect. There is no reason to doubt that it is connected with the name of the Pataria of Milan; besides the assonance of the names Cathari - Patari, an awareness of certain similarities between the two movements may also have played a part in Italy.[1]

Florence was the centre for one of the two Cathar churches attested in Tuscany by the treatise "De heresi catharorum in Lombardia".[2] The second must be sought in the region of "Roman Tuscany"- Rainerio Sacchoni called it the *ecclesia de Valle Spoletana*. Here, within the sphere of the Patrimony of St.Peter, within reach of the papal power, dualist doctrines had taken very deep root. At the end of the sixties of the twelfth century, if not earlier still, the Florentine Diotesalvi was preaching dualism in Orvieto, abd he had had a precursor there in the person of one Hermanninus of Parma. After Diotesalvi and his companion, Gerard, the names of further teachers are recorded in the town, and against all of them persecution proved ineffective.[3]

The growth of heretical movements in Italy and elsewhere was undoubtedly encouraged by the serious blow to the authority of the Church inflicted by the Papal schism of 1159. It is well known that this was the outcome of a new conflict between the Imperial and the Papal powers, provoked by Frederick Barbarossa's ambition to rule Italy.

In Lombardy, where the urban Communities were constantly fighting among themselves, they had a hard fight to preserve their independence in the face of the dangerous expansion of Imperial power. The city of Milan was destroyed in 1162, and began to rise from the ruins in 1167 as the inhabitants gradually returned. The new Archbishop Galdinus restored ecclesiastical authority, but from the outset he had to deal with the Cathars. The sect had even begun to spread and flourish in the capital of Lombardy. The connection between heresy and the previous disruption of political and ecclesiastical life is only too obvious; the heretics took advantage of the situation and preached quite openly in the city. The tireless sermons Galdinus preached against them, literally to his last breath, were of no avail.[4] Buonaccorso, who himself had been one of the leaders of the sect in Milan, said graphically that "towns, suburbs, villages and castles" were crowded with heretics.[5]

Meanwhile on the other side of the Alps the spread of heresy was fostered to some degree by the troubles and disturbances in the wake of the long fight between the ruling house of Toulouse and that of Barcelona, for hegemony over the southern provinces of present-day France. The family of the Counts of Toulouse tried to extend its suzerainty over vast territories west of the Rhône, where the name of Provincia still recalled the days of *Narbonnensis prima*. The name of Languedoc *(Lingua Occitana)* was not in use until the end of the thirteenth century, when this region was already firmly incorporated in France. The situation was very different in the twelfth century. Since the end of the Carolingian era the gradual splintering of the civil power had led to the emergence of larger and smaller domains in the region, whose rulers took to themselves, along with the hereditary tenure of the old official dignities, regal rights and revenues. The most powerful of these southern nobles, like the Counts of Toulouse, the Viscounts of Béziers (of the house of Trencavel), the Counts of Foix, Comminges, the Viscounts of Béarn, gathered their own courts about them and took little account of the interests of the King to the north of the Loire, whose name barely appears in their charters. The rulers of these southern domains also acted independently of each other, however, and the nobles in their fee were only loosely bound to them; the towns, too, attained considerable independence, especially such cities as Toulouse or Narbonne. In the South, feudal relationships did not develop to such an involved degree, and did not play such a constructive part in the formation of the body politic, as had been the case in the North, in France proper. The anarchy which reigned in Languedoc was indeed a considerable hindrance to the efforts of the house of Toulouse for unification.

The most serious rival to the rulers of Toulouse was the house of

Trencavel, which acknowledged their suzerainty. These Viscounts of Albi and Nîmes tried to become the masters of the "lower" regions of Languedoc; they succeeded one after the other in getting a hold on Carcassonne, Razès, Agde and Béziers. At the end of the twelfth century, however, the Viscounty of Agde and that of Nîmes were in the hands of the house of Toulouse, which had long ruled the County of St.Gilles with the important town of the same name on the Petit Rhône.[6] The Counts of Toulouse soon tried to extend their rule to the east of the Rhône as well, to lands only loosely attached to the western Empire. Already in the eleventh century they claimed a hereditary right to the County of Provence, but here their policy later came up against the energetic expansionist policy of Catalonia. Like those of Toulouse, the Counts of Barcelona followed their own aims, and from the tenth century onwards had vassals on the other side of the Pyrenees. The first military clash between the two came in the first quarter of the twelfth century. The union of Catalonia and the Kingdom of Aragon (1137) provided a more powerful basis for new aggression against Toulouse after the middle of the twelfth century, and the conflict began to assume European proportions. The struggle between Toulouse and Barcelona was intertwined with the rivalry between the Capets and the Plantagenets. King Alphonse II (1162-1196) came closest to the goal of uniting under one house the regions on both sides of the Pyrenees, linked as they were by language and culture - from the Ebro to the Garonne and as far as the Alps in the east. The Viscounts of Béziers, like those of Narbonne, owed him allegiance and the Counts of Foix also recognized him as their overlord. In the Pyrenees, besides the County of Foix, Alphonse II extended his influence over the Viscounty of Béarn and gained the County of Roussillon by succession, until at last only the County of Comminges held out. East of the Rhône, however, the Catalonian-Aragonese power did not spread further; the territory of Provence remained divided between the rulers of Toulouse and of Barcelona, as settled by the peace concluded in 1125.[7]

Long-drawn-out campaigns, ending only in 1185, took up most of the reign of the lord of Toulouse, Count Raymond V (1148-1194). The countryside of Languedoc bore the painful marks of the havoc caused by the bands known as "routiers", "Brabançons", "Navarrais", "Basques" and so forth, who flourished all over what is now France, in the twelfth century.[8] The feudal lords of the southern provinces could not dispense with these mercenary forces since the loose nature of feudal bonds in this part of the country meant that they could not rely on military support from their vassals. The Church also suffered from this evil of the times, which presented an insoluble problem. These armed bands out for loot were naturally attracted by the treasures of churches and monasteries, which were usually helpless in the face of violence. These rapacious and destructive attacks were accompanied by cynical blasphemy and the desecration of sacred ground, and it is understandable that in the eyes of the ecclesiastical authorities the ravages of these godless bands merged with the actions of the heretics. Indeed the constant threat to safety and public order throughout the countryside disturbed the normal functioning of the ecclesiastical administration and thus indirectly facilitated the spread of heresy. From the moment the Church began to organize a more effective fight against sectarianism, the "routiers" were named together with the heretics in the constitutions promulgated. Those who maintained robber bands in their employ were to be punished in the same manner as those who gave their protection to heretics. Indeed the nobility did not merely play a passive role; it was not infrequently the powerful nobles themselves who organized attacks on monasteries, and they were the prime initiators of violence against monks and the clergy. The wealth of the Church aroused the avid greed of the nobility everywhere, and Languedoc was no exception. Complaints by the ecclesiastical institutions of wrongs done to them were a constant phenomenon, and the excommunication even of the most highly-placed offenders did nothing to help. In Languedoc there were many families of the lesser aristocracy whose wealth had been whittled away, divided between too many heirs.[9] In towns and villages there were men of noble birth often forced to live in poverty, and turning naturally to service with these mercenary bands, as a way of living by their arms. It is not legacies to see that the wealth of the Church, constantly augmented by lagacies and donations, was particularly tempting to these lesser nobles whose wealth was inexorably slipping from them. In Languedoc it was among these lesser noble families that the Church came up against the fiercest opposition when the Gregorian reform brought with it demands for the return of all lands and dues which were originally her right. It can readily be assumed that in the fertile plains of Languedoc the demand for the return of profitable tithes to the Church was particularly unpopular. This of course helped to prepare a very receptive public for the heretical preachers who urged that tithes and other dues should not be paid to the clergy.[10]

There were even more subtle causes for this ready inclination to heretical doctrines, to be sought in the different atmosphere conditioned by the whole historical development. The Roman province of Narbonnensis prima had very close ties with Italy, and the Latin culture was far more deeply rooted there than in the other regions of Gaul.

The continuity of urban life was not interrupted here even in the centuries of violent upheavals, from the decline of the Roman Empire, to the great offensive of Islam during the first half of the eighth century. Although seriously hindered in their development, the cities preserved their vitality even later, despite the long-drawn-out crisis caused by the decay of the Carolingian Empire. Ultimately, from the end of the eleventh century onwards, economic life here, too, began to work free from the burdensome fetters of feudal duties and the narrow isolationism of local circles. The old Roman roads came alive again with the bustle of lively trade, providing a market for the growing production of crafts and agriculture. The constant stream of pilgrims encouraged trade and helped to swell the great September fairs in St.Gilles, which at first were not even equalled by the famous fairs of Champagne. The ancient urban centres lying on land and water routes grew both in population and in economic importance. Narbonne, which had been an important trading port in Roman times, again took its place in a wider maritime network, as did Montpellier and St.Gilles. At first, of course, these towns were entirely subordinated to the interests of Genoa and Pisa, whose unbridled expansionism made them masters of the western Mediterranean for a time. At the end of the twelfth century things took a turn for the better and the coastal towns of Languedoc were now able to develop independent relations with the towns of Provence and on the Spanish coast, and secured a significant place in trade with the Levant.[11]

Economic prosperity found expression in the way of life of the great aristocratic families. At this period the ruling circles in Languedoc, as in Poitou, Gascony and Provence, united by language and culture, felt no ties with the harsher North of France. Their family ties, political interests and culture were bound up with the ruling Christian houses beyond the Pyrenees, and like them they tended in some ways to imitate the subtle luxury and cultured refinement of the Arab courts. The cultural climate on both sides of the Pyrenees was thus similar, and all the more so since Catalonia, then as now, belonged to the same linguistic region. At the courts of this southern aristocracy the poetry of the troubadours flourished, with its astonishing formal richness, and revealed to the western world the hitherto unknown sphere of the cult of the sentiments. Thus the Occitan regions of southern France, with Catalonia beyond the Pyrenees, became the home of lyric poetry; nothing came out of this region, however, to compare with the heroic epics of the north, the *chansons de geste*. This southern society, unlike that of the French North in both language and culture, had become more tolerant in matters of religion. The Jews formed considerable colonies in the towns; they were not persecuted and their learned rabbis could pursue their scholarly studies.

In this atmosphere the preachers of heretical doctrines found more favourable conditions than elsewhere, and the measures taken against heresy were certainly never pursued with the fanatical cruelty which marked the religious attitudes of the region between the Rhine and the Loire.

In those northern regions with their strong Germanic tradition the date at which the secular arm began to punish the heretics can be given exactly (as we have seen). In 1022 thirteen people proved by a Church synod to be heretics were burnt at the stake in Orleans.[12] This event aroused feelings everywhere, all the more since the victims came from high up in the ecclesiastical hierarchy. The manner of punishment for erroneous doctrinal views had no precedent in the memory of living witnesses. Indeed it was the first instance known in western Europe since the end of antiquity. In the West developments had been more complicated than in Byzantium. There, as we have seen, the suppression of heresy was the common duty of both state and Church, and in theory at least their sphere of competence was clearly marked even if in practice the delimiting line was not always observed to the same degree.[13] In the West the continuity of the Roman State was broken, and long centuries passed before Roman Law again became the subject of careful study and began to influence the life of society to any degree.

Certainly knowledge of Roman Law did not entirely die out in the West even in the Dark Ages, nor did ancient ecclesiastical traditions. At the end of the Roman period St.Augustine offered the soundest justification for employment of the secular arm to punish heretics, when Church discipline had proved insufficient. For him, nevertheless, the goal remained the conversion of the heretics and their return to the Church. From this point of view even the stern measures of secular justice wear the guise of charity, and in this conception there was no place for the death penalty. After the time of St.Augustine the idea that the secular arm might strike against heresy recurred in the writings of the Church Fathers, the decisions of oecumenical councils, and other texts going to make up the official canonical tradition. In addition to this material, the collections of ecclesiastical law also contained some imperial constitutions which, unlike canon law, established the death penalty for unrepentant heretics. These late Roman legal compilations were not unknown in the early Middle Ages, but for a long time there had been no practical opportunity to put the measures they contained into effect against heretics. The occasional aberrations in doctrine which occurred were effectively dealt with by ecclesiastical disciplinary measures, and there was no reason to call in the secular arm.[14]

It was therefore most unusual that the heretics of Orleans were condemned to die at the stake in 1022. The sentence was not based on any valid law, but was a prompt measure taken to thwart those whose heretical doctrines threatened to disturb public order. The closest parallel for infliction of the death penalty by burning was to be found in the Germanic legal tradition, where it was used to punish the casting of spells and poisoning.[15] The popular attitudes put heresy into the context of such crimes as these. The revival of heretical teaching from the early eleventh century onwards gave the secular authorities more and more frequent occasion to resort to this cruel death, which gradually became accepted in the West as the customary, if not the only, form of death prescribed for those convicted of heretical beliefs. From time to time voices were raised in protest among the dignitaries of the Church. The more enlightened believed that the sword of the secular arm should not take the lives of those mistaken souls who were rather to be won back to the Church through patient persuasion.[16] There was still considerable uncertainty about how to deal with heretics in the ecclesiastical courts, the bishops often being unable to decide what to do. At times, however, the crowd took the law into its own hands. Living in the constant threat of natural catastrophes, sickness and wars, the crowd instinctively turned against those whose rejection of the true faith might bring down the wrath of God on the whole population. In the twelfth century, when the heretical movement was gaining in strength, the ecclesiastical authorities looked on with displeasure when the crowd lynched heretics. After the people of Cologne had burned some Cathar heretics in the year 1143 or thereabouts, the provost Everwin of Steinfeld noted that this had happened "against our will" *(nobis invitis)*. Face to face with the growing threat of Catharism, however, even the clergy (at least in the German and Northern French regions) preferred to take a hard line. The edicts of the Council of Rheims against the heretics, in 1157, strongly suggest the Germanic judiciary approach, including the accusatorial form of trial, *ordalia*, and the use of branding as a punishment. On the other hand, the attitude of St.Augustine was still the guiding force in educated Church circles; the commentators on the decree of Gratianus never failed to stress that the purpose of punishment for heretics was not revenge, but the charitable attempt to save lost souls. Roland Bandinelli continued to apply this principle later, when as Alexander III he became Head of the Church. In 1162 he devoted great attention to an appeal addressed to him by a group of "Popelicans" from Flanders. He not only ordered the Archbishop of Rheims to investigate the affair carefully, but called for all the documents in order to study the case himself. We do not know what the final outcome was for these "Popelicans", who seem to have been people of good social standing. But it is certain that the Pope's efforts to ensure that investigation conformed to canon law could not change the course of developments in northern France, which followed the lines set by the Council of Rheims in 1157. In adopting the practice of burning heretics at the stake the clergy both in France itself and in neighbouring German areas were coming into line with the policy of the secular governments and the feelings of the common people. One of the decrees of the Rheims Council of 1157 already hints at the death penalty for heresy. A few years later Eckbert of Schönau explicitly approves of the baptism by fire "administered" to the Cathars by Bonn and Cologne. In 1172 Henri, Archbishop of Rheims and brother of the French King Louis VII, had burned at the stake a cleric who had been convicted of heresy. Eleven years later, under his successor the Archbishop Guillaume, a number of people, common and well-born, and including many women, were accused of heresy in the County of Flanders. The case against them was brought by the Archbishop together with the Count of Flanders, and it was on their behalf that the property of the executed heretics was confiscated. This event was not the only one of its kind. The flames shot up in many places in Flanders and in other regions under the jurisdiction of the See of Rheims. Even at the end of the twelfth century protesting voices were heard from the University of Paris, but nothing the scholars might say could affect the course of events. This form of inhumanity was already deeply rooted in France proper, but a sharp turn of events was needed before it became accepted in the southern provinces as well.[17]

The burning of Peter de Bruis, which took place under exceptional circumstances, was so far the only reliably documented case of the kind in the South.[18] The Cathar preachers in Languedoc were threatened neither by imprisonment not by lynching at the hands of a fanatical crowd. In the town of Lombers there was a public disputation in 1165, attended by crowds of listeners, at which they argued with Catholic bishops[19] openly calling them "wolves" and "hirelings". Two years later (1167) the Cathars held a Council in St.Félix de Caraman and the network of heretical organizations gradually spread throughout the southern provinces... Ten years later (1177) the Chapter General of Cîteaux was informed by the Count of Toulouse himself that conditions in the Church of Languedoc were critical. The Count gave an accurate description of the new heresy: the heretics did not believe that man was created by God, they did not believe in the resurrection of the body, and - worst of all - they "introduced two principles".[20] Even in Lanquedoc there was no longer any doubt that the new heresy was fundamentally dualist, although at the disputation in Lombers (1165) and again in Toulouse (1178) the Cathar speakers denied the true nature of their doctrines.[21]

Being unable to rely on effective help from the nobles in his domains - corrupted as they were by the "plague of infidelity" - Count Raymond V called in his letter upon the King of France to take military action against the southern heretics. The idea of a Holy War against the heretics was not alien to the minds of the Roman Church at that time. The justification could be found in the decree of the Bologna monk Gratianus (c.1140), which formed the basis of canon law, and in the works of his commentators. As we have seen above, the canonists of Bologna were faithful to the conception of St.Augustine, but they nevertheless admitted the possibility of armed intervention in cases of obstinate resistance or where there was great danger. Gratianus backed this attitude by a selection of quotations from the Church Fathers which seem in the context to justify both a holy war against the heretics and the application of the death penalty for incorrigible heretics.

Even Bernard de Clairvaux leaves us in no doubt that the Church must inevitably resort to the sword of the secular arm in order to defend herself.[22]

By a cruel irony of fate the first suggestion of a crusade against the South came from a ruler of the house of Toulouse. For the time being a special mission was sent to the County of Toulouse, its significance enhanced by the armed escort provided by Raymond V and his vassals. The mission was headed by the Papal legate, Cardinal Peter, accompanied by the Cistercian Abbot of Clairvaux and other dignitaries of the Church (1178). A remarkable trial then took place in Toulouse, at which the Catholic prelates were forced to argue with two leading Cathars in the vernacular, since the latter knew no Latin. It was not until three years later (1181) that a real crusade in the South was organized, though small in size and ephemeral in effect. It was a forceful example for the future, however, that this military expedition against the castle of Lavaur was led by a Papal legate.

Meanwhile the Third Lateran Council gave the idea of a crusade against the heretics firm legal ground in 1179, in a statute similar to that permitting a "holy war" against the infidels. The relevant article gives no precise definition of the nature of the incriminated doctrine, but it is explicitly directed against the heretics of the Languedoc, called "Cathars", "Paterines" or "Publicans".[23]

The penalties set by the Council for the activities of these heretics did not apply to Peter Valdes, a merchant of Lyons who chose the path of apostolic poverty in 1176 and went about preaching tirelessly with a group of disciples. It is noteworthy that like St.Francis of Assisi later, Valdes came of the rich merchant class. The Waldensian movement was one of the signs of a religious awareness conditioned by the growing social importance of the laity during this period of rapid development of municipal freedoms. The driving force behind the movement was the realization of the duty to bear witness to the Christian life, a mission to which all who felt the vocation should submit their lives entirely. They lived as wandering preachers, in apostolic poverty, completely liberated from the cares of daily life. Valdes was not aiming at the reform of the Church, at restoring the original forms of Christianity; he did not attack the ecclesiastical hierarchy as did the monk Henry or Arnold of Brescia. He only desired permission to preach, for himself and his disciples, in order to contribute effectively to repair the urgent lack of spiritual instruction for the people. The Waldensians wished to work in complete harmony with the Church hierarchy who taught the true faith and offered the means of salvation through the sacraments. They only refused obedience where the commandments of the Church conflicted with those of the Scriptures, for it is better to obey God than man. Here the break between the Waldensians and the Church began. At the Third Lateran Council Pope Alexander III accepted Valdes' vow of poverty with approval, but refused to grant either him or his disciples the right to preach independently. Their refusal to submit to this rejection automatically drove the Waldensians into opposition to the Church, and as early as 1182 Peter and his group were excommunicated by the Archbishop of Lyons. Even after this, however, the true followers of Valdes did not depart essentially from the doctrines of the Church and were sharp in their rejection of those of the Cathar sect. Indeed a number of articles in the famous confession of faith submitted to Valdes soon after the Lateran Council are clearly directed against the Cathar heresies. Yet when they were cast out of the Church the "poor men of Lyons" shared the same fate as the Cathars. Forced to leave Lyons, they moved into the southern provinces, where Languedoc, in particular, afforded a relatively safe refuge for heretical sects.[24] The Waldensians gained ground in Languedoc but could not displace the Cathar sect which had become established there much earlier. They kept themselves distinct from them - and indeed Waldensian preachers, like Durand de Huesca (Osca) of Aragon, engaged in fierce polemics with the Cathars. A valuable record of these doctrinal arguments is provided by the *Liber Antiheresis*, written by Durand some time in the early nineties of the twelfth century.[25] The book also provides precious evidence that there were already Cathars in Languedoc who followed the radical dualist belief, based on a symmetrical opposition of two principles. With his accurate knowledge of the Cathar sect and its doctrines this Waldensian author is clearly superior to his more learned contemporary, the Parisian theologian, Alain de Lille.

The Waldensians met the Cathars on the Italian side of the Alps, too Some of the exiles from Lyons went as far as Lombardy and strengthened the ranks of the supporters Valdes had gathered there himself on his earlier journey to Rome. The Lombard branch of the Waldensians came into contact with native heretical groups like the fraternity of the *Humiliati* and the *Arnaldists* who carried on the tradition of Arnold of Brescia.[26]

The steady growth of the sectarian movements ultimately gave rise to common action by the supreme secular and spiritual authorities together. The Emperor Frederick I. Barbarossa, and Pope Lucius III, meeting in Verona in 1184, agreed on measures to deal not only with the heretics of specific areas, but on a general scale. The Papal decree *Ad abolendam* gives the names of the sects to be proscribed, and at the same time gives the first definition of the relevant heresy. It distinguishes between those who assume the right to preach without permission of the Church, and those whose teachings on the subject of the sacraments conflict with ecclesiastical tradition. The first category covers the Waldensians and related groups, while the latter clearly refers to the Cathars. The initiative for this action taken in Verona came from the Pope, while the collaboration of Frederick Barbarossa was in keeping with the political situation of the time and with the Emperor's relations with the Pope. Acting as the defender of the faith, the Emperor put the heretics under the Imperial ban *(bannum)* with all its consequences: exile, confiscation of property, houses razed to the ground, and prohibition of holding public office. There is as yet no hint that these measures were based on the theory of lèse-majesty in Roman law, nor were the legal measures proclaimed in the Papal bull any sign of a change in the legal practice hitherto followed. Legal proceedings against the heretics were still within the traditional bounds of the Carolingian Sendgericht. It was a significant new feature, however, that special sessions, separate from those of the general Sendgericht, were to be reserved for the trial of heretics. This special measure corresponded to the greatly increased number of cases of heresy investigated at this time. Thirty years later the Fourth Lateran Council (1215) repeated what were essentially the same measures.[27]

It was significant that even while the Pope and the Emperor were meeting the Cathars held their assembly in Verona itself. Towards the end of the twelfth century the sect displayed an increase of activity in a number of Italian towns, and the number of its adherents was clearly growing. In Prato an edict of the Emperor Henry VI punished heresy by imprisonment, confiscation of property and the destruction of the heretics' houses, with the result that a large number of victims took refuge in Florence. In Ferrara and Modena the city authorities themselves took action against the heretics.[28] The heretical movement, however, was able to take advantage of the growing tension between the urban communities and the ecclesiastical hierarchy. The communities of northern Italy gradually appropriated the executive power in public affairs, previously in the hands of the bishops for the most part. Now they began to try taking over various feudal rights which were formerly reserved for the bishop, the chapter or monasteries, in the towns and their rural districts. They also tried to limit the privileges freeing the clergy from fiscal duties and removing them from the judicial power of the municipality. This led to sharp conflicts between the secular powers and the Church. The bishops repeatedly issued interdicts and excommunications and the municipalities retaliated by exiling the bishops and forbidding the people to have any relations with the clergy.[29]

In the Italian towns heresy went hand in hand with opposition to the Church, and it is not always easy to draw the line between them. This was the state of affairs in northern and central Italy, and in the second half of the twelfth century we already find references to the existence of Cathars in the south of the peninsula too. From a later source (Tractatus de hereticis) it appears that a Cathar bishop held office in Naples at a time when the sect was only just beginning to gather supporters in Lombardy in the North - the convert who was to become Bishop Marcus, and his two companions.[30] It is too short a reference, however, to allow of very definite conclusions, nor is it confirmed by any other testinomy. We do know, on the other hand, that there were many followers of the Cathar heresy in Calabria towards the end of the century. The southern Italian mystic Joachim de Fiore found them there when he began to preach. In his commentary to the Apocalypse, completed somewhere around the year 1200, he devoted much space to the strange sect of unknown origin, called *patareni* and various other names, and divided into *Perfect* and *Believers*. His testimony is highly instructive, and throws an interesting light on the way in which the Cathars gained their adherents, a proselytism in which social aid to the poor played an important part. This author also makes an interesting reference to heretics establishing relations with the Saracens.

Joachim de Fiore was fully aware how dangerous the Cathar heresy was, and what a threat to "different parts of the world".[31]

At the end of the century (1196) the Emperor Henry VI again called for joint action on the part of the imperial and the papal authorities, the spiritual and the secular arm; his sudden death not long afterwards put an end to all his bold plans.[32] The decisive role in the campaign against the heretics now fell to the new Pope, Innocent III (1198-1216), in whom the theocratic ambitions of the Roman Papacy reached their highest point.

It was an important reinforcement of papal theocracy when the Roman concept of the state was applied to the Church; the bull issued by Innocent in 1199, *Vergentis in senium*,[33] described heresy as the crime of lèse-majesty, quite in the spirit of the constitutions of the Roman emperors. The penalties set by them were also repeated in the Pope's decree, down to the severest consequences. Confiscation of the property of heretics and those who gave them protection was extended to the descendants of those convicted, even if they were of the Catholic faith. Innocent III had acquired his knowledge of the Roman imperial constitutions when he was studying in Bologna. Respect for the traditions of Roman law was also instilled into him by his teacher of canon law, Huguccio of Pisa. In the spirit of Christian charity which characterized the canonical tradition Innocent III softened the severity of justice in its application; those who became reconciled to the Church were given the possibility of recovering their civil rights, according to the bull *Vergentis in senium*, and even their property might be restored to them.

The validity of this papal decree, originally directed against Viterbo, was extended the following year (1200) to apply (though in a milder form) to Languedoc and to far-away Bosnia,[34] whence reports arriving in Rome gave warning of the rise of dualist heresy.

[1] See *Amann*, Pataria, DTC XI/2, 2243 - 2246; *Borst*, 250.

[2] For the Cathar church in Florence, see *Dondaine*, Hiérarchie, II, 299 ff.

[3] On Orvieto see *Fumi*, Orvieto, 52 - 81. The beginnings of heresy are described in the Vita S.Petri Parentii, ASS Maii V, 86 f. For the date see *Maisonneuve*, 148.

[4] Vita S. Galdini, ASS Aprilis II, 595BC.

[5] Bonaccursus, PL 204, col. 778.

[6] See *A.Molinier*, Languedoc, in: Grande Encyclopédie XXI, 905 f.; *Vaissète - Molinier* III, 852 ff.; *P.Gachon*, Histoire de Languedoc, Paris 1921, 50 ff.

[7] See *Higounet*, Grand chapitre.

[8] See *H.Géraud*, Les routiers au douzième siècle, in: Bibliothèque de l'École des Chartes 3 (1841 - 1842) 125 - 147; *Griffe*, Aventure, 119 ff., 139 ff.

[9] *Guiraud*, Inquisition I, 327 f.

[10] *Griffe*, in: Cathares en Languedoc, 232 ff.; id., Aventure, 189 ff. The preaching of the monk Henry persuaded people to refuse to give the priests *oblationes, primitias, decimas*. Ex gestis pontificum Cenomanensium, in: *Bouquet*, Recueil XII, 554.

[11] See *A.Dupont*, Les cités de la Narbonnaise Première depuis les invasions germaniques jusqu'à l'apparition du consulat, Nîmes, 1942; id., Les relations commerciales entre les cités maritimes de Languedoc et les cités méditerranéennes d'Espagne et d'Italie du Xème au XIIIème siècle, Nîmes 1942; *C.-M.Higounet*, Le milieu social et économique languedocien vers 1200, in: Cahiers de Fanjeaux, fasc. 2, 20 f.

[12] See Chapter IX, p.103 f.

[13] See Chapter II, p.37; III, p.47; VIII, p.98.

[14] See *Maisonneuve*, 36 ff.

[15] See *Ficker*, Todesstrafe, 181 f.; *Havet* I, 501 f.

[16] See the answer sent by the bishop of Liège, Wazo, to the bishop of Châlons, Roger II, Anselmi Gesta episc. Leod., MGH SS VII, 227 (cf. Chapter IX, n. 9, above). The attitude of this important supporter of ecclesiastical reform in the West here comes remarkably close to the views of Theodore of Studios (see Chapter II, p.37, above) whose efforts to make the church independent of the secular power bear a strong resemblance to the aims of the western reform movement in the eleventh century.

[17] See Maisonneuve, 99 f. For the events in northern France see Frédéricq I, Nr. 46, 48 - 55.

[18] Havet II, 572, 578, n. 2.

[19] Mansi XXII, col. 157 - 168; Hefele - Leclercq V 2, 1006 - 1010.

[20] See *Thouzellier*, Languedoc, 19 f.

[21] Ibid., 21 f. Evidence concerning the early period of the history of the Cathar sect in western Europe frequently refers to these new heretics as *Arians*. This was no novelty in the West; the application of the concept of Arianism to contemporary heretical beliefs is attested as early as the middle of the eleventh century. See *R.Manselli*, Una designazione dell'eresia catara: "Arriana Heresis", in: Bull. Ist. stor. ital. per il med. evo et Archivio Muratoriano 68 (1956) 237. The term also appeared in connection with the teachings of the monk Henry. It was easy in the first instance to associate these new forms of heresy with Arianism, the prototype heresy. This meaning for the term *Ariani* in the anti-heretical polemics of the twelfth century is aptly described by *Y.Congar*, "Arriana haeresis" comme désignation du néomanichéisme au XIIe siècle, in: Rev. des sciences philos. et théol. XLIII (1959) 449 - 461. Catharism was associated not only with "Arianism", but with other early designations as well. See Sigeberti Continuatio Aquicinctina ad an. 1183, MGH SS VI, 421, quoted by Manselli, art. cit., 241:"...quidam dicunt *Manichaeos*, alii *Catafrigas* (i.e. Montanists), nonnulli vero *Arrianos*, Alexander autem papa vocat eos Paterinos" (my italics. ML). Manselli shows how at the turn of the twelfth to thirteenth centuries the terms Manichaean and Albigensian became more frequently applied to the Cathars, while the older term Arians was falling into disuse.

22 See *Maisonneuve*, 58 - 91; reference to Bernard de Clairvaux, 105 f.

23 Toulouse 1178 - sources: letter from Henry, Abbot of Clairvaux (Audite, coeli..., PL 204, col.235 - 240); letter from the Papal legate, Peter of Pavia, Cardinal of St.Chrysogonus (PL 199, col.1119 - 1124). *Griffe*, Aventure, 90 - 111, gives a comprehensive analysis.

Lavaur 1181: see *Griffe*, Aventure, 126 ff.

The 27th canon of the Third Lateran Council (*Mansi* XXII, col. 232; *Hefele - Leclercq* V 2, 1106 f.; cf. *Thouzellier*, Languedoc, 23 f.) was directed against the "routiers" as well as against heretics. *Griffe*, Aventure, 115 ff., considers that the ruling on action by the temporal power referred only to these armed bands.

24 For the beginnings of Waldensianism see *A.Dondaine*, Aux origines du Valdéisme. Une profession de foi de Valdes, in: AFP 16 (1946) 191 - 235; *Thouzellier*, Languedoc, 24 f.; *Selge*, Die ersten Waldenser I.

25 Liber Antiheresis: *Christine Thouzellier*, Controverses vaudoises - cathares à la fin du XIIe siècle, in: Archives d'Hist. doctrin. et litt. du moyen-âge XXVII (1950) 137 - 227 (giving the text of Durand's Book II according to the Madrid MS 1114 with parallel passages from the second version of the Liber Antiheresis, MS Paris, Bibl. Nat. lat. 13446); cf. *Thouzellier*, Languedoc, 60 f. Summa of Alain de Lille (Alanus de Insulis): ibid., 81 ff. The complete text of the Liber was edited by Selge, Die ersten Waldenser II.

26 See Chapter XIV below.

27 Verona 1184: *Förg*, 22 ff.; *Maisonneuve*, 151 f.; *Thouzellier*, Languedoc, 45 ff. On the edict of the Fourth Lateran Council against the heretics (*Mansi* XXII, 986 - 990) see *Förg*, 32 ff.

28 *Schmidt I*, 63 f.

29 See *L.Salvatorelli*, L'Italia comunale, Milan 1940, 436 f.

30 Tract. her., 308 f.

31 For Joachim de Fiore see *Thouzellier*, Languedoc, 110 f. For Joachim and the Cathar heresy, see ibid., 114 f. (with quotations from the work In Apocalipsim). The reference in Tract. her. 309,12-14, to a Cathar deacon named Illarius, concerns Calabria. Cf. *Borst*, 239, n. 40. Dupré-Theseider believes that Anselm of Alexandria was speaking of conditions in his own times as though they obtained for the earliest period of Catharism in Italy. He considers it possible that the Calabrian deacon referred to was the *diaconus maior* attested in Sicily at the beginning of the fourteenth century. See *E.Dupré-Theseider*, Le catharisme languedocien et l'Italie, in: Cathares en Languedoc, 304, 314, n. 12 (cf. Chapter XVIII below, n. 39). There are no grounds for the author's hypotheses, but as he rightly adds, there is no evidence for the existence of a separate Cathar church established in Calabria or in Sicily at all. Of the frequent references to Sicily in connection with Catharism in the sources Dupré-Theseider comments that the name did not necessarily always mean the island only, but the whole "Kingdom of Sicily" (304). - In 1264 a Cathar bishop is attested in the region of Naples; he came from Languedoc and had previously lived for a long time in Piacenza. Ibid., 311 f.

32 *Thouzellier*, Languedoc, 138.

33 *Maisonneuve*, 156 f.; *Thouzellier*, Languedoc, 145 f.

34 Ibid., 155 f., 158 f.

Chapter XIII

THE BEGINNINGS OF HERESY IN BOSNIA

This far-off Balkan country began to emerge into the fuller light of history precisely at the end ot the twelfth century, when the rapid disintegration of Byzantine power after the death of the Emperor Manuel I (1180) brought with it a fundamental shift in the balance of forces in the peninsula. The Serbian state under Stephen Nemanja (Nemanya)freed itself from the burdensome tutelage of Byzantium,and a little later Bulgaria regained independence after a century and a half. The Empire itself rapidly lost what it had won in Hungary, and as early as 1183 Hungarian and Serbian armies penetrated well into Byzantine territory. It would appear that the ruler of Bosnia, Ban Kulin, also took part in this invasion. His name had already cropped up earlier, during the short period when Byzantium ruled over Bosnia. After the death of Manuel, when the Magyar-Croat state rapidly took possession of its lost territories again, Bosnia, too, came under the sovereignty of the Hungarian rulers. Feudal allegiance laid certain obligations on Kulin in his relations with the Hungarian king, his overlord, but did not prevent him developing friendly relations with the republic of Dubrovnik, whose merchants enjoyed the protection of the Bosnian Ban and important privileges in his territory.

Indeed even the Bosnian bishopric, of which we have already heard in the previous century, now came under the jurisdiction of Dubrovnik. There is no evidence, however, that at this time the maritime metropolis intervened at all effectively in the affairs of its subordinate diocese in the mountainous hinterland of Bosnia. Christianity in this inaccessible part of the country had certainly retained many distinctive features, and there can be no doubt that the Slavonic rite was still in use there.[1] Church life had not been affected by the stern system of Gregorian reform, and it was not until after the installation of Pope Innocent III that unusual circumstances led the Roman Curia to intervene directly in this outlying region.

The papal legates - Innocent's chaplain John de Casamaris and the sub-deacon Simon - then held a regional synod in the see of the Archbishop of Bar, on the shores of Serbian Zeta (Dioclea). This appanage was then in the care of Nemanja's eldest son Vlkan, under the sovereignty of his brother Stephen, Grand Župan (Zhupan) of Serbia.[2] The two brothers began to contact the powerful Roman Curia, which was to play an important part in their political plans.[3] It was apparently his desire to secure the favour of the powerful Pope that led Vlkan to write to Rome after the arrival of the papal legates in his country (1199) and draw the Pope's attention to the spread of heresy in neighbouring Bosnia. The accusations were levelled at Kulin himself, who according to Vlkan had succumbed to heretical doctrines along with his wife and his sister (the widow of Miroslav, Prince of Hum) and other members of his family. The epistle goes on to declare that the Ban of Bosnia had led over 10,000 Christians astray in his lands.[4]

Innocent III had already been informed from other quarters that the *Cathar* heresy had penetrated into Bosnia. Bernard, the Archbishop of Split, had just driven a large number of these heretics out of Split and Trogir, and the Ban of Bosnia was accused of giving them his open protection.

We are told of the existence of heretics in Split during the time of Archbishop Bernard, and of their contacts with Bosnia, in the later chronicle of Split by Thomas the Archdeacon. He refers to two brothers of Apulian origin, citizens of Zadar, who were skilful painters and goldsmiths and were also learned in Latin and Slavonic books. These two brothers, Matthew and Aristodius, spent most of their time in Bosnia, where their artistic skills seem to have been in demand. They were however the victims of heretical fallacies, which they not only believed themselves, but also taught to those around them. Bernard of Split discovered this as soon as he became Archbishop (1200). He put the brothers under anathema, confiscated their property and expelled them from the town. Not until later, when they had solemnly abjured their heretical beliefs, was the ban of excommunication lifted and their confiscated property returned to them.[5]

Intervention against the heretical movement in Bosnia was not only a matter for the Papal See, but also for the King of Hungary, the feudal overlord of the Ban of Bosnia. Innocent III had sent a letter to King Emmerich[6] already in 1200, calling upon him in accordance with the bull *Vergentis in senium* not only to drive the heretics out, but to expel Kulin himself from Bosnia and from the whole of the Hungarian kingdom, and to confiscate all his property, unless the Ban took energetic measures against the heretics in Bosnia.

The ruler of Bosnia went to great lengths to put things right in the eyes of Rome. He sent a mission there, which included the Archbishop of Split, Bernard, and the Archdeacon of Dubrovnik, Marinus, as well as Bosnians. At Kulin's express request the Pope then sent his legates to Bosnia to investigate the true state of affairs.[7] At a place called Bilino Polje (Polye) a solemn ceremony took place in 1203, a valuable document from which has come down to us.[8] In effect a solemn vow was taken that the regulations and usage of the Church of Rome would be strictly observed. In the presence of the Papal legate John de Casamaris and Ban Kulin this pledge was given by the superiors *(priores)* who represented the *fraternitas* of the Bosnian *Christiani*. This term - in its Slavonic form *krst'jane* (krstyane) - had long been used in Bosnia to denote *monks*.[9] The text which has survived clearly refers to the monasteries *(fratrum conventus)* and their property. On the directions of the Papal legate monastic life in Bosnia was to be regulated by fixed rules, as was the case in Western Europe. The rules referred to in the document concern the furnishing of monastery churches, the liturgy, fasts, the celebration of holy days and not least, the separation of the sexes in monastic establishments. The elementary nature of some of the regulations shows how chaotic monastic conditions must have been in this very neglected ecclesiastical province.

Of great significance is the regulation concerning the election of an abbot for the whole congregation; the term *magister* is used here. It appears that hitherto the decision rested with the Ban, who is here called the protector *(patronus)* of the monks of Bosnia. In future the superiors of each monastery were to elect their "magister" themselves, after consultation with the monks, and the election was to be confirmed by the Pope alone.

The Bosnian *fraternitas* was also to be given its exact place in the hierarchical structure of the Roman Church, as the monastic orders and congregations of the West had. Through their representatives present at Bilino Polje the monks also abjured the "schism" of which they had been accused, and proclaimed the Church of Rome to be "their mother and the head of the whole Christian community". The Bosnian *Christiani (krst'jane)* also forswore the use of this name, hitherto arbitrarily appropriated to themselves, and promised to be known in future as "brothers" *(fratres)*.

It is remarkable that they were not called upon to abjure any heretical doctrines, although there can be no doubt that it was the *krst'jane* - the monks - of Bosnia who were reproached with heretical opinions. This follows unquestionably from the Pope's epistle to King Emmerich in 1200. There we read that Ban Kulin gave his protection to heretics, exposing himself and the whole of his people to their wicked influence, and that at the same time he respected them as the most orthodox Catholics, referring to them by the name *Christiani*. The terms in which the Bilino Polje document is formulated certainly fundamentally restrict the extent of the accusation. The monks pledged themselves not to accept heretics in their midst in future, and pledged the entire wealth of their monasteries to guarantee that they themselves would not stray into the paths of heresy. In this connection the Bilino Polje document refers explicitly to "Manichaeans and other heretics", while in their letters Pope Innocent III and his legate John de Casamaris use the names "Cathar" or "Patarene" for the Bosnian heretics.

Two of the Bosnian *krst'jane* present at the ceremony in Bilino Polje, Lubin and Dražeta (Drazheta), then accompanied John de Casamaris to the Hungarian court, where in the presence of the King, they swore to abide by all that had been pledged, as well as to obey all commands laid on them by the Church of Rome in the future. Kulin's son undertook to pay a considerable sum of money to the royal treasury and the Roman Curia should his country ever again give protection to heretics.

Lubin and Dražeta left their signatures on the Bilino Polje document, as two of the seven representatives of the Bosnian *krst'jane*. The first place, before them, went to the signature of one Dragič (Dragitch). These three names, Dragič, Lubin and Dražeta, occur again in the list of members of the heretical Bosnian church, preserved in Batalo's fourteenth century Gospel. This could hardly be mere coincidence. Everything points to the fact that the infiltration of dualist heresy into the monasteries of Bosnia, clearly attested by this document of 1203,[10] was not an ephemeral phenomenon, but had lasting consequences. Before twenty years had passed the problem of heresy in Bosnia again began to worry the Church of Rome, and with much greater urgency. In 1203, however, things seemed to have been satisfactorily dealt with, and Innocent III could turn his undivided attention to the acute danger represented by the heretics of the West.

[1] On the early history of Bosnia see *Ćirković*, Istorija, 44 ff. On the beginnings of ecclesiastical organization there see ibid., 50 f.

[2] On the papal legates' mission to the territory of Dioclea (at the request of Vlkan) and Innocent's recognition of the Archbishopric of Bar, see *Dvorník*, The Slavs, 97.

[3] See Chapter XVI below.

[4] Innoc. III, Ep. II, 176 (PL 214, col. 725 - 726; *Theiner*, Mon. Slav. I, N° 10, 6; *Smičiklas*, CD II, N° 310, 334). See the quotation in *Mandić*, Bogomilska crkva, 65, n. 202, where the text of Vlkan's letter is corrected according to Reg. Vat. IV, fol. 185v. For the date of the letter, see ibid.

[5] For Innocent's reference to the heretics expelled by the Archbishop of Split, see n. 6 below. On Matthew and Aristodius see Hist. Sal., 80. Cf. *S.Radojčić*, Majstori starog srpskog slikarstva, Belgrade 1955, 11 f.

[6] Innoc. III, Ep. III, 3 (PL 214, col. 871 - 873), *Potthast* 1142. For the heretics expelled from Split and Trogir by Archbishop Bernard see col. 871.

[7] See the epistle of Innocent III dated November 21st, 1202: Innoc. III, Ep. V, 110 (PL 214, col. 1108 - 1109), *Potthast* 1768. *Smičiklas* III, N° 11, 14 f.

[8] Text in *Mandić*, Bogomilska crkva, 435 - 436.

[9] E.g. *krst'janin* Radochna (Radokhna), whose name is scratched on a block of stone in Kulin's church. See *Miletić*, "Krstjani", 45. For various meanings of the word *krst'janin* see ibid., 49 f.

[10] *Rački* (Bogomili i patareni, 394) had already concluded after analysis of the Bilino Polje document that "Patarenism both in the East and in Bosnia early found its way into monasteries". It is still a matter for debate, however. Some authors see the *fraternitas* referred to in the document as an orthodox monastic organization of either the eastern (Basilian) or western (Benedictine) type (*Miletić*, Krstjani; *L.Petrović*, Kršćani bosanske crkve, Sarajevo 1953). At the other extreme it is suggested that this referred to an organized dualist sect (*Kniewald*, Vjerodostojnost; *Mandić*, Bogomilska crkva). According to Mandić, Innocent III dealt with the Bosnian *krst'jane* in the same way, after they were received back into the church, as he did with the Italian *Humiliati* after their reconciliation (op. cit., 136). *Ćirković* (Istorija, 57 f.) is perhaps too cautious when he sees in the document no more than evidence of "the first contact between Christianity and heresy" on Bosnian soil.

Chapter XIV

THE CULMINATION OF THE FIGHT AGAINST HERESY IN WESTERN EUROPE

Innocent III directed his main efforts to putting down heresy in northern and central Italy and in Languedoc. In the cities of Italy the heretics turned to their own advantage the constant friction between the laity and the Church. Conflict between the bishops and the city authorities was at its sharpest in the first half of the thirteenth century. The towns tried to restrict the size of church lands and their privileges and immunity, and demanded taxes and other duties from church institutions. This brought them into conflict with the papal Curia, which intervened sharply to defend the *libertas ecclesiae* from this threat; one excommunication followed another and repeated interdicts created an atmosphere of oppressive tension. These measures provoked opposition in various forms, getting round the prohibition of public worship, and on the other hand led to boycott of the clergy by the laity. When the state imposed by interdict lasted too long the ecclesiastical hierarchy became alienated from the people, and this created the ideal conditions for sectarian propaganda. Heretical elements continued to appear openly in urban life. Nor were they a negligible factor in the internal conflicts set in motion by the sharp shifts in social stratification. The number of citizens with full rights grew with the irresistible rise of the popular element; in the towns they became freemen and as craftsmen could attain considerable prosperity. This caused a significant shift in the balance of power in the community, giving rise to sharp conflict between the *gente nuova* and the old aristocracy. The nobility itself was often divided into hostile factions, so that some of the nobles would take the side of the people against the rest of the aristocracy. The antagonism between the different social classes and groupings would break out on the most varied occasions, and was reflected particularly strongly in relations with the Church. The heretics naturally found sympathy and protection with whichever party happened to be opposed to the local bishop or to the Pope at any given moment.[1]

Developments in Orvieto at the beginning of the thirteenth century show clearly what the forces were which clashed in dramatic conflict, filling the narrow streets of Italian towns with their clamour, below the grim towers of the houses of the nobility. In Orvieto the nobles defended the independence of the town against the growing influence of the Pope, for that independence was the guarantee of the power they had enjoyed hitherto. The common people, on the other hand, would have liked to see the privileges of the aristocracy abolished, and so took the side of the papal Curia. Here and there a noble family went over to this side, too, serving individual interests stronger than those of the group. Their opposition to Rome earned for most of the nobles of Orvieto the name of heretics, and indeed they became the natural allies of the sectarians. Taking root in Orvieto very early, they became bolder as time went on. At the beginning of the pontificate of Innocent III the town found itself in conflict with the Curia, over who was to govern Acquapendente and the region of the Valle di Bolsena. The interdict and the long absence of the bishop, called to Rome, left the field open to the heretics; there was a Cathar preacher *(doctor Manichaeorum)* active in the town, one Peter Lombardus. The situation was to be taken in hand by Peter Parenzo, whom Rome placed at the head of the town. His biographer stresses, characteristically, that he paid his church tithes, unlike the godless Romans, Tuscans and others. The new rector applied with all harshness the measures against heresy laid down by civil and by canon law. The penalties inflicted on the high-born patrons of heresy were drastic - imprisonment in chains, public flogging, exile, heavy fines, and many houses razed to the ground. Peter perished at the hand of conspirators, and the Church made a martyr of him. Confiscation of property and exile awaited many of nobles for their part in the conspiracy, yet a few years later (1209) Acquapendente again roused fierce conflict between the town and the papal Curia, and the former even became for a time the ally of Emperor Otto IV against the Pope.[2]

In the immediate vicinity of Orvieto Innocent III came up against stiff opposition in Viterbo. Here the community did not hesitate to give high office to a well-known "heresiarch", John Tiniosi. A papal epistle of 1205 designated precisely the false doctrine which had seduced many in that town - belief in the Devil (Lucifer) as the creator of all visible things.

The Pope's call met with active response from the authorities in several towns of the Marches and in Tuscany, but in the north Verona remained a notorious hotbed of heresy which was widespread not only among the people but even among the clergy. In the towns of Lombardy the podestas, consuls and councils of the towns contravened the privileges of the Church in many respects and called down upon themselves the full fury of the Pope's anger. His epistle to Modena (1204) is packed with complaints about various injustices inflicted on the Church by the present podesta of the town, increasing even the injuries it had to suffer form his predecessor. Similar accusations of interfering with the *libertas ecclesiae* issued from the Throne of St.Peter addressed to many other towns - Novara, Cremona, Parma, Piacenza. These disobedient towns were to be made to feel the consequences of their attitude where they felt it most; the Pope ordered the goods of their merchants to be seized, while none were to trade with the towns or have other relations with them. In some cases Innocent III threatened to move the bishopric to another town or to divide the diocese up between two neighbouring bishoprics. Again excommunications thundered, with bells ringing and candles guttering, and again divine service was not allowed to be held. ... The interdict proved a two-edged weapon, however. In Bergamo, for instance, the fact that this abnormal state of affairs lasted so long only meant that the heretics were bolder than ever in propagating their faith in the town.

At the beginning of the second decade of the thirteenth century a new cause of conflict between the Emperor and the Papacy added yet another source of hostility between Innocent III and the towns of Lombardy. Otto IV had originally been the Pope's protêgê, and in return for his support had handed over to Innocent III all his feudal rights south of the Alps. In 1209 Otto came to Italy for his coronation and immediately took steps to deal with heresy there. He drove the Waldensians and other sects out of Turin and in Ferrara confiscated the property of those who supported heresy, and had their houses torn down. Turning against Rome, he then drew some of the north Italian towns to his side.

Innocent III punished Bologna for siding with the Emperor, and threatened to move its famous schools elsewhere. He also addressed sharp accusations to Alexandria, but his greatest vehemence was directed against Milan, which at that moment was in the anti-papal Ghibelline camp for a time. Innocent's well-known epistle defined three fundamental points on which the papal Curia based its hostile attitude to this city: the protection given to heretics there, the contempt for the *libertas ecclesiae*, and the fact that the city had broken off its alliance with the Roman Curia.[3]

The case of Milan shows with great clarity how the fight against heresy was intertwined with the other aspects of papal policy in Italy. In Milan itself, as in other Italians cities, the spread of heresy was largely due to internal shifts in the structure of urban society. The popular elements, organized in the religious society *credenza di S. Ambrogio*, were irresistibly gaining the decisive influence in the community, a great wave which threatened to sweep away the varied accumulation of feudal rights as time went on; history was moving towards the formation of a single civic law. This development threatened the supremacy of aristocratic privilege, just as it threatened the privileged position of the Church, and the nobles, disturbed by this threat, gathered round the archbishop of Milan. The civic community, identifying more and more with the successful rising popular class, thus found itself in a position of hostility towards the Church in Milan,[4] which also explains its tolerance of heresy. The followers of heretical sects - Cathars and Waldensians - could count with support from this quarter, and not from the Milanese city nobles traditionally allied with the Church. The favourable attitude of the authorities towards the heretics had nothing to do with the external political situation, in which Milan was for a time engaged in the anti-papal coalition. In fact their attitude towards the heretics did not really change even after the city returned to the papal (Guelph) camp.[5]

Innocent III put all the weight of his influence over the Christian world to bring the rebellious capital of Lombardy to heel. It was to be isolated from the rest of the world not only by ecclesiastical disciplinary measures, but by the seizure of merchandise and by forbidding citizens of Milan to hold the office of podesta or rector in other cities. The papal epistle to the Milanese, however, culminated in an undisguised threat that a crusade might be sent against them as it had been against Languedoc. It was autumn 1212[6] when this threat was uttered, and by that time war had been devastating the southern regions of France, beyond the Alps, for several years.

Here we must go back to the years which immediately preceded the tragedy of Languedoc.

The problem of dealing with heresy in Languedoc was fundamentally different from the one which faced the papal Curia in France itself. In the North the heretics were persecuted with excessive diligence by the local bishops, who were fully supported by the secular authorities. In 1204 Gui Parê, archbishop of Rheims, convicted some heretics of Catharism; they were afterwards burned at the stake in the presence of Count Robert de Dreux and other secular gentlemen. Among the victims one Nicolaus is explicitly named, a painter famed throughout France.[7] Chronicles tell of the burning of Cathars elsewhere too, notably in Troyes, which belonged to the archdiocese of Sens. When Innocent III intervened in the persecution of heretics it was in order to force the local prelates to observe strictly the correct legal procedures. It was in the archdiocese of Sens that he had more than once to restrain the ruthless ardour of the despotic bishop of Auxerres, Hugo de Noyers. He often went through doubtful cases himself, and refused to condemn as heretics even cases where the suspicion was very strong. In most cases the accused came from La Charité, a town on the Loire where heresy remained deeply rooted for a long time. Innocent III himself was aware of this, as we can see from a letter he wrote to the new bishop of Auxerres (1208), referring to the large number of people in La Charité who had relapsed into heresy again. Under the pretext of carrying on trade they were making contact with leading members of the sect and bringing them into the town. The papal epistle designated these heresiarchs as *consolatores*, that is to say, people who could administer the *consolamentum*.[8] There is thus no doubt that these were Cathars, referred to by contemporary chroniclers in northern France as "Popelicans" or "Bulgars".[9]

From the outset Innocent III himself had to deal with the growing danger of heresy in Languedoc. There was only Montpellier to be relied on as a bastion of Catholicism. From the point of view of Rome the whole situation in the south had worsened at the end of the twelfth century, when the younger generation of the ruling houses of Toulouse and Barcelona came to power. Alphonse II, who had issued a very strict edict against heretics in his lands (1194), died in 1196. The "King of the Pyrenees" himself undid the work of his lifetime when he divided his domains between his two sons before he died. The younger, Alphonse, inherited the County of Provence with Millau and Guévaudan in Languedoc, which were part of the Barcelona domain proper; Peter, the elder brother, was given the throne of Aragon and the County of Barcelona as well as the County of Roussillon and the fiefs on the other side of the Pyrenees. Alphonse the younger did not renew his father's stern measures against the heretics, but his brother, now King Peter II, made the terms of the edict of Alphonse II even more severe in 1198, introducing the penalty of death at the stake and confiscation of property for followers of heresy who were caught after the short period of grace allowed them to leave the country. The edict remained limited in effect, however, to the areas under Peter's direct rule, south of the Pyrenees. In Languedoc neither the Count of Foix nor Vicomte Raymond-Roger de Trencavel followed his example, although both owed allegiance to the King of Aragon.[10]

Meanwhile two years before the death of Alphonse, Raymond V had died (1194) in Toulouse, a ruler who had done so much to get a crusade sent against his own country. His son Raymond VI immediately came into conflict with the Church for injustice done to the monks. The Abbots of St.Gilles continued to have good reason to complain bitterly about him, and St.Gilles is only one name among many in the list of estates which the son of Raymond VI had to return to the Church, as having be- confiscated by his father, once the war was lost. Conflict between the ruler of Toulouse and the Church of Rome did not stem only from such seizures and acts of violence, however, such things being of daily occurrence in Languedoc as elsewhere. The main source of growing tension was the question of the number of heretics and their supporters in the county, already reaching vast proportions. It was of course as impossible for Raymond to drive them out as it was for him to disband his mercenary gangs. The new Count of Toulouse did not merely tolerate the heretics from necessity, however; he openly showed them favour. It was well known that when he met "good people" (i.e. the "perfect" of the Cathar sect) he paid them the ritual honour offered by all "believers". There can be no doubt that there were "good people" in his immediate entourage. When he left for battle he took one of them to accompany him, so that in mortal need he could receive the *consolamentum* at his hands. It was certainly not without malice that he once invited the bishop of Toulouse to his palace to listen to a Cathar preacher. He is said to have shown his disrespect for the clergy on many occasions, but this was nothing unusual in his time; there is even a story that he egged his clown on to parody the celebrant at Mass. Yet many years later, when his son wanted to clear his father's reputation, he produced evidence of many generous donations by Raymond to ecclesiastical institutions, and proof that he was devoted to the Catholic faith.[11]

The mentality of the age and of the social class to which Raymond VI belonged, and some features characteristic of the South, will help us to understand the apparently insoluble contradictions in the personality of this "son of eternal damnation" as a monkish chronicler called him. Raymond-Roger de Foix, too, attacked in similar terms of hatred by Peter of Vaux-de-Cernay, must also have been very close to the heretics. His sister Esclarmonde received the *consolamentum* at the hands of the Cathar bishop in his presence, in the castle of Fanjeaux. Raymond-Roger's other sister was a Waldensian while his wife Philippa was the head of a Cathar *domus hereticarum*. The Count himself, however, held himself apart from these heretical groups.[12]

The young Vicomte of Béziers and Carcassonne, Raymond - Roger de Trencavel, later to become a tragic victim of the crusade, was himself a Catholic, but according to Guillaume of Tudela he was "too young" and therefore his vassals did not respect his authority and extended their protection to heretics in their "castles and towers" on their own account.[13]

At the end of the twelfth and beginning of the thirteenth centuries the sects in Languedoc felt very safe under the protection of the powerful nobles. Three or four decades later the records of the Inquisition bring back to us this irrevocably lost time of freedom. There was no need as yet for the heretics to hide in lonely farms or in the forests. The Waldensians met openly, even in churches, to read the Scriptures and sing hymns. The Cathar "perfect", known as "Christians" or "good people", appeared boldly in towns and villages, and their friends sheltered them openly in their homes. The "believers" gathered to hear their sermons, which were sometimes delivered by women. There is more than one note of "believer" women kneeling respectfully three times before one of these "heretical women" and after the triple *Benedicite* and the customary ritual prayer, receiving the blessing that God may make her "a good Christian" and "bring her to a good end". Those of the Cathar believers who attained this "good end", i.e. received the *consolamentum* from the "hands of a good Christian" before death, could be buried in one of the "heretical graveyards" mentioned in the Inquisition records. The Cathars of Languedoc at this time lived openly without fear, and without fear buried their dead. In 1203 a deceased nobleman of the Cathar faith was buried in a "graveyard of heretics" in the diocese of Toulouse; he was accompanied by a procession bearing lighted candles, and by all the inhabitants of the fortified town *(castrum)* both high-born and low.[14]

There was a widespread network of deacons in Languedoc at this time, each in a certain locality. Cathar houses *(domus haereticorum)* gave shelter to members of the sect travelling from afar. These houses resembled monasteries; in them the "Christians" (the perfect) lived a communal life under one of their number. Their diet was regulated by the ascetic rules of the sect and contained no flesh in any form (except fish) and not even eggs or cheese. Visitors brought them the only foods permitted: bread, wine, fruit and nuts. In these houses the Cathars prayed together, passed round "the bread of holy prayer" before their common meal, made their common confession and performed all the religious practices of their church. Not only men chose this communal life; there were communities of women, including women of high birth, who had received the *consolamentum* and thus renounced family life. No doubt their families provided these houses with their modest needs, however. In the Lauragais district the Cathar women took girls of impoverished noble families into their houses, giving them free board and education. The girls were brought up in the Cathar faith, so that when they finally received the *consolamentum* they would swell the ranks of the "perfect".[15]

Obviously the Cathar church had become a regular institution in Languedoc, a state of affairs which was laid at the door of the local bishops. Guillaume of Puylaurens,writing his history of the Albigensian Wars somewhat later, showed more understanding of their difficult situation. He was not ignorant of the fact that the bishop of Toulouse lived in considerable poverty in his residence, since he received no share of the tithes of the parishes under him. Nor could he carry out his visitations, because it was not possible to move safely about the countryside, at the mercy of attack from armed bands, without a military escort. Another reason why the Bishop of Toulouse - like other bishops in Languedoc – was powerless, derived of course from the fact that they had gradually lost their temporal jurisdiction. Even so Guillaume of Puylaurens blames the bishops of the time for their inertia.[16] Their contemporaries - Innocent III and his plenipotentiaries in Languedoc - were even sharper in their condemnation of the bishops. From the outset of his pontificate innocent III sent legates to this threatened region for ten years, mainly chosen from the Cistercian order. They were given extensive powers which were gradually increased as moderate measures proved ineffective. Raul and Peter of Castelnau fulfilled this function from 1203,and their mission was strengthened later by Arnold Amalric, the Abbot of Cîteaux. They did not hesitate to apply their disciplinary powers which now included the right to suspend unsuitable priests and remove bishops from office. In this way Foulques, who had been a professional *jongleur* before he became a monk and later abbot of a Cistercian monastery, became bishop of Toulouse.[17] The legates clashed sharply with the archbishop of Narbonne, Bérenger II, who had his work cut out to defend himself in Rome.[18] The complaints brought against the bishops of Languedoc by the Papal Curia were undoubtedly exaggerated in some respects. The diocese of Albi, whose bishops retained their temporal power, was administered with due care under the long episcopate of Guillaume Peyre (1185-1227). He was not a ruthless fanatic, and managed to spare his diocese the catastrophic consequences of the tragedy hurled down upon the regions of Languedoc by the pitiless action of external forces.[19]

Apart from questions of discipline there was the urgent task of instructing the people in the faith, as an effective antidote to the influence of the heretics. The legates sent by Innocent III met this need by tireless preaching, and also by arranging public disputations between Catholics and heretics, before a jury made up of people from both sides. Their work received a new impulse with the arrival of the Spanish bishop Diego of Osma. Under his direction the legates travelled about the countryside without pomp and ceremony, going barefoot from town to town like the apostles of old. Innocent III gave this new policy his full support, and numbers of Cistercian monks and abbots joined the legates to work as preachers in the different districts, living in simple poverty. Diego himself, with a young companion (later to become St.Dominic), remained in Prouille where they founded a convent for girls of impoverished noble families and thus paralysed the influence of the Cathar houses on the education of these girls. There were other successes, too, attending the work of Diego, the legates, and their Cistercian assistants, but taken all in all the results were far from proportionate to the effort expended. At a famous disputation in Pamiers, in the castle of the Count de Foix in 1207, Diego and the bishop of Toulouse, Foulques, with others by his side, won a number of converts to the Catholic faith; it is significant, however, that they were Waldensians, those who were still faithful to the teachings of their founder and were estranged from the Church only because of their demand for freedom to preach. This group of converts from Waldensianism, led by Diego's compatriot Durand de Huesca, found its mission somewhat later in the body of the Church as the society of *Pauperes Catholici* with the explicit task of preaching.

Unlike the Waldensians, the Cathars in Pamiers remained obdurate. This was indeed typical of the situation throughout Languedoc at this time when the fight against the dangerous sect was proving fruitless.[20] Yet the number of real members of the sect ("perfect") was relatively tiny and even the numbers of the "believers" should not be overestimated. The numbers who sympathized with the Cathar sect without actually accepting their faith, however, were much greater.[21] Heretical propaganda had much in common with the criticism of the Church and with the satire of ecclesiastical life found in the tendentious poetry of some of the Occitan troubadours, in their *sirventes*.[22] We know, too, that in Languedoc the ground had been prepared for Catharism by the activities of Peter de Bruis and the monk Henry, whose attacks on the Church had fitted in so well with the material interests of the local aristocracy.[23] The influence of the Cathars, however. was more profound and more lasting, for the sect was not the following of a single vivid personality, but a body witn certain fixed forms of organization which guaranteed continuity. The pure morals and ascetic life of the "good people" were a constant and powerful influence on their surroundings and gained them real converts to the faith even in aristocratic circles. The favour shown to these heretics by the noble families of the south was not merely the result of purely material calculation in the power game because of the sect's opposition to the Church, but had its strong emotional reasons.[24] The Cathar sect had such a strong influence in these noble families primarily because of the women who showed boundless devotion to the faith. Undisturbed in its growth in Languedoc for many years, heresy had become firmly interwoven in the whole fabric of upper-class life there. The greatest eloquence and the most suggestive example of Christian poverty were of no avail here. Only violence could uproot this "weed of heresy".

The Cathar sect had of course its followers among all classes of the people.[25] The Cathar preachers met with a great response among the villagers who were for the most part no longer serfs. We later hear of them freeing heretics from the hands of the secular authorities, or helping them to escape.[26]

In the towns, as far as we can form an idea of the situation from our sources, things were not simple. The rise of plebeian elements growing rich on trade and usury was a threat to the positions of the old patriciate in Toulouse at the turn of the twelfth to thirteenth centuries. Bishop Foulques in his ruthless attack on usury as well as heresy found support both among the patricians and among the craftsmen. On the other hand, the Cathar propaganda found a hearing among these same elements of the population. Among the names of people persecuted by the Inquisition in Toulouse, noted by the Dominican Guillaume Pelhisson, there is more than one craftsman. His chronicle of the Inquisition in Toulouse leaves us in no doubt, however, that the patrician class played an outstanding role in the heretical movement in Toulouse, and the records of the Inquisition confirm this.[27]

Toulouse was an urban centre of the southern type, resembling the towns of Italy both in its appearance, with the towers of the patrician family houses,and in the dynamics of its social life. As in the Italian centres, the reaction to heresy was bound up in various ways with the antagonisms between different social groups and interests, but the sources do not allow us to trace all these relationships in detail.

The activity of the sect was centred not in the cities but in the countryside, in small fortified communities.[28] It was not an accident that the bishop of the Cathar diocese of Albi did not reside in that town but in the nearby "castrum" of Lombers, ruled in condominium by some fifty families of the lesser nobility. Such fortified localities *(castra)* were the most frequent place for Cathar gatherings, for here they felt most secure. To uproot the sect it was first necessary to bring the Languedoc nobility to submission by the use of force, and rob the sect of its powerful protection.

The Pope's letters of 1204 had already contained an urgent appeal for secular action against the heretics of the South, offering the same indulgences for those taking part in a war against the heretics as those given to warriors taking part in the crusades to the Holy Land. For the first time Innocent III gave the French king the right to annex the vast territory under the rule of this heretical southern aristocracy. King Philippe-Auguste, however, refused to acknowledge the Pope's right to deal with matters of feudal allegiance.[29] Nevertheless in 1207, after the Cistercian mission in Languedoc had begun to disintegrate on the death of Raul and Diego, the idea of a crusade became of real concern. The sword must be used where spiritual weapons had proved useless. This was the message of Innocent III's letter calling upon King Philippe-Auguste and his powerful vassals, upon all the feudal lords of France, and referring darkly to the miseries of war which would be the only way to bring the followers of "heretical infidelity"[30] to their senses. War was already in the air over the Occitan south, and the violent death of the papal legate Peter of Castelnau at the beginning of 1208 only hastened the moment of its unleashing.

We learn the story of the crusade[31] against Languedoc from chronicles which sometimes present us with the same event seen from different angles, at least for half of the period concerned, when eye-witnesses each gave their own fresh account. The bias of the Cistercian monk, Peter of Vaux-de-Cernay, does not detract from the accuracy of his facts, but gives his whole picture a horrifying sharpness. He makes no attempt to subdue the hatred with which the whole campaign was carried on, or to pass over in silence the inhumanities committed in the course of it. The same events pass before our eyes again in the Occitan chronicle in verse by Guillaume of Tudela, born on the other side of the Pyrenees. His "Song of the Crusade" as it is called today leaves us in no doubt about the orthodox views of the author, a cleric and *jongleur* at the same time; he saw heresy as the cause of the misfortunes now befalling the southern provinces, but he felt no fanatical hatred for the heretics. More than once his mind drew back in horror at the scenes of incredible cruelty, and there are moments when he openly expresses humane compassion. Time did not allow Guillaume to bring his chronicle further than the year 1213, and an anonymous author continued it; he was clearly on the other side, although by no means a heretic. He was convinced of the justice of the cause of the Count of Toulouse. Thanks to this unknown author we have an invaluable insight into the attitude of the South, the victim of aggression, at least for the years 1213 to 1219. How things developed further is told by the later chronicle of Guillaume de Puylaurens. The author did not begin his work until the war had been over for twenty years (1249) and finished it (1272) only a year before his death. From this distance in time his view of the war was calmer and in a sense more objective. Although the level is not maintained throughout, and the composition is not well balanced, this work contains much valuable material and clarifies much in the situation of Languedoc in the years immediately preceding the war.[32] From these chronicles, supplemented by the evidence of documentary material, we can evoke a vivid picture of this long war between the "North" and "South" of France.

After hearing of the murder of his legate Innocent III placed Raymond VI, accused of complicity in the crime, under the anathema of the Church and personal interdict, and excommunicated him. At the same time he invoked secular sanctions against the Count of Toulouse; his vassals were released from their pledge of loyalty to him and his lands were "handed over" to Catholic lords to take for themselves.[33] These measures were based on the authority of the Church Fathers and the decisions of the Third Lateran Council.[34] The justification of war against the heretics was established in the works of the Bologna canonists and sanctioned by the Third Lateran Council. Innocent III turned to this extreme measure after all others had failed, and not even then did he leave the excommunicated Raymond VI without the possibility of reconciliation with the Church and restoration of his rights. The Pope's attitude was thus in the spirit of the canonical tradition, but further developments no longer depended only on his decisions. Once war had been let loose it went its own way, and no theoretical approach could foresee what that would be.

From the outset it was an unwelcome complication that King Philippe-Auguste obstinately refused to head the military campaign, and his absence inevitably meant that the power of the papal legates in Languedoc increased. In that they commanded the crusaders in person, they largely evaded the Pope's authority, and often acted independently regardless of his instructions. Matters soon began to get out of hand, and go far beyond the original purpose of the crusade. Far from the scene of fighting, Innocent III could not prevent the "Holy War" from turning into a rapacious campaign of devastation. His repeated attempts to insist on the careful observance of the law came up against the fanatical obstinacy of the papal legates and in spite of himself the Pope found himself carried along by the will of the radical forces.[35]

Throughout northern France the fiery call of the Cistercians from their pulpits, rousing the people to battle, met with great response. It was indeed a most tempting proposition: it was not far, while for far less risk the same indulgences and the same material advantages could be acquired as were offered in return for the much more difficult and dangerous "pilgrimage" to fight the Saracens in the Holy Land or beyond the Pyrenees. No doubt the postponement of debts and remission of usurers' interest were a strong inducement to the debt-ridden French nobility and prelates to go into battle against the "enemies of Christ" under the banner of the Cross.[36]

The South was unable to unite its forces against this overwhelming numerical superiority. The political anarchy reigning there served them ill in the face of mortal danger.[37]

It was in any case Rome's intention to attack each of the "enemies of the Church" separately, as the Pope's instructions to his legates explicitly set forth: they were not to begin outright by attacking the Count of Toulouse, but to use skilful trickery to restrain him from taking action while the other supporters of heresy in the South were being dealt with.[38] Raymond VI was unable to collect sufficient independent determination to take a firm stand himself and bring down on his own head the odium of being a protector of heretics.[39] He therefore accepted the severe terms laid on him and received absolution at the humiliating cost of public repentance in the cathedral of St.Gilles. In the end he joined the army of the crusaders which set out in the early summer of 1209. The first to feel the hurricane of destruction bearing down from the North were the lands of Vicomte Raymond-Roger. Béziers was the first goal of the army under the papal legate Arnold Amalric. Despite its fortified position the town could not withstand the brutal attack made on it by the lowest elements in the crusading forces. In the confusion the people sought refuge in the great monastery, where they all perished - priests, women and children among them: "the Cross could not save them, nor the altar, nor the figure of the Crucified".[40] Immediately on the fall of Béziers the crusaders appeared before the walls of Carcassonne, and in the heat of summer the besieged town suffered intolerably from a lack of water. Vicomte Raymond-Roger, "a man of great heart", refused the offer of his life and caught in ambush, died in prison.

After the fall of Carcassonne the conquered territories had to be administered, to form them into a reliable base for further operations. The choice fell on a sixty-year-old nobleman from the Isle-de-France, Simon de Montfort, Earl of Leicester. This rough soldier, fanatically devoted to the cause of the Church and surrounded by a legend of miraculous invulnerability, was at the same time ruthlessly furthering the power of his own house He had no scruples about taking over Carcassonne,[41] but he was in a difficult position there from the start. While it is true that many fortified places gave in without a fight, after the horrors which followed the conquest of Béziers and Carcassonne; the heretics fled to the mountains for safety and Raymond-Roger de Foix submitted to the legates; nevertheless after their obligatory forty days had passed the French nobles returned home. Simon was left alone with a handful of soldiers in the heart of hostile country. He needed military reinforcements and financial support from the Christian world, and turned to Rome with an urgent appeal for help.

Meanwhile Raymond VI had deserted the crusaders and before long openly broke with the legates. The town of Toulouse and the Count himself refused to surrender the known and suspected heretics named in a list presented by Arnold Amalric. Refusal to comply with this unrealistic demand served the legates as a pretext to excommunicate the Count again and deprive him of his rights. The seven castles he had given as a guarantee were to pass to the Roman Church, along with the County of Melgueil. Raymond VI set off for Rome, and the town of Toulouse sent a delegation to the Pope too; the results of their complaint are presented completely differently by the Song of the Crusade and the History by Peter of Vaux-de-Cernay. The letters of Pope Innocent III himself, however, show clearly that he did not approve of his legates' arbitrary actions, and ordered them to call a council to go carefully once more into the question of the accusation of heresy brought against the Count of Toulouse and that of his complicity in the murder of Peter of Castelnau.[42]

The legates showed little interest in punctilious observation of legal forms; their aim was to break the power of Raymond VI and not to reconcile him with the Church. It was therefore necessary to reduce the city of Toulouse, the principal bastion of resistance in the South. Bishop Foulques of Toulouse played a dark role in the execution of their plans; he gathered round himself the elements in the City who favoured the crusaders, forming them into an armed fraternity. Toulouse showed the utmost willingness to come to terms with the legates, but refused to desert Raymond VI and therefore - in direct contravention of the Pope's instructions - was again put under the anathema of the Church.

At the beginning of 1210 the war took a turn highly unfavourable to Simon when the nobles around Béziers and Carcassonne, secretly instigated by the King of Aragon, took up arms again; the Count of Foix also showed open hostility. The arrival of fresh reinforcements relieved Simon's position, however, and allowed the war to go forward. Minerva, with its apparently impregnable natural defences, surrendered after a month's siege. The description given by Peter of Vaux-de-Cernay eloquently conveys the tense and horrified expectation of the citadel when the crusaders took it over in the summer of 1210. The banner of the Cross and the banner of Simon de Montfort flew over the conquered fortress: *vexillum Christi, vexillum comitis.* Guido, Abbot of Vaux-de-Cernay entered a house where the "perfect" were gathered; his persuasion met with unequivocal rejection, and the Cathar women, when he passed to their gathering, were even more obdurate. After another vain attempt at conversion Simon de Montfort had many heretics burned, both men and women - there were almost a hundred and forty of them, and they threw themselves into the flames.[43] The crusade against the South of France brought with it the inhuman practice of burning heretics alive, and in numbers never seen before.[44] Constant new reinforcements kept up the tempo of war, and meanwhile the legates deliberately hindered any possibility of Raymond clearing himself. In the end they even won the Pope over to their side, and drove Raymond into such a situation that he could only declare open resistance. If he had really met the demands made upon him by the Council of Arles in 1211 he would have enslaved himself and all his subjects. Publication of these articles rang out everywhere as a call to fight to the death. Rejection of the ultimatum of Arles was followed by the excommunication of Raymond, this time confirmed by the Pope himself.

The crusade continued with the siege of the important fortress of Lavaur, and its fall was accompanied by renewed scenes of horror. The barbarous murder of Lady Giraude, beloved of all, left a deep mark on the minds of her contemporaries;[45] she was thrown into the castle well. Beyond the town walls the crusaders, in the words of Peter of Vaux-de-Cernay "burned innumerable heretics with immense joy". *Cum ingenti gaudio,* horrifyingly sincere words which recur again and again in Peter's account of similar happenings.[46]

For a time the situation favoured Raymond VI. The Count of Foix defeated a large contingent of crusaders and the Vicomte of Béarn and the Count of Comminges sent soldiers to help Raymond. Bishop Foulques was forced to leave Toulouse and German crusaders besieged the town in vain. With their departure the numbers of the crusaders fell considerably and Simon de Montfort lost many of the fortresses he had previously conquered. Not until fresh reinforcements arrived did his situation begin to improve again; these new forces were the response to the ceaseless agitation of preachers in northern Prance and in the German lands. They helped Simon to regain his lost positions and take more important fortresses,[47] and at the end of 1212 he proclaimed a statute for the conquered lands, in Pamiers. The rough conqueror who now ruled Raymond's lost territory began systematically building up a hereditary domain for his house in the South. The Roman Curia had its share of the spoils in Languedoc, too; the Pope added the County of Melgueil, formerly the property of Raymond, to the patrimony of St. Peter, later renting it out to the bishop of Maguelone. Innocent III also carefully noted the receipt of three denars annually from each house in the conquered territory, collected by Simon de Montfort on behalf of the Holy See. This was of no little significance from the legal point of view; it was the visible symbol of the supreme temporal power which Innocent III took to himself in Languedoc as elsewhere in the Christian world, in the spirit of his theocratic conception.[48]

Simon de Montfort, whose power increased so rapidly, had a dangerous rival in Arnold Amalric. Like Simon de Montfort, the Abbot of Cîteaux was not concerned only with the *negotium Christi*. The powerful legate was not content with the archbishopric of Narbonne, but regardless of Simon's claim took to himself the title of Duke of Narbonne as well, a title which was traditionally held by the ruling house of Toulouse.[49]

This newly formed constellation of power in Languedoc was a serious threat to the interests of the house of Barcelona on this side of the Pyrenees, and King Peter II never recognized the seizure of the lands belonging to his vassal, Raymond-Roger de Trencavel. Not only did he refuse to accept Simon de Montfort's oath of allegiance on behalf of these lands; he secretly instigated the local nobility against their new lord. Once more, however, divergent interests prevented united action in defence of the independence of the South at the decisive moment. At the moment when Raymond of Toulouse, with the help of his allies, finally roused himself to effective resistance, the King of Aragon marched off to fight the Saracens in the Iberian peninsula (1211).[50] A year later, when he was free to turn his attention to Languedoc again, the catastrophe of the South was practically complete. Raymond VI was left with Toulouse and Montauban alone, and much of the lands of Peter's vassals the Count of Foix, the Vicomte of Béarn and the Count of Comminges was also in the hands of Simon de Montfort. Only effective diplomatic steps could change matters; on the urgent request of Toulouse the king sent a mission to Rome, where the ground was already favourably prepared by his recent participation in the decisive victory over the Arab Almohadas. The mission from Aragon was very successful in Rome: At the moment when Simon de Montfort was so near his goal, Innocent III took a clear stand against the expropriation of Raymond's lands so long as he had not been given the chance to clear himself, in accordance with the Pope's earlier instructions, and his guilt legally proved. Innocent also instructed his legate Arnold Almaric to see that peace was soon restored in Languedoc, since the *negotium fidei* had been completed with a sufficient degree of success and the whole Christian world now needed to gather its strength for the fight against the infidels in the Holy Land and beyond the Pyrenees. Yet not even the supreme authority in the Church could arrest the forces which had now been set in motion. Before the Pope's letters sent at the beginning of 1213 had reached their destination, the council called together by Arnold Amalric had vehemently rejected Raymond's defence and thus decided to fight on until he was completely defeated.[51] Innocent III gave way again, in the train of events he could no longer control. There was nothing left to the king of Aragon but open hostilities now that Rome was no longer on his side. In September 1213 the victor of Las Navas perished in the unlucky battle before the walls of Muret. Raymond's defeat was now complete, and the victorious Simon de Montfort prepared to rule the lands of Toulouse on the east bank of the Rhône too. At the same time, however, papal diplomacy set to work in Languedoc, and Innocent III took the initiative again. The defeated Raymond VI surrendered unconditionally to the Pope - and not to Simon de Montfort! - along with his son, and with all the lands belonging to both of them (April 1214). The new papal legate, Cardinal Peter de Benevento, gave him absolution. The town of Toulouse sent hostages and was reconciled to Rome. In the same way the legate reconciled Raymond's allies with the Church and accepted as their guarantee fortresses whose garrisons now rendered allegiance directly to the Pope. The castle of Foix itself was now in the hands of the Church. All this was against the interest of Simon de Montfort; with the help of another papal legate from northern France he began a campaign to defend his own interests, but he did not succeed in getting from Peter of Benevento all of Raymond's fiefs. It began to look as though things in Languedoc would be settled in accordance with the wishes of the Curia, when an unexpected factor appeared; the Capets in the North took action. Philippe-Auguste who had hitherto kept apart from the campaign in Languedoc, now sent his son Louis to the South. The French army, together with Simon de Montfort's men, passed unopposed through Languedoc, and all the towns opened their gates to the royal prince, accompanied by Simon de Montfort and Cardinal Peter of Benevento. Once again barefaced seizure by force prevailed over all legal scruples. In this changed situation Peter the legate could not prevent the walls of Toulouse and Narbonne from being knocked down, nor hinder Simon de Montfort's seizure of the city of Toulouse and the castle of Foix. In the end the Pope himself empowered Simon to assume temporary control over all the lands conquered. The final decision on the fate of the fiefs of the house of Toulouse was to be taken by a general council, and in the struggle for the duchy of Narbonne which now broke out quite openly between Simon and Arnold Amalric, Innocent III took the side of his legate against Simon.[52]

The first act of the tragedy of Languedoc closed in Rome, against the background of the Fourth Lateran Council (1215). The second author of the Song of the Crusade gave dramatic dialogue form to this last attempt to save the family lands of Raymond VI.[53] The Count of Toulouse had nobles and prelates to speak for him, but Simon de Montfort had much stronger forces behind him, against whom not even the Pope could maintain his moderate policy.[54] The Count was deprived of his fief, which fell to Simon. The papal decree legalizing this handover stressed the rights of the French king, to whom Simon de Montfort was to pledge his allegiance. Beaucaire and Nîmes were left to Raymond's son, as well as that part of Provence which under the title of a Marquisate had been held by the Toulouse family for almost a century. Innocent III could not foresee that this domain in Provence would become the base from which the Count of Toulouse and his son would again enter the battle successfully.

The younger Raymond left Rome a little later than his father, and the two met in Genoa. Their arrival in Provence marked the turn in the fortunes of war. Marseilles, Avignon and other towns, as well as the nobles of Provence, enthusiastically rose to fight behind the Count of Toulouse and his son. In Languedoc Beaucaire opened its gates to young Raymond, and in the end Simon de Montfort had to raise the vain siege of the town and pit all his forces against Toulouse itself. There was fighting in the streets and the people fell back before superior numbers. Large numbers of hostages were taken and the town paid large sums in ransom. The walls were torn down and many of the houses damaged Yet immediately afterwards Raymond VI was given a tremendous welcome when he returned to his town. People of all classes, both men and women, worked to rebuild the walls; lights flared all night long and musical instruments played, while girls sang *chansons et ballades*.[55] Simon laid siege to Toulouse again, but early in the summer of 1217 he was killed by a stone catapulted from the walls.

His son Amalric was given every support by Pope Honorius III as well as receiving help from the French king, but in three years his situation was desperate. He wanted to hand over to Philippe-Auguste the right to all the lands conquered by his father. Raymond VII, whose father already lay dead, buried without the rites of the Church in a forgotten spot on the outskirts of Toulouse, began to negotiate with Amalric. No agreement was reached, however, and finally Amalric handed over to Louis VIII the rights to all the land seized by his family during the Albigensian war. The new French king had to regain these lands by the sword, however, and in 1226 the decision was taken in Paris to start off another campaign against the South.

The Pope confirmed the French king's right to Raymond's lands, and money for the new campaign was provided by a new tithe of all Church revenues in France. A great army of crusaders swept southward, its passage marked by the burning of heretics at the stake. An epidemic broke out among the troops, however, sudden death bearing off even the king, Louis VIII. For a time the South rallied effectively, and then the exhaustion of war on both sides led to negotiations for peace, in Meaux (1229). The strength of the South had been drained away, and Raymond VII was forced to accept hard terms including not only the obligation to protect the rights and privileges of the Church and to take stern measures against heresy, but also the loss of any possibility to take up arms to resist again. Two thirds of his lands were taken over by the king, as were the Trencavel lands, and in addition the betrothal of Raymond's daughter and the Prince Alphonse, as yet a child, paved the way for the rest of his lands to merge with those of the French crown in the near future. After Raymond had signed the peace his allies followed suit.

The treaties of Meaux and of Paris formed an important step forward in the policy of uniting the whole of France under the Capets. This foreshadowed the final lot of the Cathar sect, which was far from having been destroyed by the war itself. Many Cathar lives had been lost when towns fell, trapping a number of heretics together; elsewhere, however, they managed to hide in time and when the storm had passed over the Cathar communities quickly returned to normal.[56] There were even places which escaped the horrors of the war altogether, and there the members of the sect lived in calm safety as in the years before the crusade began. In the Abbey of Sorèze there was a perfect symbiosis and the monks and the local priest lived on friendly terms with the heretics.[57]

The Cathar sect in Languedoc was able to renew its strength during the years when the cause of Almaric de Montfort seemed lost for good. The papal legate Conrad de Porto, who was seriously disturbed by this renewal of heresy, called an ecclesiastical council to meet in Sens; transferred to Paris, the gathering was unexpectedly broken up by the death of Philippe-Auguste. Conrad de Porto lost all hope of rooting out heresy in the Occitan regions and asked the Pope to recall him. A letter from the legate has been preserved, dating from his stay in Languedoc, in which he informs the archbishop of Rouen, Theobald, of the activity of the heretics in the South.[58] A Cathar dignitary by the name of Bartholomaeus, from Carcassonne, came to the diocese of Agen. Bishop Vigorosus of Bathona, whose residence was in Pojors, gave up his see to Bartholomaeus and moved to the Toulouse region. According to Conrad's information, this latter was the deputy of the heresiarch called by the Albigensians their "Pope", to whom they turned for advice as to their supreme authority. This Cathar anti-Pope was to be found somewhere in the neighbourhood of Hungary, between Bulgaria, Croatia and Dalmatia. This geographical localization can be said to point to Bosnia, in all probability. As we shall see below, later reports did in fact give this country as the seat of the supreme head of the Cathar sect.

Conrad de Porto's letter was written in 1223. At this time the Cathars had leisure to settle several doubtful questions such as the disputed boundary between the Cathar bishopric of Toulouse and that of Carcassonne. To this end a copy of the records of the assembly in St. Félix de Caraman in 1167 was made; at this gathering of the heretics the geographical boundaries of the Cathar bishoprics had been laid down according to those of the Catholic dioceses. It was clear from this earlier record that the region of Razès, which was the cause of the dispute, belonged to the church of Carcassonne. Nevertheless a gathering of the Cathars in 1225 decided in favour of a compromise, and Razès was made a separate bishopric.[59]

Evidence has also come down to us of disputes between different groups of Cathars in Languedoc at this time, on matters of principle; these groups depended on different mother churches in the East. This subject was touched on by Durand de Huesca in a new polemical work written many years after his conversion to Catholicism, when this former Waldensian preacher was at the head of the society of the *pauperes Catholici*. This treatise can be dated 1222 - 1223.[60]

Such spiritual leaders of the crusade as Arnold Amalric had no time for written polemics, or even to work against heresy by their preaching. They were too concerned with leading military campaigns and making sure of the gains of war. Even during the turmoils of war, however, the one-time faithful companion of Diego, the Castilian Dominic, remained tireless in his enthusiasm for preaching. Gradually he gathered a band of friars round him, with a clearly defined mission. Equipped with thorough theological training they were to counteract the influence of the heretics' preaching by their own assiduous work in this field, while their strictly moral lives were to vie with those of the Cathars and Waldensians. Thus there came into being a new monastic order which was not completely approved in Rome until the successor of Innocent III was elected, but which by the time its founder died in 1221 was known in most of the countries of Europe.

The Church of Rome had already gained valuable helpers in the work of systematic teaching of the faith, in the members of the fraternity of the *Pauperes Catholici*; there was also another group of converts from Waldensianism, led by Bernard Prim. In the fraternity led by Durand, and later in that of Bernard, too, the chief part was played by the clerics, while they were helped by laymen who were particularly gifted for preaching. They followed the tradition of the Waldensian "schools", proceeding from Bible readings to dogmatic interpretation and at the highest stage going on to learned disputation, often pitted against heretics. The anti-Cathar polemic in Durand's Liber contra Manichaeos develops the basic themes covered by the author in his Waldensian period; his arguments were directed solely against the Cathars, even after his conversion, and not against the heretics whose faith he had formerly shared. Holding themselves apart from all the clamour of war, the *Pauperes Catholici* and Bernard's fraternity carried on their peaceful work among the Waldensians, still resembling them both in their apparel and in their strict poverty.[61]

Far greater historical significance than that of the modest activities of these converts from Waldensianism or of the Italian fraternity of the *humiliati*, accrued to the work of Francis of Assisi. In 1208 his small fraternity took the pledge of absolute poverty and humble devoted service to their fellow-men. In the midst of prosperous urban life there appeared a handful of people who found the whole purpose of their lives in conscious return to the simple commands of the Gospel. Like the conversion of Peter Valdes earlier, this was a sign of the mood of the time, an expression of the longing which never quite died out in western society. The secret of the extraordinary success of the idea in this case, however, lay in the unique charm of the Italian saint's personality. Against his will and against his original intention, Francis of Assisi saw the enormously swollen ranks of his followers forming a firmly organized monastic order. It was at this time that he met Dominic, who adopted for his own monks the vow of absolute poverty preached by Francis. Both orders did in fact live on the alms they received. The work of both Dominicans and Franciscans lay in the towns, where their spacious new churches were filled by crowds who gathered to hear them preach. Both these "begging orders" were especially fitted to work directly among the people and perform the task that neither the uneducated parish priests nor the existing monastic orders, too shut off from the everyday world, could fulfil.But now they had to face the strong menace of heresy.

While in the south of France war had begun to destroy the heretics and their protectors, in northern Italy the Pope's threats had no permanent effect. Both Cathars and Waldensians were very active throughout Lombardy. Some time during 1214 - 1215, Yvo of Narbonne passed through all the towns of the Po valley; everywhere he was given letters of recommendation to Cathars in the next town along his route. This cleric, who had fled his country under suspicion of heresy, had no scruples about accepting the hospitality of the Lombard heretics, drank the ***nobilissima Paterinorum vina*** with them, and pretended everywhere to be devoted to the faith. Not until his travels took him to the foothills of the Alps, to Friuli, did he rouse the suspicions of a Cathar bishop, Peter Gallus, in Gemona. By this time, however, Yvo had found out many of the heretics' secrets, and during his wanderings had seen at first hand their effective activity. They sent students to Paris from all over Lombardy and Tuscany, to acquire there the knowledge of logic and theology necessary to be able to justify their faith and rebut the attacks of Catholic polemists. The Cathars often appeared in markets, trading to get money for their church, and at the same time skilfully infecting with their beliefs the people among whom they lived and boarded.[62]

The heretics continued to play their part in internal conflicts in the Italian towns. In Brescia things went so far in 1224 that there was an armed conflict. The nobles who protected the heretics attacked their opponents from their fortified towers, and churches were burned down and looted. Pope Honorius III ordered their towers to be destroyed completely or partially, according to the extent of each family's guilt.[63] The families concerned declared that their *civitas* had for long been divided into two hostile camps, and that if the heretics were on their side, then they had been given protection as coming under that set of interests, and not because of their heretical opinions. To a large extent this was certainly the case. There is nevertheless no doubt that the heretical movement was still very strong in Brescia and that it had followers and protectors among the local nobles and among the city authorities. This tradition had survived in Brescia since the days of the unfortunate Arnold, whose work had done so much to prepare the ground for Catharism.[64]

Not only the Cathars, but the Waldensians, too, drew on the heritage of earlier heretical trends in Italy. The Lombardy Waldensians, no doubt influenced by the ***humiliati***, deserted the ideal of Valdes, that of wandering preachers living in absolute poverty and apart from their families. This in itself distinguished them markedly from their fellow-believers beyond the Alps and the differences went even deeper. In Lombardy the donatist trend gradually prevailed, which made the validity of the sacraments dependent on the moral qualities of the priest administering them. Like the Arnaldists - and perhaps in imitation of them - the Lombardy Waldensians set up their own "order of the priests of Christ". Their consistent Biblicism, rejecting everything that runs counter to the evidence of the Scriptures, led this branch of the Waldensians to break off their links with the Church completely. The Pauperes Lombardi thus came to share the view of the Cathars that the Church of Rome is not the Church of God.

The split between the Pauperes of Lyons and those of Lombardy was complete in 1205. The latter, excommunicated by Valdes himself, found a leader in John de Ronco, who came from Piacenza, and took their name from him: the Runcarians. The efforts to reunite, made at a conference in Bergamo (1218) only showed up the profound differences between the two trends, each following its own path. The Pauperes Lombardi were active in their missionary work north of the Alps, in Germany, where the Waldensians had already begun to penetrate (from France and Italy) before the split developed.[65]

Waldensianism and Catharism were firmly rooted in Milan, which remained the refuge of heretical sects in spite of all threats.[66] The article requiring the expulsion of heretics, incorporated into the statute of the Milan community in 1211,[67] seems to have been without effect. Developments in Milan continued to be favourable to the heretics. From 1214 the aristocratic families began to "secede" and in 1221 the archbishop also left the city. The community of Milan was now made up of plebeian elements, and the government of the city was in their hands.[68]

Both Pope Honorius III and the young emperor, Frederick II, took an unfavourable view of this situation. Not even the life and death conflict which later broke out between the Hohenstaufen ruler and the successor to Honorius could affect the Emperor's basically negative attitude towards the "popular" movement and towards heresy, which was widespread in Milan and elsewhere. Frederick's Italian policy was directed to strengthening the remnants of episcopal power in the communities throughout the country. Milan, to make things worse, was still the city of rebellion and heresy in the eyes of the grandson of Barbarossa. Frederick's laws made no basic distinction between these two concepts, regarding both as crimes against the Emperor. In the laws issued to combat heresy by Frederick the Roman idea of *laesa maiestas* was applied far more thoroughly than in the earlier edict of Alphonse II of Aragon (1194). Heresy, which attacks the very essence of divine majesty, is now defined as the most serious form of offence against the majesty of the ruler. Frederick could only apply this concept to the full in his Kingdom of Sicily, where the Church was subordinate to the sovereign state power and therefore heresy was a crime against the state. In the versions of the Emperor's laws against heresy which were formulated for the Empire (1232, 1238, 1239) definitions are not nearly as sharply worded as in his Sicilian code (Constitutiones regni Siciliae, 1231).

The traditions of Roman law, which the conscientious endeavours of the Bologna legists had revived, thus gradually formed the legal conception of heresy accepted by the highest authorities of the western Christian world, first by Innocent III and thirty years later by Frederick II. In his measures against heresy, however, the Hohenstaufen emperor combined Roman legal theory with the local tradition, long accepted in Germany, of inhuman punitive practice. This was not yet evident in Frederick's Frankfurt constitution of 1220, but four years later in Catania, in Sicily, he issued a constitution affecting the towns of Lombardy which set the penalty of death at the stake or the cutting off of the tongue for unrepentant heretics (1224).[69]

Emperor Frederick was thus the second secular ruler to give the burning of heretics legal grounds, the first being Peter II of Aragon (1198).[70] The Church immediately seized on these new secular laws as an effective weapon against the ever present threat of heresy.[71] In 1227 Pope Honorius III called upon the towns of Lombardy to incorporate into their statutes the existing and future anti-heretical laws of Frederick, King of Sicily and Holy Roman Emperor. This call, which clearly includes Frederick's "Sicilian" constitution of 1224, was repeated soon afterwards by Honorius' successor. Gregory IX (1227-1241) despite his great age displayed tenacious energy in his fight to strengthen the power of the Papacy in the Christian world, and from the outset struck hard at heresy. According to his constitution of February 1231, heretics condemned by an ecclesiastical court were to be handed over to the secular arm for "appropriate" punishment. Those led by fear to show signs of repentance were to be allowed to live - but in prison for life. In content and in the terms used this papal constitution agrees with that issued by Frederick II in Ravenna in 1232. The neutral expression *animadversione debita* used in both texts, as well as in the 1229 decision of the Council of Toulouse, undoubtedly refers to the death penalty. This is equally true of the notorious constitution of the Roman senator Annibaldo (1231).[72] This decree against heretics and their patrons in the Eternal City repeats practically word for word the sanctions against heretics pronounced as early as 1207 by Innocent III in Viterbo, valid for the whole of the Roman patrimony. It was from here that the penalty of confiscation of property and destruction of houses which sheltered heretics was taken. These two constitutions issued at the same time (1231) by the Pope and by Rome respectively were then sent to the bishops as binding "statutes of the Holy See"; judicial authority over the heretics rested in their hands. From the outset Gregory IX made use not only of the secular clergy, but also of the two new "begging" orders, whose monks were constantly among the people and in the best position to combat heresy effectively. In the time of Honorius III the Dominican order had in fact been established explicitly to fight heresy as well as to help reform church life. The powers now granted the order by Gregory IX authorized the *Fratres Praedicatores* to preach anywhere, in open spaces as well as in churches. Preaching directed against heretical beliefs, however, went hand in hand with *inquisitio haereticae pravitatis*. In the Roman constitution of Annibaldo the term *inquisitores* still had its literal meaning of "searching out" heretics.[73] Effective aid in this task was given by lay societies too, the Militiae Jesu Christi which were then being formed in Italian towns. The Dominican monks appeared everywhere their services were urgently required, usually sent by their superiors or by the Pope himself. They were to give their services to the local bishops who alone, at first, had the right to judge heretics. Things very soon developed in such a way that this judicial function, too, passed to the monks of the Inquisition. This happened first in Germany (1231). The episcopal courts did not cease to exist, but they were pushed into the background. It was now that the form of trial *per inquisitionem* was used in cases of heresy; introduced into ecclesiastical jurisprudence by Innocent III, it had not as yet been used in dealing with heretics.[74] Certain procedures used only in cases of heresy developed within the inquisitional form of trial: the names of prosecution witnesses were kept secret, while the accused was not as a rule allowed to call witnesses for the defence, or to take a lawyer to defend him. The only way to avoid severe penalties was to make a voluntary confession during the set time *(tempus gratiae)* before the trial began. Once he appeared before the Inquisition judges the accused was at a great disadvantage, robbed of all the basic safeguards provided by common law. The Inquisitor, of course, was not only a judge; his fundamental mission was the defence of the faith and the salvation of souls which had gone astray. This spiritual mission was taken to justify the use of various means to the given end. The efforts of the judges were aimed at obtaining a confession from the accused. The system of leading questions was worked out to the point of virtuosity, but often the victim was brought to submission only after prolonged and variously aggravated imprisonment in the justly feared prisons of the Inquisition, or later by the use of instruments of torture. Physical torture, the use of which was taken over in the West from Roman law, was in sharp contradiction to the canonical tradition. It does not seem to have been an easy matter everywhere to call upon the secular organs for help in this respect, as the bull *Ad extirpanda* (1252) laid down for Italy. On the other hand, any Inquisitor who himself was present when an accused heretic was being tortured would be committing an irregularity; as we shall see, however, a way out of this difficulty was found in time.[75]

Formally the Church never took the responsibility for the death penalty, which was inflicted by the secular arm. The customary plea that the victim's life should be spared was of course pure formality, and to comply with it would have meant the Church anathema on the person concerned.

The mass burning of heretics in the thirteenth century was primarily the work of the crusaders, before the Inquisition tribunals had been established. Even then, as we shall see, at first the fanaticism of some Inquisitors who cared nothing for legal forms claimed many victims. Beside this the Inquisition itself arranged two largescale executions which helped decisively to break the Cathar heresy in the South of France and in the North of Italy (1244, 1278). In the normal practice of the Inquisition, however, there were relatively few cases of death for obdurate heretics or those who relapsed.[76] Life imprisonment was the usual punishment for confirmed heretics; it was not seen as a penalty but as a form of penance. Nevertheless the long torments of prison, especially in its worst forms, could hardly be regarded as a lesser evil than the shorter torment of death at the stake. The property of those handed over to the secular arm, or imprisoned for life, was confiscated, and even that of dead heretics whose bones were ordered to be exhumed and burned. In this way whole families were beggared and reduced to despair. Of the more moderate forms of "penance" set by the Inquisition tribunals that of being forced to wear a visible cross sewn on to the clothing was particularly invidious.

In Inquisitional practice it was not rare to allow various alleviations or commutations of the penalty inflicted, and acquittals were also known. It cannot be denied that some of the Inquisitors tried to maintain a just procedure. The very character of the Inquisition itself, however, as an extraordinary form of jurisdiction with proceedings in secret, facilitated arbitrary action and the abuse of juridical powers. Making it a duty to denounce everything connected with heresy the Inquisition encouraged informers. Everywhere the tribunals created an atmosphere of insecurity and terror.[77]

It was not easy to force the Italian urban authorities to accept the extremely strict measures demanded by the Pope and the Emperor; the towns wanted above all to remain independent of both. They did their best to refuse the new laws against heretics and even if these measures found their way into the statutes this by no means meant that they would be put into practice. In the Apennine peninsula as everywhere else, there was strong opposition to the institution of the new Inquisition tribunals, whose activities ruthlessly cut across the whole of urban life. The efforts of Gregory IX and his assistants from the monastic orders were only partially successful in Italy, and even then it was often an ephemeral success.

Of the towns of Lombardy, Brescia had already incorporated into its city statutes (1230) articles binding the podesta to pronounce a permanent *bannum* on heretics, as well as to seek them out, put them in prison and ultimately – after they had been condemned by the bishop – to inflict physical punishment. Frederick's "Sicilian" constitution of 1224 was included in the statutes of Brescia word for word. This was due to the efforts of the Dominican prior Guala Ronio, who was elected bishop of Brescia in 1230.[78] This Dominican monk was close to Gregory IX and was very influential in the affairs of Lombardy. In Bergamo, however, he came up against obstinate opposition from the *podesta*, de Mandella, a citizen of Milan who released a number of imprisoned heretics, on his own authority. An interdict and other sanctions were pronounced against Bergamo. Meanwhile in Mandella's native city the flames leaped up, for the first time since the terrible auto-da-fé of the heretics of Monteforte, two centuries earlier. The normal penalties for heresy pronounced by the authorities of Milan as early as 1228 (exile, confiscation of property and destruction of houses) against heretics and their patrons, failed to satisfy the Roman Curia for long; the latter demanded the strictest application of the new papal statutes. Thus in the end heretics were bitterly persecuted in Milan, as we are told in a brief record in the chronicle for 1233. There is no doubt that here, too, the Dominicans showed their initiative; at any rate the Pope thanked the archbishop of Milan for the help he had given them. The secular authorities had given in to the Church. The *podesta* of Milan, Oldrado Tresseno "had Cathars burned, as was his duty" and an inscription on his statue recalled this worthy act for future generations. Yet even the persecution of 1233 failed to have a permanent effect. It was forced on the city by the constant accusations of heresy brought forward by the Emperor against Milan as against other towns of Lombardy which were hostile to Frederick II. The peace which the Emperor and the Pope made in 1230 robbed the Milanese of the support of the Roman Curia and made them all the more fearful of the catastrophe which might befall them.[79]

That same year (1233) a cruel blow was struck at heresy in Verona, where Giovanni da Schio was master of the situation for a while. A monk, he was elected *rector* of the town by the people of Verona, and succeeded in reconciling the Guelph and Ghibelline parties, putting an end to hostilities. He had sixty heretics burned in the amphitheatre, men and women of the upper classes in the city. Soone afterwards "frater

Ioannes" was taken prisoner by his enemies, however, and his authority collapsed. The measures against heresy remained ineffective and for a long time Verona and the surrounding countryside offered the Cathars their safest refuge.[80]

The general dislike of the monks' Inquisition sometimes broke out into violence. In Piacenza, for instance, the Dominican monk Roland of Cremona was attacked as he preached in front of the church, barely escaping with his life, and one of the monks with him was killed.[81] In 1239 the Dominican monastery was attacked by armed men in Orvieto, and cruel penalties were inflicted on those who took part, mostly members of the urban nobility in Orvieto. Nevertheless it took the Inquisition three decades to root the "evil" out in this town.[82]

In Florence all efforts to get rid of heresy failed for a long time. There the Cathar sect found its safest refuge in the castles and towers guarding the lands of aristocratic families in the surroundings of the city. The measures against heresy were incorporated into the city statutes, but their execution was skilfully avoided. The heretics always succeeded in finding shelter; it was difficult to punish such large numbers of accomplices, and nobody bothered about the Pope's anathemas and writs of excommunication. In Florence Ruggiero de Calcagni the Inquisitor failed notably.

The arrival of Peter of Verona, the Milanese Inquisitor, wrought a change in the situation. A Dominican monk credited with miraculous powers, he was not only a man of moving eloquence, but of energetic action. He joined forces with one party among the Florentine nobles, against the powerful families who were protecting the heretics; he also found support among the common people. The city authorities gave permission to purchase more land so that the area in front of the church of Santa Maria Novella could hold the vast crowds who came to hear Peter preach. Sharp conflict broke out, however, when Ruggiero de Calcagni brought the main patrons of the heretics before the tribunal of the Inquisition. The *podesta* whose authority was thus infringed upon took energetic steps to protest to the Inquisitor. The tension reached a climax when the monks called the people to listen to their sermons against the head of the city, in two of the churches. They were interrupted by supporters of the powerful Barone family, and armed horsemen dispersed the congregation. Peter of Verona, in front of the church of Santa Maria Novella, inflamed the hatred of the people against the intruders, and placed himself at the head of the armed fraternity he had given rise to in the city, The enemy retreated, but were caught and perished before they could reach the other side of the Arno and take up the defensive.[83] Opposition to the Inquisition was drowned in blood, and the Patarene heresy in Florence never recovered from the blow. Elsewhere, however, it was far from extinction. Peter of Verona, the victor of Florence, was shortly afterwards murdered on the way from Como to Milan (1252); Rainerio Sacchoni, a former member of the Cathar sect converted by Peter to the Catholic faith, was also intended to be a victim. Another Inquisitor from Milan, the learned Dominican Moneta of Cremona, energetically foiled a murderous attempt on his own life.[84]

The happenings in Florence (1244-1245) were already under the pontificate of Innocent IV (1243-1254). At first the conflict between the Papacy and the Hohenstaufen ruling house, which was rising to a climax, prevented Innocent IV from concentrating his attack on the heretics in Italy, until after the death of Frederick II (1250). On his return to Italy he organized Inquisition missions and repeatedly put pressure on the podestas and municipal councils to give the Inquisition all the help they could. His famous bull *Ad extirpanda* (May 1252) laid upon the podestas and rectors the strict obligation to organize the search for heretics and to carry out their punishment. It was at this time that the organization of the Inquisition in northern and central Italy was improved by dividing the separate regions between the Dominicans and the Franciscans.[85]

To the north of the Alps attention was focussed on the central Rhine region in the early thirties; here the persecution of the heretics reached unheard of proportions.[86] Immediately on taking office Pope Gregory IX entrusted Conrad of Marburg with the duties of the Inquisition here (1227); his contemporaries looked upon him as a saintly man of Old Testament severity. Years earlier "Magister Conrad" had put his eloquence at the service of recruitment for the crusade which was then being organized. In Thuringia, whither he came as confessor to the young Countess Elizabeth, he showed himself capable of carrying out the principles of church reform energetically, in the proper award of church benefices. He was given full powers to carry out this task and at the same time the Pope empowered him to persecute the heretics, to reform the morals of the German clergy, and to carry out visitation of monasteries (1227). With the death of Count Louis (1227) Conrad's office at court came to an end, but not so his stern influence over the young Elizabeth. The future saint, who had from the outset been enthralled by the Franciscan ideal of poverty and became a member of the "third order" of St.Francis, spent the last years of her life in the most devoted service of the poorest of the poor. There is nothing to suggest that Conrad himself was a member of one of the "begging" orders; he seems to have been a secular cleric, but while the absolute poverty he professed brought him close to the Franciscans, his fanatical zeal in the persecution of heresy in no way lagged behind the ardour of the Dominicans.

The Pope's mandate of 1227 charged Conrad of Marburg with the functions of Inquisitor in the sense in which Annibaldo's decree of 1231 understood them. Independently of Conrad two laymen began a heresy-hunt in the Rhineland on their own account, in 1231. Conrad Dorso and his companion John, the latter marked by physical infirmities, were prepared to burn a hundred innocents to make sure of one heretic. The chronicler of Worms called them "imperfect judges who know no mercy". In fact they did not pass judgment themselves, but only handed over to the secular courts people whom they declared to be heretics. They not only swayed the crowds, but won over the bishops and the secular authorities, tempted by the vision of wealth accruing from confiscation of the property of those victims who were rich. They themselves were indifferent to material gain, like Conrad of Marburg, who included them in his mission in 1231, thus legalizing their work. Conrad also called on the help of some Franciscan and Dominican monks. In the autumn of 1231 Gregory IX sent him the new statutes and explicitly gave him judicial powers. The Pope s instructions to Conrad reveal clearly the fundamental principles of Inquisitional procedure. In time the same powers were given to Dominicans in Friesach, Strassburg, Würzburg, Regensburg and Bremen, and thus the judgment of heretics in Germany became a matter for people specially trained and chosen for the task; the normal episcopal courts began to lose their importance. The "Ravenna" constitution issued by Frederick II (1232) making death the legal penalty for heresy throughout the Empire gave official sanction to what had long been the actual practice in Germany. The actions of Conrad and his companions, differing considerably from the valid canonical regulations, showed the marked influence of German tradition. They shaved the heads of those who admitted heretical beliefs out of fear; whoever denied his guilt was burned at the stake, and the only hope was thus to admit everything even if the accusations were absurd. The most fantastic inventions were current at the time in German circles, and there is no objective description of the Cathar faith such as we find in later Inquisition handbooks. Instead there are tales of "Luciferians",[87] turning the dualistic faith upside down as if the sect expected the restoration of Lucifer, once upon a time unjustly cast down from Heaven. We have seen a similar travesty of the Bogomil faith in the pamphlet by Euthymius of Acmonia in the eleventh century. As in Byzantium, so in thirteenth century Germany secret devil-worship was attributed to the sect, but here the facts disappeared completely beneath the fantastic descriptions of obscure rites. On the basis of what Conrad reported Gregory IX reproduced in his epistle of 1233 all the ridiculous inventions of the day - which indeed were no novelty.[88]

Although in the times of Conrad of Marburg physical torture was not yet used to force confessions of guilt, the accused were nevertheless subjected to considerable mental torture. If the victim confessed in order to save his life, he was forced to name his accomplices, and thus false denunciations brought ever new victims before the Inquisition judges. Nobody dared to raise his voice in defence of the suspects and nobody was safe before the ever-widening waves of denunciation. Pitiless jurisdiction struck at all classes without distinction. At the beginning of 1233 Gregory IX appealed for a crusade against the protectors of the heretics in the Rhineland. Sure of the support of Rome, Conrad did not hesitate to attack the highest in the land, but his fate was sealed when he called the Count of Sayn to appear before him. The accused did not comply, but instead defended himself successfully against Conrad's accusations, before an ecclesiastical council in Mainz (1233). The latent conflict between Conrad of Marburg and the Rhine episcopate now broke into the open, and the Pope was informed of the way Conrad had been carrying out his mission. Before anything could be decided in Rome, however, Magister Conrad perished at the hand of assassins, and shortly afterwards his assistants Conrad Dorso and John also met a violent death.

The assembly called by King Henry VII in 1234 and the church synod which followed it restored a measure of security, and in the future the treatment of heresy assumed regular forms in Germany. Nevertheless the reign of terror which had lasted three years did not fade from the minds of the people of the Rhineland; the chroniclers wrote of a storm of destruction, and of the countless victims of the flames, which "began at that time to gain power over mortals".

In spite of the unusual nature of Conrad's activities they fitted in with the atmosphere of the time, when the violent suppression of heresy had become a common occurrence. The age-old tradition of the Church, regarding even the sternest justice in the light of Christian charity, had markedly given way before the harshness of the times. The highest dignitaries of the Church of Rome no longer held back before the shedding of blood, as one of Gregory IX's epistles explicitly reveals.[89] It concerned the situation in northern France, where similar things were happening in the thirties of the thirteenth century to those which had horrified the Rhineland.

Here, too, in the *regnum Franciae*, the main initiative came from one outstanding personality. Robert the Dominican[90] was a learned and experienced preacher, like Conrad of Marburg, but unlike him had a heretical past behind him, unkindly preserved in his nickname "Bulgarus". In 1233 Pope Gregory IX sent this fanatical convert to the very centre of heresy on the Loire. In La Charité, for long the home of heretics, all the efforts of the local clergy had been in vain because the nobility gave the heretics powerful protection. The infection had already spread to the neighbouring regions of Champagne and into Flanders. It was essential that the Dominican Inquisition take a hand, and Robert the "Bulgar" received extraordinary powers from the Pope. He was of course supposed to act in concert with the local hierarchy, but his excessive zeal soon gave rise to serious complaints that he was violating the bishops' rights. In spite of this, Gregory IX again empowered Robert and other Dominicans to carry on the Inquisition; their mandate covered the whole kingdom of France, but Robert concentrated on Champagne and Flanders. His fanatical cruelty was responsible for great numbers of deaths - Matthew of Paris mentioned fifty people burned or buried alive by Robert in the course of three months, in his Chronicle. His contemporaries were not so shocked by this inhumanity, of course, for north of the Loire it was by no means exceptional. The people hated the wealthy merchants and bankers of Flanders, and certainly viewed with sympathy the stern measures taken by the Inquisitor against usurers. The conflict between Robert and the French bishops became so sharp, however, that it finally led to his undoing. The accusations against him could be substantiated by innumerable cases in which he had flouted canon law, including the illegal punishment of people to whom absolution had been granted, or who were in the process of appealing to the Pope. Robert was now in disgrace, but nothing is reliably known of his fate.

From about the middle of the thirteenth century the Inquisition in northern France (as in the South) was for a time in the hands of the bishops. The accepted practice in dealing with heretics was later codified in various collections of customary law, and finally the long accepted practice of burning heretics alive was also given its legal basis there.[91]

The most memorable act of Robert the monk was the mass execution of 183 heretics, men and women, in Mont Wimer (Mont-Aimé) in 1239. The nobles and prelates of Champagne and Flanders were called together to watch the terrible scene;[92] except for one "perfect" of the Cathar faith, the victims were "believers" who received the *consolamentum* at the last moment.[93] Yet the *Ecclesia Franciae* was not rooted out even then, and large numbers of Cathars found refuge in Italy.

Meanwhile the clouds of despair lay low over the South. The songs of the troubadours could still be heard, but the lyric tones of love and courtship had given place to sorrow and bitter hatred of the French and the clergy. The war had robbed the troubadours of their background, destroyed the little courts where they made their living, and scattered their public. Their secular, aristocratic outlook meant that most of them were far from the spirit of the Cathars, yet the gloom that now descended on them brought them closer to the persecuted sect.[94]

Once the war was over the whole strength of secular and ecclesiastical power was turned against the Cathars.[95] On the orders of Louis IX his officials in the territories now handed over to the French crown were to encourage the denunciation of heretics by offering rewards of money. Raymond VII, too, in an attempt to keep the terms of the peace treaty, offered rewards to informers in his territory, and sent armed units to search for heretics. An ecclesiastical synod held in Toulouse in 1229 agreed on a number of measures intended to keep check on the orthodoxy of the people and especially that of the younger generation. It was forbidden to the laity to have a copy of the Bible in the vernacular in their possession. In every parish a committee was to be set up, composed of one priest and two or more laymen, to carry out systematic searches for heretics.[96] The synod of Toulouse reserved the right to judge heretics to the bishops alone, but even in this matter there was to be a fundamental change very soon.

In the spring of 1233 the Dominicans were given orders by Rome to carry on the Inquisition in southern France according to directives which were identical with those given to Conrad of Marburg at the end of 1231.[97] In the painful situation in the South the activity of the Dominican Inquisition roused particularly bitter hatred. From the outset the Inquisitors behaved with the utmost ruthlessness, aptly described in a letter sent by the consuls of Narbonne to Nîmes:[98] with no regard for the statutes of canon or secular law, innocent people were deprived of their property or left to die in prison. The judges questioned simple people so cunningly that however they answered they were bound to fall into a trap. This letter was not simply an expression of the bias of the leading figures in the *Bourg* of Narbonne, whose people had come into violent and dramatic conflict with the Dominican Inquisition and the archbishop, who were supported by the *City*, a conflict sparked off by their sympathy towards heresy.[99] The Dominican Guillaume Pelhisson who himself was active in the work of the Inquisition gives an even darker picture of the way it worked in Toulouse.[100] He records quite openly a number of episodes which aptly illustrate the activities of the Dominicans there and of Bishop Raymond de Falga. The heretics gave the *consolamentum* to an old woman who lay seriously ill. Hearing of this the bishop of Toulouse lost not a moment; cunningly quesrioning the old woman, he proved her heresy, and handed her over to the *vicarius* of the Count of Toulouse, who had her cast into the flames, bed and all. The bishop and the Dominicans then returned to the meal prepared for them after solemn Mass, and "gladly ate, giving thanks to God and St.Dominic".[101]

This happened in 1234. Three years earlier - before the Inquisition had been introduced into the South of France - the Dominican Roland of Cremona, later active in Piacenza, had two heretics exhumed, one after the other, in the *Bourg* of Toulouse, and their bodies burned. One was a canon of the monastery of St.Sernin, who had received the *consolamentum* in secret, and the other a Waldensian dignitary. Their bodies were dragged through the streets "for the honour of the Church of Rome", in Pelhisson's words. The spontaneous candour of this Dominican chronicler is an excellent illustration of the mentality of these men for whom the idea of a ruthless battle against heresy blotted out everything else. At the same time Pelhisson provides evidence of the fierce opposition aroused by their ruthless cruelty in the embittered atmosphere of the South.[102] In Albi in 1234 two Dominican Inquisitors were almost killed by an angry crowd when they tried to exhume and burn the remains of a woman who had died a heretic. The *capituli* of Toulouse forbade the people to comply when called before the tribunal of the Inquisition, and ultimately drove the Inquisitors from the city. All the Dominicans left Toulouse, together with the bishop (1235); but not for long. Raymond VII was in an exceedingly difficult position, and was forced to agree to the return of the Dominicans and the Inquisition (1236).[103]

While the towns were heavy with the horror of Inquisition trials, the "good people" found refuge in the countryside, in the castles of nobles, or in cottages deep in the forest. They undertook long journeys to administer the *consolamentum* to the dying and bring spiritual sustenance to their faithful followers. They carried the New Testament hidden under their cloak; when news of their arrival passed round, the "believers" would come from far and wide to hear them preach and to receive the "bread of holy prayer", which they piously preserved for years. There was much greater danger in all they did now than in the years before the war.. The "perfect" could not carry on without the devoted help of their "believers". They needed experienced guides to help them round dangerous places and bring them safely to reliable friends. It was often necessary to pay an armed guard or to reward those who had provided shelter; to have a supply of money to hand for such expenses, regular collections were made among believers and supporters of the sect. There were improvised collections, too, when a ransom had to be found for a fellow Cathar in prison. A significant source of funds were the bequests made by "believers" who received the *consolamentum* on their deathbeds. The maintenance of communications between the "good people" and the "believers" was a task of primary importance, and there were accepted terms for the people who performed this kind of service; the records of the Inquisition show us "believers" serving as *ductores, receptatores, questores, nuncii.*[104] Only in some out-of-the-way country places was the sect left alone for a few years after the war. But even in Sorèze these idyllic conditions came to an end with the arrival of a new parish priest at the beginning of the forties. The ruthless persecution of the heretics began there, too.

Yet in spite of all the persecution the Cathar church in Languedoc maintained its coherence and its continuity. By this time, though, all their activity centred round a single point. In 1232 Raymond Perelle had been asked by the Cathar bishops to provide a refuge for the "good people" at his seat on Montségur in the county of Foix. It was here that the leaders of the sect in Languedoc were now usually to be found; they formed a kind of collegiate administration.[105] Important meetings were held in Montségur and it was there that the rank of new members of the Cathar hierarchy was confirmed. There was a "great house" in the castle, for religious services. From near and far the followers of the faith looked to this citadel as the only firm pillar on which their security rested.

The activities of the Inquisition in Toulouse, which had been interrupted in 1238,[106] was taken up again with renewed intensity in 1241. The extreme severity of the penalties inflicted on the heretics and their protectors aroused violent hatred on all sides. Even the Count of Toulouse found it impossible to accept the methods of the Dominican monks, obedient only to their Provincial and evading all control by the local bishops. Raymond VII was therefore constantly complaining to the Pope, and urging the bishops of Languedoc to take the initiative themselves and carry out the Inquisition in their own dioceses. He had nothing against the Dominicans working for the Inquisition, provided they acted as ordinary monks, under the orders of the local hierarchy. He was even willing to give them help and protection.[107] Yet almost immediately a murderous attack cost the lives of two Inquisitors and several of their assistants, in one of his castles (1242). Later confessions recorded by the Inquisition allow us to reconstruct the horrible May night in Avignonet. At a given moment the gates opened to admit soldiers from the Montségur garrison, reinforced by local people. The seneschal of the Count of Toulouse, Raymond d'Alfaro, was in league with them and led them to the room where the Inquisitors were sleeping.

There is no evidence that Raymond VII himself knew anything about this plot; indeed, the murder made his situation even more difficult just at the moment when he had decided to take action.[108] Two years after young Trencavel, son of the unfortunate Vicomte Raymond-Roger, had made an unsuccessful attempt to win back his family's lands, the Count of Toulouse made up his mind to make another attempt to fight free of his humiliating position. At the beginning of 1242 he was negotiating with Hugo, Count of the March (in the North-West of the Massif central), who was plotting against the French king and had secured the help of his son-in-law, King Henry of England. A great anti-French league was in the making. Raymond VII secured the help of the Count of Comminges and Roger Bernard II of Foix, and among other supporters was the young Trencavel, now in exile in Catalonia. The English king did in fact come to the help of Hugo and Raymond, but suffered a defeat. Raymond VII with Trencavel and others of his allies penetrated far into the lands annexed by the French king in 1229. Raymond entered Narbonne and again took up the title of Duke, hereditary in his family. The Count of Foix deserted him, however, and in the end he had to surrender to superior forces (Peace of Lorris, 1243).[109]

The records of the Inquisition reveal the reactions of the people of the South to the news of what had happened in Avignonet. Many openly rejoiced at the deed, which they hoped would mean the end of the terror of the Inquisition.[110] Their hopes were dashed, however, when the military action of Raymond VII failed. The Dominicans themselves, shocked by the Avignonet murders, sent a request to Rome to be relieved of their mission, but the new pope, Innocent IV, refused to grant it. The regular machinery of Inquisition tribunals was again set in motion, ever more exactly regulated by papal bulls and decisions taken by the legates sent by the Pope to settle problems on the spot. The pontificate of Innocent IV was by far the most important stage in the formation of legal and juridical norms for Inquisition practice. A number of provincial councils in the South of France also contributed much in this direction. The first of these, held in Narbonne at the end of 1243 or beginning of 1244 gave detailed answers to Inquisitors' questions, in the form of twenty-nine articles.[111] Armed with new instructions, the Dominican Inquisitors had the remains of deceased heretics publicly burned in many places, and called before their tribunals anyone they accused or suspected of heresy however highly placed in society.[112] But military action was necessary, to strike at the heart of the sect.

The siege of Montségur began some time in May 1243. The garrison was made up of very brave and seasoned fighters, and for many months they defended the little plateau where the castle stood, at the top of the steep mountain. After the last major attack the defenders' position became hopeless; the besiegers were exhausted, too, and capitulation was agreed at the beginning of March 1244. The conditions were relatively favourable; the warriors of Montségur were to be allowed to pass out, and even to keep their arms, only having to make a confession later to an Inquisition tribunal. This meant that the murder in Avignonet, which half the garrison had been involved in, would no longer be held against them. The Inquisition was richly rewarded, however, by the number of victims surrendered after the fall of Montségur. There was no mercy shown to heretics who would not recant, and none of the "perfect" of Montségur would have bought his life at that price. In spite of the sufferings of the long siege the moral strength of leading members of the sect was not only as firm as ever, but exercised a magnetic effect on those around them. The Inquisition records recall the ferment of the last attack, when many a mortally wounded soldier recieved the *consolamentum* from the hands of the "good people". Not all of the members of the garrison who survived took advantage of the conditions offered them; reports speak of some twelve who received the Cathar spiritual baptism at the last moment and thus voluntarily chose death. About 210 or 215 men and women were taken from Montségur to the town of Bram, and burned at the stake probably after a summary trial. Montségur no longer existed for the "Church of God"... The three or four "perfect" who escaped secretly, as determined by the rest of the community, hurried to their brethren to pass on the news of the last moments of the Cathar citadel.[113]

The last fortified place where the sect could still take refuge fell in 1255; this was the little castle of Queribus, whose ruins still point to the sky on a peak of one of the Corbières mountain ridges, on the borders of Languedoc and the County of Roussillon.[114]

A few years before the tragedy of Montségur stern action was taken against the Cathar sect on the other side of the Pyrenees,in the northern part of the peninsula. Catharism seems to have crossed the Pyreness from the South of France rather late. The stern edicts of the Kings of Aragon at the end of the twelfth century were directed against the Waldensians, not the Cathars, of whom there is no mention at that time. Even at the beginning of the thirteenth century Raymond VI knew of no "good people" on the other side of the Pyrenees in whose hands he would be able to die a good death.[115] In the course of a few years, however, Cathar preachers from Languedoc appeared in the Kingdom of Léon. Persecution could not prevent the new teaching from spreading among the people of the region. In 1232 crowds were making the pilgrimage to the spot where the remains of the murdered Cathar missionary, Arnold, had been laid to rest sixteen years earlier. Miraculous healing powers were attributed to a spring which rose from the ground there, and soon a chapel was built over it. It did not help for long to have the chapel torn down and the heretics expelled. Heretical writings appeared, cleverly left lying by the roadside, among rocks and in woods, to be found by shepherds. The seed of heresy - skilfully disguised as letters from Heaven - was carried throughout the country as if by the winds. Lucas Tudensis tells us these things in his somewhat confused treatise on the Cathars.[116] This Spanish bishop was himself an ardent persecutor of the heretics.

The safest refuge for the Cathars in Catalonia was the territory of the Viscount of Castelbo, Arnold (d.1226). It is clear from confessions made to the Inquisition in Languedoc, on the other side of the Pyrenees, about the middle of the forties, that Arnold was a patron of heretics. The Cathars held their gatherings openly in the castle of Castelbo, in the presence of the Viscount himself, and of some other nobles. The *consolamentum* had also been administered there. The daughter of the Viscount, Ermessinda, was a "believer" of the Cathar sect and did not neglect to pay the "perfect" the customary "adoration". Things did not change in Castelbo under Roger Bernard II,Count of Foix, who gained control over the viscounty when he married Ermessinda.

Rome continued to watch with distress the way heresy was growing in the peninsula, constantly reinforced from Languedoc. An edict of 1233 issued by James I of Aragon organized the persecution of heretics along the lines laid down by the synod of Toulouse in 1229. On the Pope's orders the Dominicans began their Inquisition activity in the king's territory too.

In 1237 Roger Bernard II was unable to prevent the Inquisition from moving into the territory of the Viscounty of Castelbo. The tribunal condemned 45 live heretics who were thrown into the archbishop's prison in Tarragona, and 18 dead heretics, whose remains were exhumed and burned.

In Catalonia and Aragon the Inquisition, carried on by the Dominicans and the Franciscans, continued to persecute both the Cathars and the Waldensians. As late as 1269 the Inquisitors, along with the bishop of Barcelona,ordered that the bones of Viscount Arnold and his daughter Ermessinda should be dug up and burned.[117]

During the time of Gregory IX the energetic action of Rome was being felt in another region on the periphery of Christendom, at the far end of Europe. From the beginning of the thirteenth century Rome became increasingly interested in Bosnia, a country about which little was known. At that time the Bosnian countryside experienced a Dominican mission and the Inquisition and finally a crusade as well. Developments in this country followed curious lines, however, and the final outcome was very different from what the papal Curia had planned.

[1] *Volpe*, Movimenti, 97 ff.

[2] *Fumi*, Orvieto.

[3] *Thouzellier*, Languedoc, 161 - 168, with references to the epistles of Innocent III concerning the towns mentioned here.

[4] *Visconti*, 211 f.

[5] *Borst*, 131, n. 10.

[6] Innoc. III, Ep. XV, 189 (PL 216, col. 710 - 714), *Potthast 4611*.

[7] Braisne-sur-Vesle: Chronicon anonymi Laudunensis canonici, an. 1204, in: *Bouquet*, Recueil XVIII, 713; cf. *Schmidt* I, 365.

[8] *Chénon*, La Charité, 320 f.; *Thouzellier*, Languedoc, 149 f.

[9] See Chapter IX, 118.

[10] *Bigounet*, Grand chapitre, 320 f.; *Thouzellier*, Languedoc, 134 - 138.

[11] *Guiraud*, Inquisition I, 303 - 310; *Luchaire*, Albigeois, 28 ff.; *Griffe*, Languedoc, 28 f.

[12] *Guiraud*, op. cit. I, 310 - 314; *Luchaire*, op. cit., 28, 31. On Esclarmonde and Philippa see *Koch*, Frauenfrage, 50 - 54.

[13] Chanson de la Croisade I, 17: "E tut sei cavalier et l'autre valvassor / Tenian los eretges, qui en castel qui en tor."

[14] *Thouzellier*, Languedoc, 245 - 249, gives the references to Inquisition records. See also *Griffe*, Aventure, 147 ff.; id., Languedoc, 24.

[15] For Lauragais see the map by *J.Guiraud*, corrected by *Y.Dossat*, showing where the Cathar deacons were, in: Cathares en Languedoc, 26. - For Cathar houses in Languedoc see *Guiraud*, Inquisition I, 146 - 151; *Koch*, Frauenfrage, 55 - 64; *Dossat*, Les cathares d'après les documents de l'inquisition, in: Cathares en Languedoc, 72.

[16] Guill. Podiol., t. 19, 199 f. On the situation of the bishops of Toulouse see *J.H.Mundy*, Liberty and Political Power in Toulouse 1050 - 1230, New York 1954, 81 f. See also *Griffe*, Aventure, 142 ff., 178 ff.

[17] *Thouzellier*, Languedoc, 139 ff., 187, 189, 191 f.

[18] Letter from the archbishop to the papal legates, *Vaissète - Molinier* VIII, No. 124, LXXIX, 509.

[19] See *L. de Lacger*, L'Albigeois pendant la crise de l'albigéisme. L'Episcopat de Guilhem Peire 1185-1227, in: RHE 29 (1933) 272 - 315; 586 - 633; 849 - 904. *Bécamel*, in: Cathares en Languedoc, 239 - 245. See also *C.P.Bru*, Éléments pour une interprétation sociologique du catharisme occitan, in: Spiritualité de l'hérésie: le catharisme, Paris 1953, 40: "...là où le pouvoir appartenait aux évêques, l'hérésie voyait son développement limité ou rendu impossible." According to *Y.Dossat*,

Le Clergé méridional à la veille de la Croisade Albigeoise, in: Revue du Languedoc (1944) 268, [in the Albi diocese] "véritables nids d'hérétiques sont les fiefs laïques tels que Lombers. Cathares et Vaudois pullulent dans le Sud du Quercy, ...dans la partie du diocèse qui échappe à l'influence politique des évêques" (quoted by *Bru,* art. cit., 40, n. 28). Cf. also *Griffe,* Lanquedoc, 61. On the term *Albigenses* applied to the Cathars see *Christine Thouzellier,* Hérésie et hérétiques Vaudois, Cathares, Patarins, Albigeois, Rome 1969, 223 - 262 (not available to the author).

[20] On Diego of Osma and the controversies between Catholics and heretics see *Thouzellier,* Languedoc, 193 - 204. Durand of Huesca and his *Pauperes catholici: Selge,* Die ersten Waldenser I, 193 ff. Waldensianism as a less important element in the anti-ecclesiastical front in Languedoc: *Guiraud,* Inquisition I, 258 f. It was the Cathars and not the Waldensians who were heretics par excellence in the eyes of the Church, as the frequent distinction made between *heretici* and *Valdenses, heresis* and *Valdesia,* shows. Cf. *C.Douais,* Les hérétiques du comté de Toulouse, in: Compte rendu du Congrès scient. intern. des catholiques (1891) 7. For the attitude of contemporary Catholics to the Waldensians, see also *Griffe,* Languedoc, 59 (referring to Peter Sarn. I, 18, No. 18).

[21] The term *fautores hereticorum* included "une adhésion même plus ou moins lointaine aux dogmes du catharisme: ...relations de famille ou de voisinage, de rapports ... entre des concitoyens..." (*Molinier,* Église et société III, 274.).

[22] *Nelli,* Le catharisme vu à travers les troubadours, in: Cathares en Languedoc, 187.

[23] The monk Henry taught: "Episcopi et sacerdotes non debent habere honores et pecunias." Monk Guillaume. ed. *Manselli,* l.c. (see Chapter IX, n.35), 60.

[24] See Guill. Podiol., t. 19, 200 (the well-known comment by a nobleman of Languedoc).

[25] For the social stratification of the Cathars in Languedoc see Molinier, Église et société III, 269 f.; *Koch,* Frauenfrage, 13 f.

[26] See *Guiraud,* Inquisition I, 355; *Dossat,* in: Cathares en Languedoc, 98. *Griffe,* Languedoc, 57, Cf. n. 102 below.

[27] See *Mundy,* op. cit., 77 ff., 289 (n. 17).

[28] See *Griffe,* in: Cathares en Languedoc, 225; id., Aventure, 10 ff., 154 ff., 182 ff.

The record of the disputation in Lombers (1165) says explicitly that the heretics known as *Boni homines* were under the protection of the nobles *(milites)* of that *castrum.* É.Griffe rightly concludes that the purpose of the disputation was not so much to convert the heretics as to deter the nobles from joining them. (*Griffe,* Languedoc, 24; cf. id., Aventure, 59 - 67). On the *castrum* of St.Félix see ibid., 67.

[29] *Thouzellier,* Languedoc, 187 f.

[30] Ibid., 204 f. The Pope's epistle (17.XI.1207): Innoc. III, Ep. X, 149 (PL 215, col. 1246 - 1247), *Potthast* 3223.

[31] *Schmidt* I, 219 - 293.

[32] P.Meyer evaluates these narrative sources in his Foreword to the edition of the Chanson de la Croisade. Cf. *Y.Dossat,* Le chroniqueur Guillaume de Puylaurens était-il chapelain de Raymond VII ou notaire de l'Inquisition toulousaine, in: Annales du Midi (juillet 1953) 343 - 353; quoted here from *Maisonneuve,* 200, n.1.

[33] Letter of 10.III.1208. Innoc. III, Ep. XI, 26 (PL 215, col. 1354 - 1358), *Potthast* 3324.

[34] See *Maisonneuve*, 200 f.

[35] For an analysis of the policy of Innocent during the period of the Crusade in Languedoc see *Luchaire*, Albigeois; *Maisonneuve*, 218 - 233 with references to the papal correspondence.

[36] For the legal status of those taking part in the crusade see *Maisonneuve*, 207; *J.A.Brundage*, Medieval Canon Law and the Crusader, London 1969.

[37] Chanson de la Croisade I, 10, l. 195 - 201; 11, l. 221 - 225.

[38] Innoc. III, Ep. XI, 232, (Pl 215, col. 1546 - 1547), *Potthast* 3642.

[39] *Luchaire*, Albigeois, 130.

[40] Chanson de la Croisade I, 31; Ibid.,17.

[41] The action of Innocent III, confirming the right of Simon and his heirs to rule the Vicomtê of Bêziers and that of Carcassonne, instead of the legal heirs to these domains, was clearly contrary to canon law. See *Maisonneuve*, 218.

[42] *Luchaire*, Albigeois, 152 f.; *Maisonneuve*, 206.

[43] Peter Sarn. I, 159 ff.

[44] *Havet* II, 499.

[45] Chanson de la Croisade I, 71.

[46] Peter Sarn. I, 228, 233; II, 208.

[47] *Thouzellier*, Languedoc, 241 f.

[48] *Luchaire*, Albigeois, 176 f.; *Maisonneuve*, 225; *Thouzellier*, Languedoc, 243.

[49] *Luchaire*, op. cit., 188.

[50] Ibid., 179 f.; *Maisonneuve*, 219 f.

[51] *Luchaire*, op. cit., 195 - 211.

[52] Ibid., 230 - 246.

[53] Chanson de la Croisade I, 141 - 161.

[54] The second author of the Chanson de la Croisade employed a good deal of emotional exaggeration in his description of the Pope's efforts to protect the rights of Raymond VI and his son. Basically, however, he gave the correct interpretation of Innocênt's aim, as we also see it in the work of Guillaume le Breton, historian of Philippe-Auguste. See *Luchaire*, Albigeois, 248 f.

[55] Chanson de la Croisade I, 176.

[56] *Dossat*, in: Cathares en Languedoc, 72 f.

[57] Ibid., 97 f.

[58] The argument as to which of the two preserved versions of Conrad's letter is the earlier has now been definitely settled in favour of the longer one, included in a collection of letters of the Praemonstratensian abbot Gervaise (ed. *Mansi* XXII, col. 1203 - 1206). See *Thouzellier*, Un traitê cathare, 33 f. The shorter version of the letter, preserved in the work by Roger of Wendover, Flores Historiarum (see *Mandić*, Bogomilska crkva, 438 - 439) and in the chronicle of Matthew of Paris (from where Mansi printed it, XXII, 1201 - 1202), is nevertheless important for the reconstruction of the original text. It gives a more faithful rendering of the opening lines of the letter than the first version does. See *J.Šidak*, O pitanju; for a confrontation of the two versions, pp. 6 - 7.

[59] *Borst*, 12, n. 27.

[60] Liber contra Manichaeos, 138,30 - 139,1. See Chapter XVII below, p.240.

[61] On Dominic see *Thouzellier*, Languedoc, 250 - 255. On the *Pauperes catholici* of Durand de Huesca, ibid., 215 - 232; 255 - 262. Bernard Prim: ibid., 232 - 237;

62 Text: Matthew of Paris, an. 1242, MGH SS XXVIII, 231.

63 See the letters of Honorius III: Hon. III, Ep. IX, 146, 149 (MGH, Epist. Saec. XIII, t. I, Nos 264, 265, 266), *Potthast* 7346.

64 *Volpe*, Movimenti, 101 f.

65 *Selge*, Die ersten Waldenser I, 173 - 188.

66 For references to the sources see Schmidt I, 145, n. 6.

67 *C.Manaresi*, Gli atti del comune di Milano fino all' anno MCCXVI. Milan 1919, 471.

68 *Visconti*, 211 f.

69 See *E.Kantorowicz*, Kaiser Friedrich der Zweite. Berlin 1928, 139 - 143, 239 - 247. The antiheretical *Constitutiones* of Frederick II: MGH, Const. 2, 108 f., 126 f., 196 f., 280 ff. Cf. Borst, 129.

70 See *Thouzellier*, Languedoc, 134 f.

71 *Maisonneuve*, 243 f.

72 On the interpretation of the text of the constitution of Gregory IX, ibid., 245 f.

73 *Förg*, 49.

74 Ibid., 26 ff.

75 See below, Chap. XVIII, p.278.

76 See *Dossat*, Crises, 247 ff.; Id., Le bûcher de Montségur, les bûchers de l'Inquisition, in: Le Credo, la morale et l'Inquisition (= Cahiers de Fanjeaux, fasc.6), Toulouse 1971, 370 ff.

77 See *H.-C.Lea - J.Hansen*, Geschichte der Inquisition im Mittelalter, Bonn 1905, 412 ff.; *E.Vacandard*, L'Inquisition. Étude historique et critique sur le pouvoir coërcitif de l'Eglise, Paris 1907, 218 ff.

78 *Ficker*, Todesstrafe, 199; *Maisonneuve*, 252.

79 On Bergamo, see *Maisonneuve*, 252; the burning of heretics in Milan, in 1233: Memoriae Mediolanenses, an. 1233, MGH SS XVIII, 402; Matth. Paris an. 1240, MGH SS XXVIII, 197,31 - 39. See also ibid., an. 1236, 133. Cf. *Borst*, 131.

80 *Cipolla*, Verona I, 65 f.

81 *Schmidt* I, 164; *Maisonneuve*, 252.

82 *Fumi*, Orvieto, 64 f.

83 *Perrens*, 359 - 382; *Volpe*, Movimenti, 104 - 106; cf. ibid., 91, n. 1, with references to further literature and to sources.

84 *Schmidt* I, 170.

85 References to the relevant texts in *Maisonneuve*, 307 f.

86 *B.Kaltner*, Konrad von Marburg und die Inquisition in Deutschland. Prague 1882; *Ficker*, Todesstrafe, 213 f.; *Förg*, 71 - 90; *Maisonneuve*, 257 ff.

87 *E.Amann*, Lucifériens, DTC IX/1, 1047 f.

88 See Alanus de Insulis, Contra haer. l. I, c. LXIII, PL 210, col. 366. Cf. *Amann*, art. cit., 1047. For references in western sources to devil worship and obscene rites, see *Borst*, 2, n. 6.

Epistle of Gregory IX, 13.VI.1233: MGH IX/1, Epist. saec. XIII, t. I, No 537, 432 - 434, *Potthast* 9230. The unreliability of these accounts of Luciferians is admitted by *P.Alphandéry*, who thinks however that in later years, the Cathars, like some mystics, might have considered the possibility of the redemption of Lucifer. The references he gives are not sufficient to confirm this hypothesis, however. See *Alphandéry*, Idées morales, 174 - 180.

The Trier chronicle provides some real factual information on contemporary heresies in Germany. See Gestorum Treverorum Continuatio IV, MGH SS XXIV, 401. According to the letter sent to Pope Gregory IX by the Archbishop of Mainz and the Dominican monk Bernard, after the death of Conrad of Marburg, the latter opposed both the Poor of Lyons and the heresy of the "Manichees". See Chronica Albrici Monachi Trium Fontium (Aubry des Trois Fontaines), MGH SS XXIII, 931.

89 See *Maisonneuve*, 266: referring to the epistle of Gregory IX, 4.II.1234.

90 See *Chénon*, La Charité, 324 ff.; *Maisonneuve*, 266 ff.

91 The report by Matthew of Paris that Pope Gregory IX sentenced Robert to life imprisonment, must be treated with great caution. See *Y.Dossat*, L'hérésie en Champagne aux XII et XIII s., in: Mémoires de la Soc. d'agricult., commerce, sciences et arts, du dép. de la Marne LXXXIV/1969/72. For the Inquisition in northern France after Robert the "Bulgar" see id., in: Le siècle de Saint Louis, Paris 1970, 265 f. For references to French books of customary law see *Havet* II, 598 f.

92 Aubry des Trois Fontaines, l.c., 944 - 945. Cf. *Borst*, 231, n. 3.

93 See *Dossat*, art. cit., 69 ff.

94 See *Nelli*, in: Cathares en Languedoc, 196 f.

95 *Schmidt* I, 293 ff.

96 *Förg*, 43 f.

97 *Maisonneuve*, 270 f.

98 *Schmidt* I, 298, n. 2.

99 For the happenings in Narbonne see *C.Douais*, L'Albigéisme et les Frères Prêcheurs à Narbonne au XIIIe siècle, Paris 1894.

100 The chronicle of Guillaume Pelhisson, ed. *C.Douais*, Les sources de l'histoire de l'Inquisition dans le midi de la France aux XIIIe et XIVe siècles, Paris 1881 (text: 81 - 118).

101 Ibid., 97 f.

102 On the exhumations see ibid., 88. Cf. *Y.Dossat*, La répression de l'hérésie par les évêques, in: Cahiers de Fanjeaux, fasc. 6, 237. - See *Pelhisson*, 99, for Cassès, where some of the heretics escaped arrest *adjutorio rusticorum...*

103 Albi: For references see *Dossat*, Crises, 124, n. 142. Cf. *Bécamel*, in: Cathares en Languedoc, 248. Toulouse: *Dossat*, Crises, 131 ff.

104 *Dossat*, ibid., 89 ff.

105 On the "Montségur" period of the Cathar sect in Languedoc see *Dossat*, ibid., 73 f. Cf. *Schmidt* I, 300 f.

106 On the date see *Dossat*, Crises, 143 f.

107 *Vaissète - Molinier* VI, 737 f.

108 Ibid., 741. Cf. *Dossat*, Crises, 150. *F.Niel*, Montségur, la dernière forteresse cathare, assumes that Raymond VII was behind the Avignonet affair, in: *J. Duvernoy - R.Nelli - F.Niel - D.Roché*, Les Cathares, Paris 1965, 300 f.

109 *Vaissète - Molinier* VI, 718 f. (Trencavel's rebellion), 736 f. (the formation of an anti-French league), 742 ff. (Raymond VII's campaign).

110 See *Dossat*, Crises, 150.

111 Ibid., 158 ff.

112 *Vaissète - Molinier* VI, 766.

113 Ibid., 768 ff.; *Niel*, art. cit.; *Dossat*, Le "bûcher de Montségur"..., l.c., 361 ff.

114 *Niel*, art. cit., 357. Cf. *D.Paladilhe*, Les grandes heures cathares, Paris 1969, 251 f.

[115] *Thouzellier*, Languedoc, 14, n. 7; 136 f.

[116] Lucae Tudensis episcopi Adversus Albigensium errores libri III, ed. *I.Mariana*, Ingolstadii 1612, lib. III, 169 ff., 182 f., 184. Cf. *Schmidt* I, 368 ff. On the tradition of the "letters from Heaven", see *Borst*, 127, n. 29.

[117] *V.Subirats*, El catarismo en Cataluña, in: Boletín de la Real. Acad. de Buenas Letras de Barcelona XXVIII (1960) 75 - 158; id., Le catharisme en Catalogne, in: Cahiers d'Études cathares XIV (automne 1963) IIe S., No. 19, 3 - 25; see also *Maisonneuve*, 275 ff.

Chapter XV

HERESY "IN PARTIBUS SCLAVONIAE" — THE CRUSADE AGAINST BOSNIA

The very first reports of dualist heresy in Bosnia clearly pointed to a connection between the Bosnian hinterland and the ancient towns on the Dalmatian coast. Heretical teachings, already firmly rooted in Dalmatia, travelled inland along with craftsmanship and artistic traditions. The lively atmosphere of the maritime towns provided the same favourable conditions for the growth of heresy as did the Italian city states on the other side of the Adriatic. The letter of Innocent III we have already mentioned merely refers in passing to the heretics of Split and Trogir whom Archbishop Bernard expelled, and who later found refuge in Bosnia. Thomas, Archdeacon of Split, gives more details of two particular heretics whose corrupting activities in Split were firmly put down by the Archbishop. We should bear in mind that these two brothers, of southern Italian origin, Matthew and Aristodius, who spent much of their time in Bosnia, had been citizens of Zadar since childhood. The connection between heresy and this town on the Dalmatian coast stands out quite clearly in that passage of the chronicle of Split which describes the horrors of the sacking of Zadar by crusaders in 1202. Before proceeding to his account of the shocking event, Thomas the archdeacon reproaches the citizens of Zadar bitterly for having - among other things - forsaken the Catholic faith. It is his contention that practically all the well-born and influential people of the city had shown their sympathy for the heretics.[1] Here, again, we find the links between heresy and the upper classes which we have seen in Italy.

Elsewhere in his history Thomas accuses the ruler of Split himself of supporting the heretics; this was Prince Vyšen (Vyshen), who ruled in the early twenties of the thirteenth century (1221-1223). Suspicion of heresy was also attached to his successor Peter (the ruler of Hum), and there was even a sharp burst of hostility between the laity and the clergy when the latter tried to prevent the new ruler from entering the church. In the end the people of Split used force to get him in, and in consequence the papal legate Acontius put the city under anathema.[2] We cannot draw more definite conclusions from the sparse information at our disposal, as to whether the accusation of sympathy for heretics was based on fact in these particular cases, or whether it was invented for the purpose of political intrigue. However that may have been, we find here a remarkable analogy with contemporary conditions in Italian cities, where heresy - real or imagined - was a significant factor in the ferment of political life.

It appears from the account given by Thomas in his chronicle that when this disturbing event occurred the papal legate was no longer in Split itself.[3] Acontius had come to the city in connection with a mission entrusted to him by Pope Honorius III in the spring of 1221; he was to organize energetic measures against the pirates of Omiš (Omish) who were seizing boats sailing for the Holy Land under the banner of the Cross.[4] By the end of that same year he was given an even more difficult task, that of decisive action against the heretics in Bosnia.[5] The forces of the King of Hungary were to be brought into the campaign, and Honorius did in fact appeal to King Andrew II, as well as to the Archbishops and bishops of Hungary, to provide Acontius with effective support.[6]

We do not know for certain when Acontius first set out for Bosnia. The author of the history of Split speaks of the legate's visit in connection with the disturbances in the city after the ascension of the new ruler, Prince Peter. It would appear that it was from Bosnia that Acontius sent his letter placing the city under anathema. The legate was a sick man, who died soon afterwards,[7] and we only learn of the steps he had taken against the Bosnian heresy from an epistle written (15. V. 1225) after his death, by Pope Honorius.[8]

The tone of the papal correspondence in the first half of the twenties of the thirteenth century is a clear reflection of the change which had taken place in Bosnia since the time when John de Casamaris had been sent there. Then, in the first years of the pontificate of Innocent III, Rome was certainly disquieted at reports of the "Manichaean" heresy which was beginning to spread in Bosnia, where the organization of the Church was far from effective. There was as yet no clash. The Roman Curia did not carry out completely the reforms which would have been necessary in the Bosnian Church. The over-large Bosnian diocese was not divided up into several bishoprics, as John de Casamaris had proposed, nor do we know to what extent the strict rules laid on the Bosnian monks by the papal legate were actually put into practice. There seems to be no evidence of the Curia having bothered with Bosnia at all during the ensuing years. At the beginning of the thirteenth century Rome still did not seem to be aware of the true extent of the danger; in the course of the next twenty years it attained truly menacing proportions.

The heretics who have found refuge in Bosnia preach their false doctrines quite openly, Honorius III wrote to his legate Acontius at the end of 1221.[9] Rome was now determined to begin as unrelenting a fight against heresy in the Balkans as was being carried on against heresy in Western Europe, but considerable difficulties arose to hinder it. The King of Hungary, Andrew, was fully occupied with his conflict with the nobles, and allies had to be sought among the powerful prelates. We learn from the Pope's letter dated 15th May 1225 that the deceased legate Acontius had succeeded in enlisting the support of the Archbishop of Kâlocs, Ugrinus. The Pope now called upon him to organize a crusade against Bosnia. At the same time he confirmed King Andrew's donation, founding the legal claim of the see of Kâlocs to secular rule over the whole of Bosnia, including the northern regions of Usora and Soli.[10] Archbishop Ugrinus was to purge this territory of heresy by force, but actually no campaign was undertaken.[11]

It was not until several years later, when the Church of Rome was ruled by the strong personality of Gregory IX, that an energetic attempt was made to settle the Bosnian question from the very roots. Reports reaching the Roman Curia in 1232 bore witness to the incredibly primitive forms of ecclesiastical life in that out-of-the-way region, about which little had hitherto been heard. It now came to the knowledge of Rome that the bishop of Bosnia did not serve Mass in his church, nor administer the sacraments; he was quite illiterate[12] and did not even know the fundamental church rite of baptism. The most serious accusation was that of intimate contact with heretics. It was said that he had acquired office through simony, having bribed a certain heretic to support him. To cap it all the bishop's own brother was an acknowledged heresiarch, to whom he extended his protection; the bishop himself was said to reside in a village along with the followers of heretical beliefs.

Although these accusations may be somewhat exaggerated, it cannot be denied that what they say of conditions in the Bosnian bishopric are in notable agreement with the earlier document, from Bilino Polje (1203), illustrating monastic life in Bosnia at that time. Both these sources give roughly the same picture: ecclesiastical institutions at a very low level and the resulting spread of heresy, which found fertile ground in this backward environment. The earlier document spoke of heresy penetrating monastic life, and now thirty years later we hear of the disturbing influence exerted by the heretics on the bishop himself. It is clear that the sect which had cautiously worked in secret had gradually gained adherents in the country until it now felt sufficiently safe to come out into the open. The reference to a heretic to whom the bishop owed his office is the first sign that in Bosnia powerful and influential men had taken to heresy.

This account of the state of the Bosnian church has not come down to us in its original form, but indirectly, as reproduced in a papal epistle of mid-1232, addressed to the Archbishop of Kâlocs, the bishop of Zagreb and the provost of the monastery of St.Laurence in the archdiocese of Kâlocs.[13] The Pope called upon the prelates to ascertain in all thoroughness the true state of affairs in Bosnia. The papal legate in Hungary, Cardinal Jacob Pecorarius of Piacenza, then took over the investigations in Bosnia. The result was that the bishop of Bosnia was deposed and in his place the legate was to instal two to four bishops who would share the ecclesiastical administration of the extensive territory of Bosnia.[14] Again, however, this longplanned scheme of the papal Curia was not put into effect; instead the legate, acting on his own responsibility, removed the see of Bosnia from the jurisdiction of the Archbishop of Dubrovnik, and it remained for the time being directly subordinate to Rome.[15] A Dominican mission was brought into the country, and the man chosen to fill the bishop's office was Johannes Wildeshausen, a Westphalian German who had been the Provincial of the Hungarian Dominicans since 1232. His exemplary life seemed to qualify him to put into effect the reforms intended to bind Bosnia firmly to the centralized system of the Church of Rome.

There can be no doubt that these sudden changes, combined with the installation of a foreigner as bishop, were regarded with great distaste by the Bosnians, still accustomed to the ancient patriarchal way of life.[16] At this time, however, the policy of Rome towards Bosnia had the support of the Bosnian ruling house. Ban *Ninoslav*, whose name first appears in the sources in this connection, was also supported by the Pope in 1233 when internal opposition (perhaps already encouraged from within Hungary) was causing him trouble.[17]

The correspondence between Gregory IX and Ban Ninoslav that year (1233) throws a very interesting light on the church policy followed by the ruler of Bosnia at that time. A glimmer is even thrown on the preceding shadowy period of Bosnian history, the years in between the reign of Ban Kulin, of which we do not know when it ended, and that of Ban Ninoslav, of whom we do not know when he came to the throne. The favour or at least tolerance extended to the heretics by Ninoslav's predecessors[18] and at first by Ninoslav himself, helps to explain how the seed of dualist heresy brought from outside managed to proliferate so on Bosnian ground. The Pope's epistle dated at the beginning of 1234 sets before the eyes of the prelates of the neighbouring countries the distressing state of the Christian Church in Bosnia, compared to a pathless wilderness overgrown with thorns and nettles.[19]

The correspondence of 1233 referred to above shows clearly that Ninoslav and other members of his family were decidedly inclined towards the Church of Rome and that the Pope relied on them as his allies in the fight against heresy.[20] This concord between the aims of the Ban of Bosnia and the interests of the Church of Rome did not last long. In the events which now followed fast, the papal Curia had to rely primarily on the military force of Duke Koloman, whose personal aim, however, was to get Bosnia into his power. We may justifiably conclude that this Hungarian aggression, unreservedly supported by Rome, roused general opposition to the Roman Church among the Bosnians, a fact which favoured further strengthening of the "Patarene" sect there. There can be no doubt that in this out-lying region, as in Western Europe, the success of the heretical movement was largely determined by the play of political forces. The paucity of information provided by the sources does not allow us to trace the developments which must have taken place within the ruling class of Bosnia, but we are left in no doubt as to their outcome. By far the great majority of Bosnian nobles, as we have seen, were hostile to Rome. Even Ban Ninoslav could no longer find favour with the Pope. The fateful necessity to defend his own vital interests drove him to take a stand against Hungarian aggression. This brought him into the anti-Rome front as well, and thus laid him open to the accusation of backsliding into heresy again.

The papal correspondence gives us some idea of the course of events, 1234-1239. At the beginning of 1234 Gregory IX made the Prior of the Carthusian monastery of St.Bartholomew in Trisult his plenipotentiary legate for Bosnia; we learn of this appointment from the papal epistle already quoted,[21] addressed to the bishops and "all faithful Christians" not only in Bosnia, but also in the surrounding Slavonic countries, embraced by the general term *Sclavonia*. In this epistle Gregory commends the new legate to their care and calls for obedience to his instructions. All who responded to the legate's call and took up the Cross, joining the ranks fighting against heresy in Bosnia, would be given the same indulgences and advantages as those who joined a crusade to free the Holy Land.

Six epistles issuing from the Papal office in mid-October 1234[22] reflect the mustering of forces for a decisive struggle. Gregory IX called on Duke Koloman to attack Bosnia with his army. The latter, and all who took part in the crusade, received guarantees that their property would not be touched for the whole time of their absence, and the bishop of Zagreb was enjoined to see that these guarentees were observed. The bishop of Bosnia was called upon to do his utmost to stamp out heresy in his diocese and the immediate vicinity. The instructions issued by the Pope once more stressed the advantages offered those fighting for the faith in Bosnia, which were the same as those enjoyed by crusaders setting out to free the Holy Land from the infidel. The bishop of Bosnia was also given wide powers of absolution, including men excommunicated for serious crimes, so long as they responded to the call to arms and persevered in the fight.

We do not know the size of the army gathered against Bosnia. It is known that this time the campaign was not headed by Archbishop Ugrinus of Kâlocs, but by the Croatian Duke Koloman. To him - as previously to Ugrinus - the Hungarian king entrusted the government of Bosnia, and once again the Pope agreed to this donation (epistle of August 9th 1235).[23] The burden of responsibility for the conduct of this crusade lay heavily on the Bosnian bishop; already in the autumn of 1235 the Pope sent him a special letter to persuade him not to insist on retiring from office as bishop.[24] This learned Dominican, who longed for a life of abnegation away from the stresses of the world, was little fitted for the bitter struggle which was his lot in this harsh and unyielding country.

Practically the whole of Bosnia was in fact united in opposition to Rome, as the papal epistles of 8th August 1236 bear eloquent witness.[25] Among the Bosnian nobles, infected to a man with the plague of heresy, only Prince Sibislav of Usora and his mother Ancilla shone "like lilies among thorns". The Pope took them both under his protection and guaranteed their rule and all their property, with the single condition that the rights of the Hungarian king must be preserved. The Hungarian episcopacy headed by the Archbishop of Esztergom was charged with seeing that this protection was effective. Basically this guarantee was similar to that given by the Pope to Ban Ninoslav three years previously, and in both cases devotion to the Church of Rome was the essential condition. From the epistles of favour to Sibislav and Ancilla it appears that the Pope did not wish Duke Koloman's rule to extend to Usora, so that this part of Northern Bosnia would be subject only to the Hungarian king as feudal overlord. We do not know what the attitude of the Hungarian princes was to this matter.

While the papal correspondence provides us with interesting sidelights on the situation in Bosnia at the time of the Hungarian invasion, it does not tell us anything about the actual course of the campaign. In 1237 things were far from clear in Bosnia. We can deduce this from the decision taken by the Dubrovnik assembly, which forbade the people of Dubrovnik to have any trade dealings with Bosnia until the return of a mission which the republic had just sent there.[26] From the spring of 1238, however, the papal correspondence already refers to Koloman's victory, and it seems that in addition to Bosnia he was already the master of part of neighbouring Hum. Rome could now proceed to energetic measures to complete the undertaking in Bosnia, interrupted by the war. The bishopric had to be filled, vacated by the resignation of John Wildeshausen the previous spring. His successor Ponsa was again a Dominican; he was also entrusted, for the time being, with the spiritual administration of the conquered regions of Hum. The Bishop of the Hungarian Cumans was charged by the Pope not only with the consecration of the new Bishop of Bosnia, but also with the establishment of the chapter for the future cathedral in Bosnia, and the nomination of a Provost. He was also to determine, on his own responsibility, the amount of the tithes and other dues to be rendered to the see, the chapter and the rest of the clergy.[27]

It was essential for the time being, however, to secure extraordinary financial subsidies for the Church in Bosnia, and so at the end of 1238 we already find a number of orders to that end issuing from Rome. The most interesting piece of information concerns the money set aside for the building of the new cathedral, from which we learn that Ban Ninoslav himself had at one time lodged a certain sum of money with the Dominicans of Pecs, for this purpose. The papal epistle says that the Ban of Bosnia, who had turned from heresy in the past and become reconciled in the bosom of the Catholic Church, later became a backslider into heresy.[28] In all these critical years (1234-1239) this is the only mention of Ninoslav. There is no hint of whether he still held any part of Bosnian territory in 1238. Pope Gregory's epistle, dated December 23rd.,[29] tells us only that the fighting was not yet over. Bishop Ponsa, whom the Pope had made his plenipotentiary legate, received orders to recruit in Hungary warriors against "the remnants of the plague of heresy" in the diocese of Bosnia. The Pope took such warriors under his protection, along with their families and their property.[30] Any man who contributed - directly or indirectly - to the success of a crusade against the Bosnians would receive absolution of his sins in proportion to his deserts. Duke Koloman was called upon to persevere in his present efforts to "glorify the Name of God" in Bosnia. And a year later the Pope confirmed Koloman's donation of estates to the Bishop of Bosnia and the Bosnian Chapter of St. Peter.[31]

In the course of 1239 the Provincial of the Hungarian Dominicans was twice requested to send some of his monks to Bosnia, eloquent preachers and resolute missionaries. An epistle of the Pope to the Bosnian Dominicans, written right at the end of 1239, again repeats the Pope's orders to hand over the money set aside by the Ban of Bosnia for the building of the Cathedral. There is no mention, however, of the fate of Ninoslav at the time of writing.[32]

Our surprise is therefore all the greater when we read a charter delivered by this ruler of Bosnia in the early spring of the following year, in which he figures once again as the sovereign ruler of his country.[33] "The supreme lord of Bosnia, Ban Matthew Ninoslav" was in Dubrovnik with his "boyars" at that moment. He had come to visit the people of Dubrovnik as to "old friends", to make "eternal fast peace" with them, ensuring them freedom of movement as well as complete security and freeing them from the payment of tithes and other dues. He also promised his help to the community of Dubrovnik should they come into conflict with the King of Serbia. He pledged himself to this treaty in the sight of "Christ and His Mother, the Cross and the Gospels, and of St.Blaise and all the saints", a vow which contrasts strangely with the recent accusations that the Ban of Bosnia was a heretic.

The sources which have survived do not allow us to form any opinion on the causes of the turn of events reflected in that epistle of 1240. It is only quite clear that events now pursued a course which was extremely favourable to Ninoslav. The wave of Mongol invasion ravaged Hungary and King Bela IV himself was forced to take refuge on the Dalmatian islands. Ninoslav could strengthen his power undisturbed. At a time when Hungary had not yet recovered from this terrible catastrophe the Ban of Bosnia was openly on the side of the Hungarian king's enemies. In the war which broke out between the Dalmatian towns of Split and Trogir Ninoslav gave his aid to Split, which was then hostile to Bela IV. This conflict gave Bela a welcome opportunity to revenge himself on the town for its refusal to help him in his moment of greatest need. At the same time he attacked Bosnia with his army.[34] Ban Ninoslav undoubtedly fell back before the superior Hungarian forces led by the king himself. The treaty then concluded between them led to considerable strengthening of the Catholic Church in Bosnia, and with it increased Hungarian influence. The text of a charter by Bela IV has been preserved,[35] setting forth exactly the rights of the Church in Bosnia. The Bosnian bishopric already owned lands in Slavonia (Djakovo and Blizna) which had been donated by the king's dead brother Koloman. Ban Ninoslav was now forced to hand over to the Church various lands in Usora in compensation, as it were, for "various injuries and damage" done to it. Tithes were now to be paid throughout Bosnia, as was already the case in the rest of the Hungarian realm.[36] The people who lived on land belonging to the bishopric and chapter were removed entirely from the jurisdiction of the regional secular rulers ("princes and bans") and placed under that of the ecclesiastical courts. The bishop and chapter would in future collect the dues from these lands, which had formerly been paid to the Ban of Bosnia. There can be no doubt that all this was a noticeable weakening of the secular power, to the advantage of the Church.

This document also describes the landed property of the Church in the different župas (zhupas) of Bosnia as foundations of the respective churches; at that time there was already a cathedral, dedicated to St. Peter, at a place called Brdo, in the *župa of Vrhbosna*. Besides the cathedral, there was at least one church in each "župa" (zhupa), and in one of them we are told of three churches. Everything seems to point to normal development of church life in Bosnia by that time, as the chronicle proudly relates; written by Suibertus, prior of the Dominican monastery of Bodrog, it gives the story of the Hungarian Dominicans.[37] We read of the renovation of the churches in Bosnia, previously ruined and overgrown with thorns and briars. Yet this same source goes on to reveal that the heretics had burned down two of the monasteries the Dominicans built in Bosnia. Suibertus in his brief account nevertheless bears clear witness to the bitter nature of the struggle. In Bosnia the merciless machinery of the Inquisition was set in motion. The secular authorities subject to Duke Koloman were no more than tools in the hands of the Inquisition, handing over numerous victims to the stake. In Bosnia, too, the terrible flames leaped up, and more than one Dominican monk paid the price of fearful vengeance.[38]

In his chronicle the Prior did not fail to stress the unique services rendered by his Order, which undertook an almost hopeless cause, with the help of Duke Koloman, after the crusade led by the Archbishop of Kâlocs yielded no results. Suibertus is of course wrong in saying that the Archbishop of Kâlocs had already at that time performed the functions of papal legate; we know that before Koloman intervened in Bosnia Archbishop Ugrinus had undertaken no military campaign, and indeed the only ally he had lined up was a dismal failure. We have no information as to whether the see took part, later, in the work of organizing the crusade against Bosnia in the years 1234 - 1239. Ugrinus himself perished in the fighting with the Tartars (1241) and it was not until several years later that his successor Benedict appeared at the head of a new crusade, instigated by the Roman Curia.

The treaty between Bela IV and Ban Matthew Ninoslav clearly failed to provide any effective solution to the problem of the Church in Bosnia. Rome had in any case already been disturbed by the growth of the heretical movement in the close vicinity of Bosnia; as early as 1243 Pope Innocent IV had appointed Stephen, the former Bishop of Vâcz, as his legate to Croatia and Dalmatia with the express task of clearing these regions of heresy.[39] Two years later (1245) the Pope was preparing new crusade against Bosnia.[40] To this end he did in fact give the Archbishop of Kâlocs, Benedict, the powers of a legate, and in 1246 sent him the Cross in the name of which he was to lead a holy war.[41]

Two epistles sent by Innocent IV from Lyons to Hungary in the years 1246 and 1247 throw light on the ecclesiastical and secular status of Bosnia. The diocese of Bosnia, which for a time had been directly subject to the jurisdiction of the Pope, was now subordinate to the Archbishop (1247). The arguments put forward by Benedict, supported by King Bela IV of Hungary and the Bishop of Bosnia, to sustain his claim in this matter, are most interesting. There is a clear tendency to give Benedict and his predecessor Ugrinus the chief credit for what was gained by the war against the Bosnian heretics. The successes are emphasized - the large areas of Bosnia conquered, the "many thousands" of heretics carried off - but the sacrifices made by the archdiocese of Kâlocs are also stressed, for the long-drawn-out struggle had exhausted it to no small degree. Most significant of all, we learn that the positions captured in Bosnia were lost once more, for the fortresses built up against the heretics were not capable of resisting their attacks. Since there could be no hope that this obdurate land would voluntarily return to the fold of "church unity", the only solution was to carry on with the fight.[42] In effect, this was a confession of failure: all that had been done so far had failed to subdue heresy in Bosnia. At a time when the conquest of Montségur (1244) had taken the last stronghold from the Languedoc Cathars, Bosnia remained unsubdued.

The letters of Pope Innocent IV written in Lyons in 1246 and 1247 echo with the aspirations of the Kâlocs prelate to secular rule *(dominium temporale)* over Bosnia. This claim was founded on the donation of these rights by Andrew II, confirmed by Pope Honorius himself (1225) Under Innocent IV, however, there was no longer any serious thought of putting it into effect. The papal Curia was again forced to rely on the support of the local rulers, having decided that Ban Ninoslav had not really deserted the true faith, and that only political expediency had forced him to temporary alliance with the heretics.[43] There was authentic evidence of his orthodoxy, yet in spite of this the Pope asked for detailed information.[44] We are not told the results of the investigation, however.

We last hear of Ban Matthew Ninoslav in 1249, when he made a new treaty with Dubrovnik, renewing the conditions of 1240.[45] Soon afterwards, Hungarian pressure on Bosnia increased, and for many decades the country was so far from independence that, divided into two parts, it was even under foreign domination for a time. Having achieved political domination over Bosnia, the kings of Hungary lost all interest in further crusades. From the middle of the thirteenth century there were no more campaigns to wipe out heresy in Bosnia and to revive the Church there. Indeed, as we shall see, developments within Hungary itself were not such as to favour intervention of that kind.

Thus the bishop of Bosnia appointed from Rome never came back to his diocese. He remained at Djakovo (Dyakovo) in Slavonia for good, on the lands donated to the Bosnian see by Duke Koloman. The dualist sect now had a free hand in Bosnia; undoubtedly the remnants of the native clergy and monks merged with the sect, for they had long been close to the heretical movement.[46] When the local sources speak of the *Bosnian Church* they now mean the heretical church with its dualist doctrines, which writers of other nations also call the *ecclesia Sclavoniae.* The term *Sclavonia,* of course, is not used for a single entity; it expresses the ethnical character of a number of Slav lands included in the kingdom of Hungary and on its borders. *Sclavonia* could mean not only the Slavonia of today but maritime Croatia and Dalmatia, and even Istria and Kranj, as well as Bosnia, Hum, Zeta or Rascia.[47] The name was sometimes applied to one of these regions, and sometimes to a group of them.

It is noteworthy that the Dominican chronicler Suibertus connects the heretical "ecclesia Sclavoniae" with *Bosnia* and *Dalmatia.* We know that in the records of the Cathar assembly of St.Félix de Caraman in 1167, the *ecclesia Dalmatiae* is cited as one of "original churches" still in existence. It is highly probable that this heretical "Dalmatian" church is what was meant by the vague *ecclesia Sclavoniae* attested for the second half of the twelfth century by the writer of "De heresi Catharorum in Lombardia". It would be difficult to apply the term to Bosnia at this period. The first reliable evidence of dualist heresy there dates from the turn of the twelfth to thirteenth century. The sources make it quite clear that the heretics had taken refuge in Bosnia to escape persecution in the Dalmatian towns. Heresy did not disappear from Dalmatia, of course, but there is no doubt that Bosnia, at least after the crusades ceased, offered a much safer refuge for the followers of the heretical faith. It is therefore possible that the organization of the sect throughout the Bosnian and Dalmatian region was then centred in Bosnia. This could explain why Suibertus included both Bosnia and Dalmatia under the term *ecclesia Sclavoniae.*[48] Somewhat later the writer of the "Tractatus de hereticis" mentions the heretical bishop of Bosnia, and in this context clearly identifies Sclavonia with Bosnia, where there was an important centre of the heretics' organization and which, in the minds of contemporaries, was the heretical country par excellence.

This is the picture given us by the sparse references in the sources, of the course and outcome of the bitter struggle carried on by the Church of Rome in Bosnia in the first half of the thirteenth century, without success. At the same time, under Gregory IX there was an attempt to organize a crusade against Bulgaria as well, but here the fight against heresy was no more than a pretext for intervention in a region which Rome still considered its own domain and which it had recently attempted to bind closer by means of a church union. In fact, this union to which Bulgaria had agreed for a time out of political expediency was a purely external affair. Neighbouring Serbia, for which Rome had similar plans, remained outside the papal jurisdiction. The existing order in each of these Churches, that of Bulgaria like that of Serbia, did not change, nor did their measures against heresy depart from the customary practice of the Byzantine world.

1 Hist. Sal., 83.

2 Ibid., 102.

3 "Que res, postquam ad aures Acontii pervenit, *misit* et totam civitatem interdicto supposuit" (our italics ML). Ibid., 103.

4 Epistle from Honorius III to the town of Split (13.IV.1221). *Smičiklas*, CD III, No 166, 191 f.; Hist. Sal., 95.

5 Epistle from Honorius III to Acontius (3.XII.1221). *Smičiklas*, CD III, N° 171, 196 f.

6 Epistle from Honorius III. to the Archbishop of Esztergom and his suffragans. *Theiner*, Mon. Hung. I, N° 63, 31 f. *Potthast* 6729.

7 Thomas of Split wrongly gives the year 1222 as the date of Acontius' death. See *Farlati* III, 254 f. Epistle from Honorius III to Acontius, 27.VII.1223, ibid., 254.

8 *Smičiklas*, CD III, N° 216, 242 f.

9 In the epistle referred to in n. 5.

10 See n. 8. Cf. *Mandić*, Bogomilska crkva, 143 f.

11 This was already shown by *A.Hoffer*, Dva odlomka iz povećeg rada o kršćanskoj crkvi u Bosni, in: Spomen knjiga iz Bosne, 1901, 70 f.; and again by *J.Šidak*, Misija dominikanaca, 17 f. The suggestion put forward by *D.Mandić*, that Ugrinus led a crusade against Bosnia and conquered most of the territory of Bosnia, is untenable (it should be noted that he takes 15.V.1225 - the date of the Pope's epistle to Ugrinus - as the terminus ad quem). *Mandić*, Bogomilska crkva, 139 - 144.

12 This could of course mean simply ignorance of Latin (see *Ćirković*, Istorija, 60).

13 *Smičiklas*, CD III, N° 315, 361 f.

14 Epistle of Gregory IX, 30.V.1233: *Smičiklas*, CD III, N° 327, 379 f.

15 *Šidak*, Misija dominikanaca, 19.

16 Ibid., 22.

17 Ibid., 21.

18 "...progenitores tui qui fuerunt vitio heretice pravitatis infecti..." Epistle from Gregory IX to Ban Ninoslav, 10.X.1233: *Smičiklas*, CD III, No 335,

388. Cf. the epistle sent on the same day to Duke Koloman: *Smičiklas,* CD III, No 336, 388 f.

[19] *Smičiklas,* CD III, No 343, 396 - 398.

[20] See *Šidak,* Misija dominikanaca, 21.

[21] See n. 19.

[22] *Šidak,* Misija dominikanaca, 23.

[23] *Smičiklas,* CD III, No 385, 443.

[24] *Smičiklas,* CD III, No 387, 444 f.

[25] *Smičiklas,* CD IV, N°s 12 - 15, pp. 15 - 18.

[26] *Smičiklas,* CD IV, No 24, 27; cf. *Šidak,* Misija dominikanaca, 25.

[27] Epistle of Gregory IX, 26.IV.1238, addresed to the bishop of the Hungarian Cumans. *Smičiklas,* CD IV, No 50, 56 f.

[28] *Smičiklas,* CD IV, No 60, 65.

[29] *Smičiklas,* CD IV, No 63, 67 f.

[30] Cf. Chapter XIV, n. 36.

[31] *Smičiklas,* CD IV, No 87, 94.

[32] *Theiner,* Mon. Hung. I, No 307, 170, *Potthast* 10779; *Smičiklas,* CD IV, No 88, 94; ibid., No 89, 95. Cf. *Šidak,* Misija dominikanaca, 28.

[33] *Smičiklas,* CD IV, No 99, 107 f.

[34] *Šidak,* art. cit., 29 f.

[35] *Smičiklas,* CD IV, No 208, 236 - 240.

[36] It is not certain that tithes were established in Bosnia at this date, as *Šidak* believes (art. cit., 28). He rightly assumes however that this question aroused very strong opposition in the country.

[37] Commentariolum de provinciae Hungariae originibus, in: Fratris Gerardi de Fracheto O.P., Vitae Fratrum Ordinis Praedicatorum (= Monum. Ord. Fr. Praed. Historica I), Rome - Stuttgart 1897, 305 - 308. This neglected source was pointed out by *Šidak,* Misija dominikanaca, 13, 37 f.

[38] Professor *Šidak,* on the evidence of Suibertus' chronicle, denies that there was loss of life among the Dominicans in Bosnia (Misija dominikanaca, 38). He quotes Suibertus according to *N. Pfeiffer,* Die ungarische Dominikanerprovinz von ihrer Gründung 1221 bis zur Tatarenverwüstung 1241 - 1242, Zürich 1913, 146. In the edition quoted above in n. 37 (to which Professor *Šidak* did not have access) a variant is given, however, in which there is a reference to the murder of several Dominican monks in Bosnia. See 308, n.1, var. lect. C: "Qui (i.e. the Dominicans) ... mirabiliter profecerunt, *et aliqui fratres ab hereticis sunt occisi et alii* multos hereticorum et credentes eorum *ad fidem converterunt,* et multi ex iisdem, qui converti nolebant, per ministrales dicti regis Colomani ignibus *traditi sunt...*" (Our italics where the reading in C differs).

[39] *Theiner,* Mon. Hung. I, No. 347, 187. *Potthast* 11095.

[40] See *Šidak,* art. cit., Addenda, 245; cf. *Šidak,* O pitanju, 15.

[41] Epistles of Innocent IV, 3.VIII.1246: *Smičiklas,* CD IV, Nos. 264 - 265, pp. 298, 299).

[42] Epistle from Innocent IV to Bishop Ortulf, the Abbots and Priors of Cistercian and Dominican monasteries in Hungary, 26.VIII.1247: *Smičiklas,* CD IV, No. 285, 322 f. Cf. the Pope's epistle to the Abbot of the Benedictine monastery of St. Martin, 20.VII.1246: *Smičiklas,* CD IV, No. 263, 297 f.

[43] Epistle from Innocent IV to Archbishop Benedict, 27.III.1248: *Smičiklas,* CD IV, No. 305, 341 f.

[44] See the Pope's epistle of the same date: *Smičiklas*, CD IV, No.306, 342.

[45] *Stojanović*, Povelje i pisma I 1, No.12. Cf. *Šidak*, Misija dominikanaca, 34.

[46] The sources do not offer substantial proof of any connection in the development of certain institutions of the Patarene Church and of earlier forms in the organization of the Catholic Church in Bosnia. There is only a later document which seems to suggest that the author regarded the "Patarene houses" (see Chapter XIX below) as Catholic monasteries *(loca)* taken over by heretics. See the epistle from Thomas Bishop of Hvar to John Capistranus (19.II.1451): "...in locis occupatis per hereticos..." and again: "...si tot essent Fratres boni religiosi, ut occuparent loca principalia regni..." *Wadding* XII, 111 (130); Acta Bosnae, No 901, 211.

[47] See the epistle of Gregory IX, 13.II.1234 (*Smičiklas*, CD III, No. 343, 396): "...Prelatis ac universis Christi fidelibus per Carneolam, Istriam, Dalmatiam, Bosnam, Cruaciam, Serviam et alias partes Sclavoniae constitutis..."

[48] See *Ćirković*, Istorija, 354, n. 5; id., Bosnische Kirche, 554, n. 17.

If the Ecclesia Sclavoniae referred to by the author of De heresi did in fact have its centre in Dalmatia, it can be concluded that the heretical Church of Bosnia, with its close affinities to the Dalmatian region, followed at first a variant of dualist doctrine similar to that found among the *Sclavini* (i.e. the Italian Cathars who took their *ordo* from Sclavonia) by the same Lombard author. If this was so, developments must have brought about a profound change, for later evidence points to relations between the Bosnian Church and Italy, and clearly shows its doctrines to have coincided with those of the church of the Albanenses. See Chapter XIX, p.298 and 323, n.25.

Chapter XVI

THE STRUGGLE AGAINST THE BOGOMIL HERESY IN SERBIA UNDER THE NEMANJID DYNASTY, IN THE SECOND BULGARIAN EMPIRE, AND IN CONTEMPORARY BYZANTIUM

The walls of Serbian churches of the Nemanjid period still show us in their murals the great ecclesiastical councils as well as native gatherings which are well known to us through contemporary Serbian authors.[1] It is not without interest that the very first such assembly attested for the early years of the Nemanjid State was concerned with *heresy*, then penetrating into Serbia. The initiative in this timely attack on heresy lay with the Grand Župan Stephen Nemanja, here as in other fields the earnest emulator of the great Comnene rulers of neighbouring Byzantium. The whole story is told in the biography which Nemanja's son Stephen, the First-Crowned, wrote in honour of his father, revered after his death as St.Symeon.[2]

In harmony with the general character of the work, this event in Nemanja's reign is given the style of a legend, and few facts can be deduced from the narrative. References to the confiscation of houses and other property suggests that heresy had its adherents among the wealthier aristocratic members of society. Nemanja's biographer tells us that some of the heretics were burned at the stake, while others suffered "various punishments", by which he may have meant the varied physical deformities which the cruel practice of Byzantine criminal jurisdiction afflicted. Only one such case is explicitly quoted in this source, however: the leader of the heretics had his tongue cut out for "not confessing the Son of God". This reference to a leading personality, together with heretical books condemned to be burned, certainly suggests that there was an organized sect; we are told nothing of its origins, however.

There is only one passage in this text which could be taken as referring to the Bogomils; the words placed in the mouth of a daughter of the Serbian ruler, married to a heretic, who saw their secret gatherings for herself, remind us of the enemies of the Bogomils, accusing them of Devil-worship. Nemanja's biographer, nevertheless, does not place the Serbian heresy in the context of dualist heresies, but connects it with the teachings of Arius "dividing the indivisible godhead". While it is well known that in western Europe the polemists used the term "Arian" to mean, in fact, the Cathar heresy,[3] we cannot be certain that the term should be given the same significance in this Serbian text. We thus are not able to claim the emergence of dualist heresy in Serbia under Stephen Nemanja, although it can easily be imagined that Bogomil beliefs trickled through from their Macedonian and Bulgarian neighbours very early on.

It is certain that under the immediate successors to Stephen the Church of Serbia was seriously disturbed by this heretical movement which was spreading throughout the surrounding countries and even in Serbia had to be borne in mind as a latent danger. We know, too, that Nemanja's son Vlkan kept close watch on the growth of heresy in neighbouring Bosnia and that some time during the year 1199, that is to say shortly after his father's death, he urgently called the attention of Pope Innocent III to this problem.[4] We have already glimpsed the political aims of these two brothers, Vlkan and Stephen, in the light of the papal correspondence on the threshold of the thirteenth century. Their aims were undoubtedly in line with those of the Papal Curia, not in the least unwilling to seize the opportunity of binding both Serbia and Bulgaria to the Church of Rome, and thus regaining the long-lost ecclesiastical jurisdiction over what was formerly Illyricum. There was an obstacle, however, in the shape of the Hungarian state, which had its own goal of hegemony over the Balkan peninsula. The king of Hungary, Emmerich, made his military power felt in Serbia, when he placed Vlkan on the throne instead of Stephen. With Bulgarian help Stephen soon regained control of Serbia (1203) but he had to wait many years before a favourable opportunity occurred for fulfilment of his old dream, the royal crown.[5]

Meanwhile the rulers of Bulgaria, which at the end of the twelfth century had emerged from over a century and a half of Byzantine domination, had gained recognition in Rome much earlier. In 1185 two feudal overlords, the brothers Peter and Asen, led the north-eastern Bulgarian region to successful revolt against the weakening Byzantine Empire. The magnetic tradition of the Tsardom of Bulgaria first led the rebels to the old seat of Preslav, and finally gathered them round the new seat of the Asen dynasty, founded somewhat to the west. Peter and Asen chose a site which had been fortified in the days of the early Byzantine Empire, against attack from the north, on a stretch of high ground rising from the winding Jantra river. An important road passed through this valley, protected here on all sides by steep cliffs. Here the new town of Trnovo rose up, with its churches and palaces, the manors and chapels of the boyars on Trapezica (Trapezitsa) hill, the workshops of artists and craftsmen, and settlements of merchants from Dubrovnik and later from Venice too. In the eyes of contemporaries Trnovo was "a glorious city of the Tsars", a "new Tsarigrad" which overshadowed the decaying "imperial city" on the Bosporus. This attitude is seen very clearly in the work of the famous Trnovo literary school, in the last period of the second Bulgarian Empire.[6] The first Asen rulers already set out with the intention of making their city the centre of the entire administration and of ecclesiastical life, on the Byzantine pattern. In the Trnovo church of St.Demetrios Asen was crowned Tsar by the Archbishop, thus placing the dynasty in the traditional line of the Preslav and Ohrid Tsars. The militant Kaloyan (1197-1207), the youngest brother of Asen and Peter, applied to Pope Innocent III for the crown.[7] The long-drawn-out negotiations were not concluded until after an event which changed the whole situation in the Balkan peninsula: in April 1204 the capital of Byzantium fell to conquerors from the West, the Latins or Franks as they were called by the Byzantines. Innocent III was now naturally anxious to reach agreement with the Bulgarian ruler, whose growing power could be very dangerous for the "Latin" empire rising from the ruins of Byzantium. King Emmerich of Hungary was forced to give way to pressure from Rome and in November 1204 Kaloyan was crowned in Trnovo by the papal legate, Cardinal Leo. Rome had once more renewed its jurisdiction over Bulgaria, but as in the days of Boris, the Papal Curia was not to enjoy its gains for long.[8] The crushing military defeat of Adrianople (1205)[9] shook the empire of the Latins to its foundations, and saved the work of rebuilding the Byzantine Empire which had started in Asia Minor under tremendous difficulties. The Byzantines created a new centre of national and ecclesiastical life in Nicaea, the seat of both the Emperor and the Patriarch.

The powerful Kaloyan, feared by both Latins and Greeks, called himself Tsar although the Pope had granted him only the rank of king, and in the same way the Archbishop (Primas) of Trnovo was called Patriarch. In any case the Asen dynasty held a purely Byzantine view of the relationship between state and Church, and there was no hope that they would ever submit to the theocratic order towards which Innocent III was working. Recognition of the Pope's supremacy was a purely formal act, and the ecclesiastical policy of the rulers of Bulgaria followed the pattern set for centuries by the world of Byzantium.[10]

This was made clear immediately after Kaloyan's reign, when his nephew Boril (1207-1218) seized the throne. The usurper was not in a strong position, for much of the country was in the hands of other members of the ruling house who did not recognise his claim. The Bogomil movement flourished during this period of uncertainty and confusion; as in the tenth century, it was again strong in the north-eastern regions, which formed the heart of the Bulgarian Empire once more. And like Tsar Peter in his day, Boril himself took action against the heretics. This is a very different situation from that in Bosnia, where everything undertaken against heresy was arranged by Rome and executed through the papal legates. In Bulgaria Tsar Boril (like the Grand Župan of Serbia, Stephen Nemanja) called a great council before whom the preachers of heretical errors were tried and condemned. The record which has survived gives a vivid picture of the memorable Council of Trnovo in 1211.[11] The bishops, clergy and monks gathered in one of the "great churches" of the royal city, along with the boyars and "great numbers of other people". Tsar Boril, clad in a cloak of purple, took his seat at their head and directed the proceedings in person. This is a clear continuation of the Byzantine tradition, where even after the time of general oecumenical councils the emperors did not relinquish their prerogative in settling questions of doctrine and in combatting heresy. Under the Comnene dynasty, in particular, there were a number of synods against heresy, presided over by the Emperors themselves and not infrequently attended by representatives of the secular power as well. In 1117, for example, Eustratius, a disciple of the philosopher John Italus, had to abjure his Christological errors before Alexius I, the permanent synod *(synodos endemusa)* and the senate.[12] We know that the public trial of the Bogomil teacher Basil was arranged in the same manner. The record of the 1211 council, celebrating Boril's victory over heresy, was incorporated into the Synodicon for the Sunday of Orthodoxy which the Tsar ordered to be translated from the Greek at that time. The Bulgarian version of the Synodicon clearly follows the provincial Byzantine tradition of the text, which was part of the liturgy for the First Sunday in Lent, celebrated as the Feast of Orthodoxy in memory of the final victory over the Iconoclasts. The Synodicon gave solemn expression, as time went on, to a number of victories gained by Byzantine orthodoxy in doctrinal questions during the *post-conciliary* centuries. The official version, developed for the use of the church of St.Sophia in Constantinople, restricted the range of questions to certain principal points of dogma.[13] The provincial versions, in so far as they have survived, are less strict in adhering to this canon and contain additional articles anathematizing acknowledged heretical teachers of the time of the oecumenical councils (Arius, Nestorius, Eutychius and others), as well as various heretical trends and individual heretics of later centuries. Thus in the provinces of Byzantium whole passages dealing with dualist heresies found their way into the Synodicon, and in particular the teachings of the Bogomil sect, which was considered the supreme heresy ever since the eleventh and twelfth century.[14]

In one of the provincial recensions of the Byzantine Synodicon for the Sunday of Orthodoxy (Cc in Gouillard) an article on the Bulgarian priest Bogomil has been inserted, taken, as is now known, from the formula given in an epistle of the Patriarch of Constantinople, Cosmas.[15] The Bulgarian Synodicon, however, includes *all* the anathematizing articles in the Patriarch's epistle, almost word for word, except for a certain amount of rearrangement and addition. Thus the Byzantine original for most of the sixteen anti-Bogomil anathemas which follow (in Boril's Synodicon) the article anathematizing the followers of the monk Nilus of Calabria, has been found.[16] These sixteen articles anathematize not only the priest Bogomil but also his "former and present disciples, known as apostles". The heresy propagated throughout Bulgaria by the priest Bogomil in the time of Tsar Peter is here described as a combination of Manichaean (i.e. Paulician) and Messalian teachings. This definition of the Bogomil heresy is quite in the spirit of the Byzantine conception found in the writings of Anna Comnena and Euthymius Zigabenus.

In its detailed description of Bogomil's teachings the Bulgarian text clearly stresses his Christological docetism - the doctrine that the birth of Christ was only an illusion. The anathemas which follow reproduce familiar Bogomil ideas: the belief that Satan was the creator and ruler of the visible world, the rejection of the Old Testament and of John the Baptist, of the sacraments of baptism and the eucharist, of the symbol of the cross, of church buildings and the liturgy. Of particular interest is the article which refers to the part played by Satan in the conception of every human being; this links up with the view we are already familiar with, that baptism with water has no power to free the new-born child from the power of the Devil. We should remember here how in the tenth century, in the days of Cosmas the Presbyter, the sect would turn away in disgust from baptised children, who remained in their eyes the "children of the Devil". According to the Synodicon of Boril, the Devil can only be driven out by prayer and fasting. We know that fasting and intense prayer formed the main part of the preparation for the spiritual baptism which brought complete freedom from the Devil, to the Bogomils. Another article, referring to the heretical manner of baptism without water, also concerns this rite of initiation; it was performed (according to the Synodicon) only by reciting the Lord's Prayer.

It is also of interest that the Bogomils are accused here of *completely* rejecting the Old Testament and all the great figures therein. Even the books of the prophets, we are told, were inspired by Satan, just as the rest of the Old Testament was. It was only "of necessity" and involuntarily that the Old Testament prophets proclaimed what was later fulfilled in Christ.[17] (Euthymius Zigabenus said that the sect included the Psalms and the prophetic books in their canon, besides the New Testament books.) The whole of this section of the Synodicon ends with the anathema pronounced against the leader of the *Byzantine* Bogomils, Basil, who sowed the tares of heresy in the very city of Constantinople itself, in the days of "Alexius, the orthodox emperor Comnenus".

There are also more or less obvious allusions to the Bogomil sect and its teachings towards the end of the dogmatic part of the Bulgarian Synodicon.[18]

There is much more significant information, however, in the "diptycha" of the Synodicon of Boril, which contain the complete record of the proceedings of the Trnovo council directed against the Bogomils. The record ends with anathemas ramarkably consistent with those which preceded them, concerning the priest Bogomil and his disciples. Now, however, the names of some of his disciples are given: Michael, Theodore, Dobr, Stephen, Vasilij (Vasiliy), Peter.[19] We cannot determine exactly when these men lived and taught, but the fact that their names have been preserved is a significant proof that the sect maintained the tradition and preserved the memory of their teachers, beginning with Bogomil himself. We find a similar phenomenon in Bosnia.[20] The name "apostles" given to the disciples of the priest Bogomil shows the deep awareness of the sect that they had a mission, that it was their duty to return to the original, genuine Christianity of apostolic times.

Along with the anathema pronounced against Bogomil and his followers, the record of the Trnovo Council ends with the "eternal memory" of those who alongside Tsar Bosil helped to reduce the evil heresy to dust. The conclusion of the description of the assembly itself stresses the fact that before Boril nobody had convened such an "orthodox assembly". In the churches of Bulgaria "the eternal memory" of the "pious Tsar" was evoked, but not a word betrayed the sorry end of his reign.

A few years after the holding of this Council, where he proudly appeared as the protector of the true faith and used his sovereign power to punish the unrepentant preachers of heresy, Tsar Boril had to relinquish control over his kingdom, and was driven to desperate measures to defend himself in his own residence. It seems that in the end the people of Trnovo themselves opened their gates to the legal heirs to the throne of Bulgaria. Taken prisoner, Boril was blinded, and Ivan Asen II became Tsar; under his rule (1218-1241) the Second Bulgarian Empire enjoyed its greatest period of power.

At about the same time, Boril's rival, the Grand Župan Stephen of Serbia, achieved what he had long been working for. In the favourable circumstances created by the departure of Andrew II of Hungary on a crusade to the Holy Land, Stephen was given the royal crown by Pope Honorius III (1217) and became known to later generations as Stephen the First-Crowned. We know no details of his negotiations with the Roman Curia, but so much is certain: the state of Serbia did not fall under the jurisdiction of the Church of Rome. There was a Catholic Archbishop in Bar (Antibari), on the territory of Zeta (Dioclea), owing allegiance to the King of Serbia; nevertheless Stephen wanted to gain independent status for the *Orthodox* bishopric of Rascia - the heart of the Serbian state. Hitherto the bishops of Rascia had been subject to Ohrid, by this time part of the territory of the rapidly growing state of Epirus. It was in Stephen's interests to escape from the ecclesiastical and political influence of this Greek neighbour. His brother, the monk Sabbas, now set out for far-away Nicaea, and there Stephen's request for an independent archbishopric was indeed favourably received. In 1219 the Patriarch (Manuel Sarantenos) consecrated Sabbas archbishop.[21]

The new archbishop was anxious to establish an effective ecclesiastical organization, and following the Byzantine model he sent visitants (exarchs) all over the country, to see to it that the Christian faith and moral principles were observed, and to insist, particularly, on the involved ecclesiastical regulations concerning marriage. In the Serbian context this meant the consistent introduction of the sacrament of marriage, for this was still far from common among the people. The Serbian church had several difficult problems to deal with, for the customs of the people went far back into pagan times. At the same time the Church had to be on its guard against the heretical teachings being propagated on all sides. The words of Sabbas, directed against heresy, in the solemn assembly called to his residence in 1221, are significant. Both his biographers refer to this occasion, which took place in the monastery of Žiča (Zhitcha).[22] The monk Domentian has the saint exhort the faithful in general terms to preserve orthodox doctrine as established by the seven oecumenical councils, and to condemn the heresies anathematized there. The second biographer, the monk Theodosius, records that Sabbas was anxious and fearful, lest "impure heresy" should appear even among his listeners. He called upon all who might have been affected perhaps through ignorance to confess it openly. During the solemn form of worship which followed the congregation repeated the words of the archbishop, in a confession of faith so formulated as to be clearly directed against the teachings of the Bogomils. Their name does not occur in the biography of Theodosius, nor does the formula given refer to the dualist metaphysics of the sect. It does, however, proclaim respect for the visible symbols of Christianity which the dualist sects rejected; the cross, pictures of Christ, the Virgin Mary and the saints; and such outward signs as churches and sacred vessels. The clearest evidence is provided, however, by the clause which stresses the significance of the eucharist as the partaking of the real body and blood of Christ, and rejects absolutely the wrong attitude of those who considered the elements of the sacrament to be no more than ordinary bread and wine. It has been suggested that the assembly in Žiča solemnly proclaimed a Serbian Synodicon of Orthodoxy, elaborated or adapted for this great occasion. Unfortunately there is no evidence. The earliest Serbian recension of the Synodicon which can be identified dates from the time of Sabbas' immediate successor, Archbishop Arsenij (Arseniy). The version of Arsenij already contains four articles directed against native heresy - the Babun faith. This strange name is explained in the Serbian translation of the Nomocanon, where the word babuni is clearly used on several occasions as a synonym for the Bogomils. At a later date, the famous secular *Code* edited by *Tsar Dušan* (Dushan) gave measures to be taken against those who "used Babun talk".[23] In mediaeval Serbia the word *Babun* became the official term for the Bogomil sect. The name Bogomil itself, however, occurs here and there in Serbian sources. It is attested primarily in the very interesting text known as the "Commandments of John Chrysostom"[24], which accurately describes some of the fundamental features of the Bogomil heresy.

The name Babun is mysterious, nor do we find it anywhere but in Serbia. It is not until later that the written sources throw light on the matter. At the time when aggressive Serbian expansion was pressing into Macedonia, King Stephen Dušan acquired Strumica (Strumitsa) and Prilep by treaty in 1334, and immediately showered gifts on the monastery of Treskavac (Treskavats) dedicated to the Virgin Mary, in the Prilep region. This is borne out by the chrysobulls he issued to confirm and extend the property rights of this important monastery. Two of the chrysobulls refer to a farm which belonged to the monastery, "u Bogomilěch" (u Bogomilyekh = by the Bogomils) and "u Bogomili v Babuně" (v Bábunye = by Bogomila in Babuna).[25] The name Babuna can certainly be found elsewhere in the Balkan peninsula among geographical names; in Macedonia it has been given to a mountain range north-east of Prilep, as well as to a river which rises in the mountain range of Jakupica (Yakupitsa) and flows ultimately into the Vardar. In the Middle Ages, however, the name was already used to denote the whole region (župa), which covered a larger area than the valley of the river alone.[26] Babuna Mountain had hidden in its sides a number of cave-chapels where remnants of frescoes can still be seen, but it could equally well hide groups of heretics. On one of its slopes there is a valley known as

Bogomiljsko polje. Dušan's chrysobull, however, undoubtedly places the village of Bogomili or Bogomila in the Babuna *region;* it must be looked for in the valley of the river of the name, where indeed a village called Bogomila still exists today.

It would thus appear that the dualist heresy reached the faraway Serbian hinterland from Macedonia, where we can guess at the existence of an important centre of the sect, or perhaps a number of such centres, connected with the whole network of Bogomil communities throughout the Byzantine world.

We may recall the laconic comment by Theodorus Balsamon, pointing to widespread occurrence of the heresy in the provinces. This was the situation at the end of Manuel's reign. The death of this great Emperor (1180) was followed by severe disturbances which led ultimately to the temporary occupation of Constantinople and large areas of the Empire by the "Latins". This was but the beginning of the involved struggles which kept changing the balance of forces in the peninsula. The lack of stability in political and ecclesiastical life certainly favoured the spread and growth of the sectarian movement. Leading figures in the Byzantine Church in its centre in Asia Minor came to realise, sooner or later, the extent of the danger which had again sprung up in the exceptional conditions of the time. This realization is clearly behind the encyclical sent to Constantinople[27] by Patriarch Germanos II (1222-1240), to be read in all churches on Sundays and feast days in order to put the orthodox on their guard against this dangerous heresy. Any man who desired to abjure these erroneous teachings and return to the orthodox faith was to repeat before the congregation the abjuration formula whose separate articles were explained in the encyclical. The description of the Bogomil heresy provided by Germanos agrees with what earlier Byzantine sources said of it, particularly the work of Euthymius Zigabenus. Once again we read that some Bogomils thought Satan was the elder son of God. Patriarch Germanos did not know the Bogomil sect only from his reading; he had undoubtedly been in direct contact with it, as some of the references in his homilies suggest quite clearly.[28] It is evidence that there was a strong heretical movement in Asia Minor, on the territory of the Byzantine emperors of Nicaea. The encyclical of Germanos suggests that in the city on the Bosporus the Bogomil heresy was once again felt to be a serious danger. Somewhat later, evidence from a western source confirms the spread of this dualist heresy both among the Greek and the Latin part of the population of Constantinople.

The Byzantine Patriarch in Nicaea could not of course back up his spiritual authority with effective help from the secular power in Constantinople, as long as the capital remained in the hands of the Latins, along with much of the European territory of the Empire. Their sphere of influence was certainly considerably restricted when the "kingdom of Salonica", the largest of the Latin feudal domains, fell into the hands of Theodore Angelos, ruler of Epirus, whose political aims were similar to those of the ruling house of Nicaea, the Lascarids. The rivalry between Epirus and Nicaea reached its peak when Theodore Angelos was crowned Emperor (1224), after taking Salonica. This solemn event, performed by the Archbishop of Ohrid, Demetrios Chomatianos, provoked sharp protest from the Patriarch Germanos. The hostility between the two Greek ecclesiastical centres, Ohrid and Nicaea, dated from the time when St.Sabbas was consecrated first archbishop of Serbia by the Patriarch of Nicaea (1219). This act reduced the sphere of influence of Ohrid considerably, for it meant that the orthodox Serbian see of Rascia had been removed from its jurisdiction. In future the name of the Nicaean Patriarch sounded in Serbian churches, and his honour and influence grew apace.[29] Even the Bulgarian Church, formally subject to Rome since the days of Kaloyan, ultimately recognized the primacy of the Byzantine Patriarch of Nicaea. This significant change reflected the sudden shift in the balance of power following the battle of Klokotnica (Klokotnitsa) on the Marica in 1230. The power of Epirus then fell as suddenly as it had risen, and the victorious Ivan Asen II reached the peak of his power. It is a curious paradox that it was this great victory which took Asen further away from his ultimate goal, the metropolis on the Bosporus.[30] Freed from the threat of attack from Epirus, the Latins broke the terms of their agreement with the Bulgarian Tsar, who as the future father-in-law of the Emperor, who was a minor, would have acted as Regent and thus gained decisive influence in Constantinople. Bulgaria now allied herself with Nicaea in order to destroy what was left of the Latin Empire. This alliance, however, formally put an end to Bulgaria's ties with the Church of Rome. After long-drawn-out negotiations Nicaea agreed to establish an independent Bulgarian Patriarchate in Trnovo (1235). The Bulgarian Synodicon, recalling every year the *revival* of the "Patriarchate of the Bulgarian Empire", shows how lively was the awareness of continuity between the Preslav and the Trnovo empires. A long description of the event was included in the Synodicon, referring with due respect to the Nicaean Patriarch Germanos and the Emperor John Vatatzes. The years which followed were not without surprises. The Tsar of Bulgaria suddenly broke off his military alliance and went over to the side of the Latins, only to conclude a peace with John Vatatzes again (1237). The Conquest of Constantinople, put off on account of disagreement between the allies, was not to be achieved during Asen's lifetime. After his death (1241) Bulgaria declined, suffering heavily from the results of Mongol invasion while Nicaea worked persistently towards complete restoration of the Byzantine Empire.[31]

The return of the powerful Bulgarian state to the sphere of influence of the Eastern Church meant the final defeat of the Pope's efforts to restore the jurisdiction of Rome over the region of ancient Illyricum. A temperament like that of Gregory IX could not suffer such a defeat easily, however. When the Crusaders were already far into Bosnia the clouds began to gather round Bulgaria too. Early in January 1238 the Pope appealed to King Bela IV with an urgent call to arms against the "perfidious Asen".[32] In this papal epistle the Tsar of Bulgaria is not only accused of schism, destroying the unity of the Church, but also of giving protection to heretics, who "it is said" flourish throughout his lands. This accusation levelled at Asen II had serious legal consequences, for it enabled the Curia to apply all the measures for combatting heresy, by now firmly embedded in canon law, to the case of Bulgaria. We have already seen that the same spiritual and temporal advantages were given to those taking part in such a "Holy War" as to those who set out under the banner of the Cross to free the Holy Land from the infidels.

It is interesting that at the same time another epistle from Gregory IX to King Bela, about calls to take arms against the Nicaean Empire, also refers to the liberation of the Holy Land and the crushing of heresy. The immediate purpose was to bring help to the Latin Empire, which was in a weak position after the alliance with the Bulgarian Tsar broke down. To help to maintain tha Latin Empire, the Pope argued, would open the way to reconquest of the Holy Land and make it possible to crush the heresy which had infected "almost the whole of Greece".[33] The epistle repeats almost word for word the accusations levelled a little earlier about the spread of heresy in Asen's territory. In connection with the campaign against Byzantium, however, the Hungarian king is only called upon to provide safe conduct and other aid to the crusaders on their way through his country. In the case of Bulgaria Bela IV was expected to undertake the whole campaign himself. The only way to get him to do so was to offer him a sufficiently great material reward, and so in his first epistle the Pope puts Bulgaria at the disposal[34] of Bela and his allies. The same day the Pope wrote to tell the Latin Emperor Balduin II of his decision, asking him not to lay obstacles in the way of the campaign and to put at the disposal of the Papal Curia any rights that he might claim over the territory of Asen. The papal correspondence on the subject of this campaign against Asen revives the concept of *exponere* a heretical land.[35]

Bela not only accepted the offer made him by Rome, but added some conditions of his own, asking that the function of papal legate for Asen's territory should be given to him, with the power to draw the boundaries of each diocese and instal bishops there. According to Bela this right had been granted to the Hungarian crown in the time of St. Stephen.[36] Gregory IX was willing to meet this request, but in such a way as to keep control over its execution in the hands of the Curia.[37]

It is not known why the negotiations between Rome and the Hungarian king dragged on. Meanwhile the Mongol threat was drawing near in the East, so soon to lay waste Hungary and other countries. Under these circumstances nothing came of the plans for a Crusade against Bulgaria, although the Pope did not lose sight of this "schismatic" country in planning his policy. The hope of bringing Bulgaria back to Roman jurisdiction remained alive even after the death of Gregory IX, as can be seen from two bulls issued by his successor, Innocent IV, in March 1245. The name of Bulgaria is constantly recurring in the instructions sent from Rome to the Franciscan monks entrusted with missionary work in the Balkan countries. Bulgaria is also among the "custodias" of the Bosnian vicariate which provided the basis for missionary work in these regions from the beginning of the forties of the fourteenth century.[38]

We do not know how far the accusations levelled at Ivan Asen II, that he offered protection to heretics, corresponded to the facts. The aims of the Pope in the matter are quite clear, but that does not mean that the whole thing can be dismissed as invention. It is certainly probable that followers and protectors of the sect cruelly persecuted by Boril would support the Asen dynasty at the right moment, and that they would therefore be left alone during the reign of the great Tsar. This would certainly encourage the growth of the sect in Bulgaria. From the correspondence of Gregory IX we learn nothing of the real strength of the Bogomil sect there, or of its organization, and there is no native Bulgarian testimony at all. The information provided by the work of an Italian inquisitor, Rainerius Sacchoni (1250), about dualist churches in Byzantium and among the southern Slavs is therefore all the more valuable.

[1] See *N.Radojčić*, Srpski državni sabori u srednjem veku, Belgrade 1940.

[2] Edited by *P.J.Šafařík*, Památky dřevního písemnictví Jihoslovanův, Život sv. Symeona od krále Štěpána, 1868.

[3] See Chapter XII above, p.160, n.21.

[4] See Chapter XIII above, p.163.

[5] See *Jireček - Radonić* I, 210 f.; *Dvorník*, The Slavs, 97.

[6] See *I.Dujčev*, Tărnovo kato političeski i duchoven centăr prez kásnoto srednovekovie, in: Archeologija VIII (1966) kn. 3, 1 - 9.

[7] *Zlatarski*, Istorija III, 149 - 211; *I.Dujčev*, Prepiska na papa Inokentija s bălgarite, Uvod, tekst i beležki, in: Godišnik na Sof. Univ., ist.-fil. fak. XXXVIII (1942) kn. 3; *Dvorník*, The Slavs, 94 f.

[8] For the development of relations between Kaloyan, the Latins and the Greeks see *Zlatarski*, Istorija III, 211 - 269; *Dvorník*, The Slavs, 96.

[9] For the events in Plovdiv at this time see *Zlatarski*, op. cit., 222 f. In his view two parties were formed in Plovdiv, the "Bulgarian" which turned to Kaloyan and the "Latin" which looked to Balduin for protection (223). Unlike other historians, Zlatarski does not consider the possibility that the Plovdiv heretics (Paulicians - Popelicans) took part in the negotiations with Kaloyan. The relevant passage in Villehardoúin reads: "...une partie des genz, qui estoient popelican, s'en alerent à Johanisse et se rendirent a lui, et li distrent: 'Sire, chevauche devant Phynepople, ou envoie t'ost: nos te renderons la ville tote.' Quant Renier de Trit le sot, qui estoit en la ville, si dota que il ne le rendissent a Johanisse: ensi que s'en issi a tant de gent com il ot, et s'esmut a une jornee, et vint par un des bors de la vile, ou li popelican erent a estage qui erent rendu a Johanisse; si mist le feuec ou borc et en art grant part..." Villehardouin, La Conquête de Constantinople, éd. et trad. par *E.Faral* II, Paris 1939, 210. I am of the opinion that the words "une partie des genz qui estoient *popelican*" means no more in this context than "part of the people of Plovdiv". The term *Popelican* was also used for the Cathar sect which had an important centre in the town (see Chapter IX above). According to *K.Jireček*, Cesty po Bulharsku, Prague 1888, 102: "In the town itself there is the 'Pavlikianska mahala' ... on the road to Stanimaca, on the site of the old quarter of the "Popelicans" in the time of Duke Renier. There is no proof, however. that the quarter was continuously in existence." B.Primov believes that by the word *Popelican* Villehardouin meant not only the Paulicians but also the Bulgarian Bogomils and perhaps the non-Bogomil people of Philippopolis as well. *B.Primov*, Žofroa d'o Vilarduen, četvărtijat krăstonosen pochod i Bălgarija, Sofia 1949, 93 f. Cf. the same author's study, Bălgari, gărci i latinci v Plovdiv prez 1204 - 1205 g. Rolja na bogomilite, in: Izv. na bălg. ist druž. XXII - XXIII (1948) 145 - 158.

[10] *Dvornik*, The Slavs, 96.

[11] Sinodik of Boril, Palauz., Nos. 110 - 112; Drin., No. 93. For the date of Boril's council, see *Zlatarski*, Istorija III, 301, n.1.

[12] *Gouillard*, Synodikon, 207.

[13] Ibid., 287 f.

[14] Ibid., 226 - 237.

[15] See Chapter VIII above, p.96. *J.Gouillard* published the letter of Patriarch Cosmas to the Metropolitan of Larissa according to Marcianus gr. II 74 (coll. 1453, olim Nanianus 96), fol. 77v - 79v. See *J.Gouillard*, Une source grecque du Sinodik de Boril, La lettre inédite du patriarche Cosmas, in: Travaux et Mémoires 4 (1970) 361 - 374. It was written either by Patriarch Cosmas "of Jerusalem" (1075 - 1081) or by Cosmas Atticus (1146 - 1147) (*Gouillard*, art. cit., 363 f.). In the editor's opinion the anathematizing formula elaborated by the Patriarch "a dû pénétrer en Bulgarie bien avant la restauration de l' empire bulgare, c'est-à-dire dès que la lettre y eut été diffusée, surtout si l'on suppose que Cosmas a adressé une copie à chacun des métropolites des provinces voisines de Larissa. Le formulaire fut alors

inséré dans le synodikon en vigueur dans le pays" (art. cit., 373).

16 For the text of the Patriarch's letter (with French translation) see Gouillard, art. cit., 369 - 372. For the sixteen anti-Bogomil articles of the Bulgarian Synodik see: Sinodik of Boril, Palauz., Nos. 38 - 53; Drin., Nos. 45 - 60.

17 *Gouillard*, art. cit., 367 finds a difference on this point between Palauz., No. 45 and the third article of Patriarch Cosmas. This interpretation, however, is due to wrong punctuation in the Palauz. text as published by *Popruženko*. If a comma is placed after ѥже съписашѫ and the fullstop after ѡ҃ х҃ѣ is removed, the Bulgarian and Greek texts are identical.

18 On the names in Palauz., Nos. 77, 78: Alexander the coppersmith, see 2 Timothy 4,14: 'Αλέξανδρος ὁ χαλκεὺς πολλά μοι κακὰ ἐνεδείξατο. Cf. the letter from Georgios Tornices to the Metropolitan of Athens, cit. edition (see Chapter VIII above, n. 17), 215,14-15: Δέδοται γοῦν μοι σκόλοψ ἄλλος χαλκεὺς 'Αλέξανδρος - referring to Soterichos Panteugenos, who spread lies about the deceased Patriarch Theodosius. The names Audinos and Photeinos in the same article have been taken from earlier heresiological catalogues, where they follow in the same order as in the Bulgarian Sinodik (See *Gouillard*, L'hérésie, 302 f.). The other people named in Palauz. Nos. 77, 78 remain a mystery. Most scholars assume that the mention of Peter of Cappadocia, дѣдць срѣдечьскаго alludes to some Bogomil bishop (дѣдъ) with his seat in Srědec (Sryedets, now Sofia). (дѣдьць is a diminutive of дѣдъ.) This is no more than a hypothesis with no supporting evidence from any source.

19 Sinodik of Boril, Palauz., No. 111.

20 See Chapter XIX below, p.303.

21 *Jireček - Radonič* I, 217 f.; *Dvorník*, The Slavs, 98 f.

22 Život svetoga Symeona i svetoga Save, napisao Domentijan, ed. *Dj.Daničič*, Belgrade 1865, 233 - 243. - Život svetoga Save napisao Teodosije, ed. *Dj.Daničič*, Belgrade 1860, 141 - 152.

23 *Solovjev*, Svedočanstva, 37 ff. On the question of the first recension of the Serbian Synodicon for the Sunday of Orthodoxy, see *Mošin*, Sinodik, 371. Cf. the comment by *J.Šidak*, O pitanju, 10 f. The four entirely original articles against the "Babun" faith precede in the Serbian synodicon the articles against iconoclasm taken from the Byzantine model. (These are a few anathemas from the Second Council of Nicaea [787] which disturbed the symmetrical arrangement of the Byzantine Synodicon for the Sunday of Orthodoxy. See *Gouillard*, Synodikon, 2). This context explains the subject of these four articles, which deal with disrespect for holy images (including the Cross). The heretics are then accused of distorting the words of the Scriptures and at the same time calling themselves "Christians". Serbian Sinodik, 301 f.

On the Nomocanon of St.Sabbas and Dušan's Code, see *Solovjev*, Svedočanstva, 29 ff., 88 ff.

24 Ibid., 33 f. See also 25 f., on the passage entitled о оучєни вѣры, known from the collection (now lost) by "pop" Dragolj. It is a compilation of some passages of Cosmas' Sermon. *J.K.Begunov* dates Dragolj's collection at the beginning of the fourteenth century, and believes that the compilation was made from a complete Bulgarian copy of the Sermon in which the text was still close to the archetype of the work. Begunov connects this compilation with the assembly in Žiča (1221). See *J.K.Begunov*, Serbskaja kom[illegible] XIII v. iz Besědy Kozmy Presvitera, in: Slovo 18-19 (Zagreb 1969) 91 - 108.

The Serbian translation of the Nomocanon proves that the common Byzantine idea of the connection between Bogomilism and Messalianism had penetrated into Serbia. The two heresies are identified in the title of a passage by Demetrios Chomatianos which reached the Serbs through the medium of the translated Nomocanon. In the Serbian marginal glosses to the text of the Nomocanon, the Messalians are also called "the teachers of the Bogomils, that is, the Babuns" The Serbian commentator also found ideas similar to those of the Bogomils in the Marcionite articles set out in Chapter 63. See Solovjev, Svedočanstva, 29 f.

25 Text: Glasnik Društva Srpske Slovesnosti XI (1859) 134; XIII (1861) 371.

26 See the Enciklopedija Jugoslavije I, 267. *Obolensky*, Bogomils, 164 f., gives a review of ideas on the origin of the name "babuni", as does *Miletič*, "Krstjani", 97 f. Cf. *M.Budimir*, Trikleti babuni i babice patarenske, God. Ist. Društva B. i H. 10 (1959) 73 - 86.

27 Ed. *Ficker*, Phundagiagiten, 115 - 125. *Grumel*, Regestes (fasc. IV, ed. V. Laurent), No 1291.

28 In exaltationem venerandae crucis et contra Bogomilos: PG 140, col. 628CD. In vivificam crucem: ibid., col. 652CD. De imaginibus: ibid., col. 664AB.

29 *Ostrogorsky*, Geschichte^3, 356.

30 Ibid., 361.

31 *Zlatarski*, Istorija III, 354 - 418.

32 *Theiner*, Mon. Hung. I, No. 283, 159 f. Cf. also No. 284 (letter to Balduin) and No. 285 (episcopo Perusino), both of the same date as the first epistle (27.I. 1238). Also: No. 286 (letter to the Hungarian bishops). *Potthast* 10505 - 10508.

33 "...propter hereticos profligandos, quibus pene tota Grecia est corrupta." Ibid., No. 288, 162. As far as we are aware, attention has not been paid to this letter in the literature on Bogomilism to date.

34 "...terram eandem exponimus prout statutum est in generali concilio occupandam." Cf. the letter to the Hungarian prelates, N° 286, 161.

35 See Chapter XIV, 206, n. 29.

36 *Theiner*, Mon. Hung. I, N° 308, 171.

37 Ibid., N° 295, 165 f.

38 *Dujčev*, Franciscanesimo...

Note: I have been unable to consult the article by *I.Dujčev* on Bogomilism in the Second Bulgarian Empire: Quelques observations à propos des courants dualistes chez les Bulgares et à Byzance aux XIIIe-XIVe siècles, in: Studi Veneziani 12 (1970) 107 - 125. According to the author's own note (BZ 65 [1972] 190), it follows the line of his earlier work in which the close linguistic and cultural affinities between the Protobulgars and the Uiguri were put forward as the ground for suggesting that the Manichaean ideas had penetrated Bulgaria from the Uigurian regions. See *I.Dujčev*, Aux origines des courants dualistes à Byzance et chez les Slaves méridionaux, in: Rev. des Ét. Sud-Est Europ. VII (1969) N° 1, 51 - 62. In his latest study D. has construed a relationship between the flourishing of dualist heresy in Bulgaria in the XIII - XIV centuries, and the relations maintained at that time with peoples who had come from Asia (Cumans, Tartars etc.).

Chapter XVII

ECCLESIA ROMANA AND "ECCLESIA DEI"

Not relying solely on crusades and courts of the Inquisition, the Church did not underestimate the value of argument in the fight against heresy. Thorough preparation was necessary, for the heretics were unusually well-read in the Scriptures and skilfully selected Biblical quotations in support of their contentions. The Cathars were no longer at that naive level of argument which provoked the irony of Guillaume de Puylaurens when he described the disputations in Languedoc before the southern crusade.[1] By the time of the Albigensian Wars the Cathars had become very able polemists indeed. Their skill forced the Church authorities, who had hitherto based their refutation of heresy on New Testament texts only, to pay attention to certain Old Testament books on which the Cathar sect drew liberally. This extension of the repertoire of Biblical authority called upon seems to have been primarily the work of Catholic polemists who had originally come from the ranks of the Waldensians.[2] We know that Durand of Huesca was in the habit of disputing with the Cathars at the time when he was still a Waldensian. At the beginning of the twenties of the thirteenth century, when he headed the society of the "Pauperes catholici", he took up his pen again to refute the dualist teachings of the Languedoc heretics, better armed with theology that he had been in those early polemics. This time he had to hand the heretics' own treatise, revealing remarkable knowledge of the Bible in the arguments proving the insoluble opposition of two Gods and two worlds. Durand copies the text of his heretical model, linking his own polemic commentary to the individual chapters of the original, which he sometimes gives in full and sometimes in shortened form.[3]

Other Catholic authors refer only to the titles of the heretical works which sometimes directly inspired their polemical treatises. The famous book, "Supra Stella" was written in 1235 in Piacenza, one of the liveliest centres of the heretical movement in northern Italy. The Catholic author, Salvo Burci, was polemizing with a heretical work entitled "Stella". Among the Cathar works now lost was the "Perpendiculum scientiarum", quoted by the Spanish writer Lucas de Tuy, and a work by one Tetricus mentioned by Moneta of Cremona. Sometimes the Catholic writer does not quote his source, but sets out the matter of individual heretical theses which he then refutes systematically. This is the arrangement used in the anonymous Summa traditionally ascribed to Prevostin of Cremona.[4]

The form of a "Summa" of the authorities, i.e. collections of quotations from the Scriptures, served the practical needs of Catholic preachers, who were thus always ready to deal with heretics.

Sometimes the confrontation of Catholic and heretical views took the form of a dialogue. One such "Disputation between a Catholic and a Patarene heretic" was written in Lombardy, perhaps before the middle of the thirteenth century. The author, one Georgius, was a layman, like Salvo Burci.[5] As we shall see later, the Bosnian Franciscans put together a similar dialogue in the following century.

Besides the work of Georgius, with its lively impression of a real spoken discussion, learned treatises were written in Italy, arranged strictly in accordance with the scholastic system. Among them is an important work which can very probably be attributed to the Dominican martyr, Peter of Verona. It was not until a little later, however, that this type of "Summa" reached its peak in the work of another Dominican, an Inquisitor of the Milan region. A contemporary of Peter's, Moneta of Cremona was at first a professor at the University of Bologna. His "Summa against the Cathars and Waldensians" is based on thorough theoretical erudition; it is also founded on real knowledge of sectarian teachings, gained in the practice of the Inquisition and by study of heretical writings.[6]

The exhaustive work of learned argument against heresy produced by this theologian of Bologna meant that his Italian colleague Rainerio Sacchoni, an equally ardent Inquisitor, could turn his pen to somewhat different purpose. Rainerio's "Summa" is clearly of a different type, providing accurate information about heresy rather than polemical material.[7] At about the same time a short work was produced in Languedoc, too, giving a brief summary of the basic facts known about the Cathar teachings in their different variants. This "Brevis Summula" was actually a compilation of various elements, including a description of the doctrines of the north Italian Cathar churches, taken from De heresi Catharorum in Lombardia. It contrasts with the work of Rainerio Sacchoni, which is a complete and unified work based on the author's own personal knowledge. Rainerio himself had spent many years among the heretics before he became their determined persecutor. He was undoubtedly one of the "perfect", and it seems to have been the influence of such a strong personality as Peter of Verona that converted him to the Catholic faith (1245).[8] This ardent convert was already writing his "Summa" in 1250; it was copied profusely, and was to be found in all the handbooks prepared for the use of the Inquisitors. These contained texts of various kinds, both concerned with the Inquisition itself, and with the heresies against which it was directed. In time the whole of this material was arranged in systematic handbooks.[9] The best known, the "Practica inquisitionis" of Bernard de Gui, includes detailed forms for inquisitional procedure and also gives a systematic and historical review of contemporary sects, based on information from reliable sources. Handbooks of this type were used as part of collections of material provided for the Inquisitors, and were also found as separate books. Anselm of Alexandria added some texts intended for the information of the Inquisitors, to his "Treatise on heretics" written between 1260 and 1270. The historical review with which the Treatise itself opens is largely identical in subject matter with the contents of the De heresi Catharorum in Lombardia. As we have already seen, in some respects it supplements this earlier work. The further sections of Anselm's work provide much interesting information about the doctrines, ritual and customs of the Cathar sect in Italy. Yet in many instances the Dominican monk is not repeating what his colleague Rainerio Sacchoni had already collected in the way of fact. Rainerio's "Summa" was interpolated later, in a mechanical way, in Anselm's text, which in itself concentrates on those aspects which are not to be found in Sacchoni's work.

The "Summa" against the Cathars, written by the former "heresiarch" Rainerio Sacchoni, and significantly supplemented by Anselm of Alexandria, gives us a very accurate picture of the Cathar churches in northern Italy around the middle of the thirteenth century. In the main outlines, this picture has not changed since the date of the De heresi Catharorum. There are still three churches in the foreground, originally centred in village communities: Concorezzo near Milan, Desenzano on Lake Garda, and Bagnolo near Mantua. These churches had recognised names; such designations formed from personal names occurred as early as the days of Salvo Burci (1235), and again in the anti-heretical constitution of Frederick II made public in Cremona and other towns of Italy (1238). The *Garatenses* were so called after the Concorezzo bishop Garattus and the *Caloiani* after Caloiannes of Bagnolo. The name Albanenses, applied to the Desenzano church at this time, was probably also derived from a personal name.[10] At the time when Rainerio Sacchoni was writing, the first two churches were already commonly known by names derived from their localities: *Concorezzenses* and *Bagnolenses*, but the old name *Albanenses* was still in use for the third.

Rainerio knew Nazarius, Bishop of Concorezzo, who even in old age remained faithful to the doctrines he had learned as a young man "across the sea". At the beginning of the thirteenth century, however, the anonymous author of the history of the Cathars of Lombardy already noted dissension among the Concorezzenses on the subject of the Virgin Mary and that of the human nature of Christ. According to Anselm of Alexandria, later, this fundamental question of doctrine had divided the Concorezzo church into two factions. One remained faithful to the traditional doctrine proclaimed by Nazarius while the other followed Desiderius. The latter, "filius maior" in the Concorezzo church, did not acknowledge the Bulgarian secret book and did not believe in the mystical stories of the birth and nature of the celestial bodies, developed by Nazarius on the basis of the Interrogatio Iohannis.[11] Primarily, however, Desiderius believed that the mother of Christ was a real woman, and taught that at birth Christ took on a real human body, in which he died and rose from the dead. On His ascension He laid this body aside in the "earthly paradise" *(in paradiso terrestri)*, where His mother abides for ever, never having died. The Son of God will assume His human body for the last time at the Day of Judgment. According to Desiderius Christ had not only a human body while he was on earth, but a real human soul as well, and not the "divinity in place of a soul" *(deitatem pro anima)* of Nazarius' teaching.[12]

Even earlier, when Rainerio Sacchoni was still one of the Cathars himself, the Albanenses had already been divided into two schools of thought with considerable differences between them on some fundamental points of doctrine.[13] According to Rainerio this split occurred in 1230, up to when the whole of the Albanensian church had preserved the radical version of the dualist myth in its original form. Even after this date Bishop Belesmanza remained faithful to the doctrines taught at the end of the twelfth century by Marchisius de Soiano and Amizo. The great majority of the older members of the church followed Belesmanza, but fewer of the younger Albanenses. Most of the latter were won over by the bold conception elaborated by the "filius maior", John de Lugio. Rainerio Sacchoni studied his "lengthy work of ten quarternions" very thoroughly, and presents the main ideas of this new teaching in his "Summa".

Basically, however, these ideas were not as novel as they appeared on first sight; what John de Lugio did was to carry to paradoxical conclusion the parallelism of two worlds elaborated by the extreme dualist speculations of the Cathars. It was very tempting to develop the idea of an invisible world composed of four immaterial elements and illumined by an invisible sun, moon and stars. Moneta said aptly of the radical dualists that "they duplicate all concepts" *(ponunt omnia duplicia)*. The idea of the "Heavenly Jerusalem" was suggested by the New Testament itself; that was where they imagined the Old Testament prophets were when they pronounced their visions.[14] Nor was it possible to imagine those visions of the appearance of the Son of God as inspired by the evil ruler of the world. Euthymius Zigabenus had already observed a radical change in the attitude of the Byzantine Bogomils towards the prophetic books of the Old Testament. Nevertheless the original idea seems to have persevered alongside the new in Byzantium[15] and the "Bulgarian" and "Sclavonian" churches appear to have remained conservative in this matter. The Concorezzo and Bagnolo Cathars, at any rate, who derived their "ordo" from these two overseas churches, took some time to develop a more receptive attitude towards the prophets and John the Baptist.[16] Rainerio Sacchoni even noted tendencies among the Concorezzo Cathars to regard Abraham, Isaac, Jacob and the other Old Testament patriarchs more favourably.[17] The Albanenses, on the other hand, placed the origin of the prophetic books, as well as some other Old Testament books like Job and the Psalms, in "the other world". John de Lugio went even further, teaching that all the books of the Old and the New Testament had been written in that supernatural sphere, and that all the events recorded in the Bible had also taken place there. The ideas hitherto held about the "new land", the world of the Heavenly Father, thus developed into the projection of the whole of Biblical history into this invisible world.

Biblical history is really the history of the human race, with all its misfortunes, errors and sins. Placing it in the realm of the good God only brought out even more urgently the problem of the interplay of good and evil, and the question of free will. The good God has His own world, and in this world live human beings with all their faults. Evil has thus not come into this celestial world only with the invasion of Satan (Lucifer) and his angels, but has always been present there. God is not omnipotent, for His actions are constantly limited by the action of the principle of evil. Indeed some acts and statements of the good God can only be explained as inspired by His opponent. He could not form His creatures as He would, perfect, because the evil principle gave each a certain element of evil which led them into sin. The only exception is the nature of Christ, in whom this disposition to sin is neutralized by His supreme goodness. The rest of the "people of the good God" *(populus boni Dei)* succumbed to sin and were therefore condemned to suffer death. The souls of the dead descended into Hell, i.e. to the earthly world. This was not a completely new idea. The older generation of the Albanensian church regarded the world of the evil God as a place of suffering where the souls of celestial beings atoned for their sins. Their sentence was carried out when they fell from the state of eternal beatitude into the darkness of this world. It was for this reason that the second coming of Christ had no place in the eschatology of the radical dualist school of Cathar thought, and that the idea of Hell was identical with that of this earthly life. It was thus only the logical consequence of this line of thought that John de Lugio regarded the coming of Christ to earth as *descensus ad inferos*.

Together with the other Albanenses the followers of John de Lugio believed in the pilgrimage of repeated incarnation, at the end of which all souls reach final release and find their own bodies again in their heavenly home. The dark earthly world where they are imprisoned, however, would endure for ever, like the world of the good God. It should be borne in mind that the dualist metaphysics of the Gnostic groups and schools of late antiquity were based on the contrast between the eternal and invisible world of the true God and the visible world whose ephemeral existence implies that the reign of the inferior forces (or one inferior power) which created it will also end in time. This basic framework which persisted in Marcionism and Paulicianism passed into the doctrines of the new mediaeval dualist sect, and it was not disturbed until some Cathar groups elaborated the idea of the *eternal existence of the principle of evil*. At first even the radical dualists (Albanenses) thought that the material world created by the "prince of darkness" would exist only for a limited time. By the time when Rainerio Sacchoni was writing, however, even the older generation (those who followed Belesmanza) were convinced that this earth of ours "would know no end". Here, too, John de Lugio was only thinking out what had been hinted at earlier. The concept of a world without end is incompatible with the idea of its creation in time. John de Lugio therefore understood the priority of the two "creators" with regard to the things they had created not in the temporal but only in the causal sense. He illustrated this idea by the example of the sun and its rays.

Let as return once more to the Christological doctrines of John de Lugio. If the appearence of Christ on earth was nothing more or less than His descent to the underworld, it necessarily means that the whole of his life as described in the Gospels was passed in the supernatural world. In that "other, higher world" *(in alio superiori mundo)* Christ was really born, there He suffered and died, and finally rose from the dead. Even His miracles, like the sight restored to the blind man and the feeding of the five thousand, really took place but in that other world. John de Lugio was not the first, however, to place the life of Jesus in the invisible supernatural world. We find a remarkably similar belief among the Cathars of Languedoc in the first years of the crusade against the South. An unknown author who belonged to Waldensian circles, and after him Peter of Vaux-de-Cernay,[18] give us a glimpse of the extravagant world of fantasy which occupied the minds of some of the Cathar groups in Languedoc when all around them death and destruction raged. In their secret doctrines there was an "evil Christ" whose life was bound up with the "earthly and visible" Bethlehem and Jerusalem, an ordinary sinful man whose wife or concubine was Mary Magdalene. The "good Christ", however, was present on this earth only in the spirit - in the figure of the apostle Paul. He was born and crucified in "another new and invisible world". On this point their views are remarkably consistent with those of John de Lugio, although we can find no evidence that the "filius maior" of the Albanenses had any knowledge of the secret doctrines of his fellow-believers on the other side of the Alps. It is certain that their conception of the "invisible land" or the "land of the living" arose from the parallelism of two Gods and two worlds. However confused it may have been, then, it belonged to the same trend of thought as the faith of the Albanensian church in Lombardy. The myths propagated in Languedoc communities were derived from a very primitive interpretation of the Biblical text, taken literally and distorting the sense excessively. It would not otherwise be possible to deduce from the prophecy of Ezekiel (Ezekiel 23,4) the paradoxical idea of two wives of the good God and the sons and daughters He begat on them. This is the absurd culmination of the image of the "land of the living" *(terra viventium)*, i.e. the realm of the good God. It is a vision of a flourishing land with castles and towns in it, green pastures, springs of clear water, woods and an abundance of game - a land where life is passed in the pleasures of love, the table, and other purely secular pastimes. This was all in accordance with the joyous tones of the 149th Psalm: "Let the saints be joyful in glory: let them sing aloud upon their beds..."

Besides these flights of the imagination aroused by texts from the Old Testament, our attention is drawn to the idea of Christ and the Devil ("the prince of this world") as the two sons of God the Father. The similarity to the secret teachings of the Byzantine Bogomils is only illusory, however, for the whole idea runs on quite different lines. According to the Cathars of Languedoc both the sons sinned, but Christ "with His people" has already been reconciled with God. The proof given is the parable (Matthew 21, 28-31) of the two brothers in the vineyard. It is quite clear that the whole legend was inspired directly by this story in the Gospel, and that it has no connection with the Byzantine tradition of Satanael as the first-born son of God.

These examples show to what extremes the unbridled fantasy of some Cathar groups could go. We find ourselves entangled in a wild undergrowth of myths often very far from the fundamentally logical dualist system.

In contrast, the speculation of the school of John de Lugio was sometimes only too bound to a few axioms from which there was no escape. This way of thinking is well illustrated by the authentic work of the Albanensian Cathars, the *Liber de duobus principiis*.[19] This forms the common title to several short treatises all tending to prove the existence of an independent origin of evil. The argumentation is completely subordinated to syllogisms based on the incompatibility of the concepts of good and evil. The actions of each individual are determined by one or other of these two opposing forces. This line of thought is connected with the doctrines of John de Lugio by the belief that the evil principle in effect restricts the omnipotence of the good God, and by the curious view of the creation *(creatio)*. John de Lugio did not accept the idea of "creation from nothing" *(creatio ex nihilo)* and defined three categories of creation: from good towards better, from evil towards good, and from evil towards even worse. According to the Book on Two Principles the *tertia creatio* consists in the fact that God, for His own purposes, sometimes allows the principle of evil or its servants to perform acts which could not take place without His consent. Only in this *conditional* sense can the true good God be considered the origin of evil. In the choice of Biblical texts to support this line of thought the author draws on a wider range, using the historical as well as the prophetic books of the Old Testament, and the New Testament. Here, too, there is a remarkable agreement with John de Lugio.

The "Liber" is more likely to be the work of one of John's disciples than his own; the unknown author certainly based his arguments on John's system although it is not set out completely nor consistently applied.

The intellectual armoury of the author of this Book on Two Principles is not well stocked. He employs some concepts current in mediaeval philosophy but he was not well read, and direct references to secular literature are few. The proofs are arranged in the usual scholastic manner *(quaestiones, oppositiones, solutiones)* but the strictly logical approach is employed only in the first part of the treatise.

In the concluding section the author launches a direct attack on the teachings of the Concorezzo church, calling its followers Garatenses, which we know was the name derived from that of their bishop. This polemic reflects the old antagonism between the two biggest Cathar communities. When Rainerio Sacchoni was writing his "Summa", this hostility had not been overcome in spite of all the efforts made to reach agreement. Salvo Burci, who was informed by impartial Cathar circles, speaks of the great endeavours made on both sides to bridge this ruinous schism, and of the long journeys made to that purpose, sparing neither effort nor expense. There was no lack of will on either side, concessions could be made in matters of dogma in order to achieve doctrinal unity again. It was obvious that the everlasting conflicts were having a bad effect on the "believers" and weakening the whole movement at a time when external pressures were increasing. The trouble was not only one of doctrinal differences, however, but of precedence. Each of these two Lombard churches was convinced that she alone was the true and original "church of God" and that the other had split away later.[20] This conflict was too deeply rooted, as we know, going back to the initial confusion in the young Italian Cathar community caused by divisions among the mother churches "across the sea". According to Durand of Huesca (in the twenties of the thirteenth century) in Languedoc, too, three Cathar factions poured recriminations and curses upon each other, one giving obedience to the *Greek* heretics, one to the *Bulgarians*, and the third to the "Drogovetia" (i.e. *Dragovitsa*) church.[21] Unfortunately we have no way of verifying whether the author of the Liber contra Manichaeos did not distort relations between these Languedoc Cathar groups.

It is quite clear that Catholic authors were always highly content to note any conflict within the Cathar sect, and to lay due emphasis upon it. It was always welcome evidence that the sect could not be the *ecclesia Dei* it claimed to be. Objective study of the sect, however, cannot fail to reveal the sense of unity and lively awareness of integration which never for a moment disappeared among the members of these groups scattered over so vast a geographical area. Neither physical distance nor differences of dogma prevented good relations. Here again we have the reliable witness of Rainerio Sacchoni; at the time of his writing (in the middle of the thirteenth century) the schism between the Cathars of Concorezzo and the Albanenses was the only serious rift in the Cathar world. The other Cathar groups in northern and central Italy had apparently long ago overcome their feelings of alienation and reached mutual understanding. And at this time, too, the other dualist churches both in the East and the West "received each other" even where their doctrines were widely different or perhaps quite contrary.[22]

We should be careful not to see the rift between the Albanenses and the Concorezzenses in the oversimplified light in which present-day authors often place it in the broader context. The usual view is that the whole Cathar community was profoundly split between the "radical" and the "moderate" dualists. Contemporary sources do not support this view, however. It is clear that *the Albanenses and the Concorezzenes* formed the two strongest Cathar churches, with branches all over Lombardy, like the *Bagnolo* church which was, however, numerically weaker.

Rainerio speaks of the three as the three main "parties" *(partes)* or "sects", holding leading positions in the whole Cathar community.[23] We know, too, that De heresi Catharorum already cited these three schools of thought as prototypes of the different variants of dualism. The Languedoc Brevis summula uses the initials of the three groups (Albanenses, Bagnolenses, Concorezzenses) in its table of heresies, to show at sight which articles of the Cathar faith were held by each of the three. It can be assumed from what Rainerio Sacchoni says that these three Italian churches "set the tone" for the attitudes of the other groups. The doctrines of the Albanensian church were widespread among the eastern branches of the sect, too; Rainerio tells us explicitly that "practically all" of the churches "across the sea" preserved the doctrines of absolute dualism in the traditional sense, as taught by Belesmanza. Yet we have not a single hint that any of the churches either East or West was drawn into the long hostility between the Albanenses and the Concorezzenses. However, not even in dogma were the Cathar communities completely dependent on the teachings of one or other of the "chief sects" of northern Italy. According to the evidence collected by Rainerio Sacchoni none of them accepted completely the doctrines of the Concorezzo church, and he names three in Italy which in his view agreed with the church of Bagnolo "in more points" than they had in common with the Albanenses.[24]

Somewhat later we are told by Anselm of Alexandria that within the Bagnolo church itself some of the faithful followed the teachings of the Albanenses.[25] It is apparent that doctrinal differences between the "radicals" and the "moderates" were not in themselves enough to provoke open rift. The case of Bagnolo shows that followers of two opposing schools of thought could exist side by side within the one church. This aspect of the developments within the Bagnolo church also suggests that their creed did not remain changeless and that some members of the church moved away from its original teachings. We have seen, too, that developments were also taking place in Concorezzensian and Albanensian thought. Desiderius and John de Lugio each went his own way, both departing considerably from the traditional conception.

It is not without interest that both these younger men, speaking for their generation in each of the two largest Cathar churches, in some sense drew closer to official Christianity. Desiderius acknowledged the real incarnation of Christ, while John de Lugio accepted all the books of the Bible without distinction. Some authorities today see these two points as a symptom of conscious retreat from the original dualist faith. Those who argue thus believe that as the sect was pressed more and more into the defensive, it found itself forced to discard more and more of the alien metaphysics derived from ancient non-Christian sources and rousing essential opposition in those lands with a Latin civilization. The sources do not provide any support for this view, however.[26] There is nothing to suggest that these new trends in the Lombard church derived in any way from a conscious desire to adapt the dualist doctrine so as to give it more chance of acceptance by people brought up in an atmosphere of strict monotheism. There is not the slightest hint that Desiderius or John de Lugio were manoeuvring to gain or keep adherents when they preached these new and perhaps more acceptable ideas. Indeed Rainerio Sacchoni shows quite clearly that the reverse was true; the new variant of the dualist doctrine, originating with the filius maior of the Albanenses, remained the spiritual property of a relatively small circle of the "perfect". John de Lugio and his followers were not out to create confusion in the minds of the "believers", and indeed they feared lest the emergence of a new doctrinal conflict within their own ranks disturb even more profoundly the confidence already shaken by the old disagreement between the Concorezzo and the Albanensian churches.[27]

As we have in fact seen, the system elaborated by John of Lugio was only the logical development of doctrines already held by the Albanenses. It thus arose in the course of developments in the thinking within that church and there is no need to explain it by external pressures. Similarly Desiderius with his realistic view of the human nature of Christ was only taking up tendencies which the De heresi Catharorum had already noted earlier among the Cathars of the "Bulgarian" persuasion.[28]

The emergence of these new and bold variations was only one aspect of the constant fluctuation to which these doctrines were subject, based as they were on free speculation on the text of the Scriptures. As early as the tenth century Cosmas the Presbyter had commented appositely on this changeable nature, and since that early period it should be borne in mind as a permanent element in the dualist teachings. We have seen that the sect's ideas on the very fundamental question of the origin of evil, and their attitude towards the creation of the visible world, were subject to permutation; new solutions were constantly being sought. The dualist myth thus underwent change not only in details, but even in general outline, moving closer to or further from the general Christian tradition as the case might be. Differences of belief thus arose between the different Cathar groups, differences which were certainly not considered insignificant by their opponents. Catholic authors could not fail to note that compared with the Albanenses' doctrine of two principles, the ultimate belief of the Cathars of Concorezzo and

Bagnolo in a single God was not in conflict with the teachings of the Church itself. There was of course always a point beyond which Cathar thought could not go, towards Church dogma. Even in its most "moderate" forms the beliefs of the sect were quite unacceptable to the official Church, in the West as in the East. No changes in the structure of the dualist myth could touch the hard core which was quite incompatible with the very substance of Christian doctrine. This was the fundamental rejection of the world which gave rise to several tenets held by all groups within the sect. Rainerio Sacchoni recognised these *communes opiniones,* which formed the inner bond between them, quite clearly.[29]

All "Cathars" - men of the pure life - were united in their revulsion against all that binds the human being to his material body. They rejected married life and would not eat meat and other animal products because they had all come "ex coitu". In Western Europe this was the reason given from the earliest days, and the sect made no secret of it.[30] There were also deeper reasons, which only gradually emerged, and which were rooted in their dualist metaphysics. Peter of Verona came upon a secret doctrine which he described as Manichaean error: this concerned some female creatures of unknown origin who appeared in Heaven at the decisive moment in the battle between Satan and the Archangel Michael. Being with child, these women could not stand the noise and the way Heaven reeled, and dropped the fruits of their womb, from which all the kinds of animals and birds came. Thus the bodies of animals are of human origin and should not be eaten.[31] There is no doubt that this mythical explanation of the origin of the animal world is of Manichaean origin, but it has been shorn of its original significance and clumsily incorporated into the Cathar myth. The sect could have become acquainted with Manichaean mythology through literature, for the works of St.Augustine[32] were not unknown and his anti-Manichaean writings could have found their way among them. There is nothing strange in their using such a bizarre motif[33] if we remember to what extravagant lengths the imagination of the sect was sometimes allowed to go.

An even more secret Cathar doctrine came to the knowledge of Peter of Verona, connecting the prohibition of meat with the mystery of metempsychosis.[34] On its long pilgrimage towards salvation the soul passes through many bodies, not only those of human beings but animals as well, those "with blood". It is therefore forbidden to eat animals and birds but there is no objection to eating fish for they "are born of water".[35]

Not all the Cathar churches, of course, believed in the transmigration of souls. This particular secret doctrine could only develop in a group which believed in the two principles. And even there it was only revealed to the few initiates; according to Peter of Verona very few of the "perfect" alone were acquainted with it.

There was another, simpler reason for not eating meat, however, expressing the general belief of the sect but kept from other people. We first hear of it in the West from Eckbert of Schönau, who said that these heretics believed all bodies to be the work of the Devil and therefore would not eat meat even in the most dire need.[36] Everwin of Steinfeld, the first western author to write about the Cathars, declared that he could not find out why the sect rejected marriage.[37] This German Praemonstratensian was obviously unaware of the myth of the creation of the first man and woman, as later attested among the Lombard Cathars. It gives clear expression to the idea that the "opus carnale" is the order of the evil creator, the Devil, hoping thus to bind the soul to the material body for ever. This idea had already been expressed in the Bulgarian secret book, where the Devil acted from without in order to rouse physical passion in the first man and woman.[38] There can be not the slightest doubt that the sect had brought with it from the East the knowledge of the diabolic origin of marriage, as of the eating of meat and other animal foods. Originally wine was also included in these forbidden foods the true Christian should not touch. From the outset, however, we see that in the West the Cathars are most temperate in their consumption of wine, but do not refuse it on principle.[39] The sect maintained its original inexorably negative attitude to wine, on the other hand, in the eastern Christian world. This is attested by the Byzantine Patriarch Germanos II in the first half of the thirteenth century, and in a Serbian text of about the same time the Bogomils are reported as believing that "wine and women are of the Devil".

This "bitter teaching",[40] the stern rejection of the joys and pleasures of life on earth, was rooted in very ancient times. As we have already seen, tendencies to extreme asceticism had existed in the Byzantine world from the earliest times, and found expression in abnegation of marital life and in abstention from wine and meat.[41] The Church fought this conception of the ascetic life as a heresy, although it was not yet bound up with dualist metaphysics. Finally in the teachings of the Bogomils these tendencies to complete renunciation of life in the world joined with the dualist teaching that this world originated in the force of evil and was ruled by it. This doctrine, too, with its sharp contrast between God and the material world, was very ancient It had been kept alive for centuries in the Byzantine world by the Paulicians, but it went right back to the mental world of the Gnostics of later antiquity. It would seem that the concept of matter as evil was already disappearing among the Paulicians, although it had been the fundamental element in all Gnostic teachings and retained its full significance in Marcionism as well. In the Bogomil-Cathar myth matter never appears as an independent factor. Whether created by the good or by the evil God, its existence is always secondary. The cosmic drama no longer moves towards liberation of sp·ritual Light from the dark prison of Matter as in Manichaeanism, but towards the liberation of the human soul from the power of evil ruling the material world. It was of little importance for the attitude adopted towards this material world whether one or another force created amorphous matter *ex nihilo*. Views on the ultimate origin of material being might change, the decisive point was that the Devil or the evil God had made *(fecit)* this world of ours and all that it contains. This was one of the fundamental theses acknowledged by all branches of the sect.[42] We find it precisely defined already in the Sermon of the Bulgarian Presbyter Cosmas: all that we see around us is the work of the Devil; he is the creator and ruler of all earthly things; it is he who gave the command to lie with women, to eat meat and to drink wine.[43]

Here we come up against the insuperable barrier. It was certainly possible to credit the good God with bringing the first man to life or separating the four elements, but this made no difference to the outcome. The world still remained in the power of the Devil, whose treachery always perverted the intentions of the good God in His acts of creation. John de Lugio even linked the creation of the world and the whole of Old Testament history with the good God, but from the earthly standpoint this was nothing but a fata morgana from another world, and it was only there that all these things had really occurred. The world in which we live seemed to John de Lugio, too, as the domain of the evil principle and the prison of the human soul, atoning here for its supernatural sin. The negative attitude towards the world was thus unchanged even in this variant of the Cathar doctrine.

In the same way the recognition of the reality of the incarnation of the Son of God did not affect the negation of the material body. Not even Desiderius attributed eternal life to the body of Christ - in his eyes it was as ephemeral as the bodies of ordinary people. At His second coming the Saviour would make His last appearance in human form; His human body would lose its significance then, however, and so it would desintegrate and return to the dust like the bodies of all human beings and animals.[44] This refusal to glorify the body of Christ meant fundamentally that the very dogma of the Incarnation of the Son of God was made light of, the very foundation stone of Christian theology. This then robbed of their significance both the sacraments of the eucharist and of baptism, which are profoundly bound up with the Incarnation. In the West the attitude of the heretics to these and to the other sacraments of the Church remained as negative as in the Byzantine world. The attitude of the sect towards *baptism* was somewhat more involved in the West, however. The arguments against the baptism of children which had been taken over from earlier local heresies were still accepted; baptism cannot bring salvation to little children who cannot yet understand and therefore cannot be instructed in the faith before baptism. They are simply not capable of the faith without which there is no salvation. Unlike the members of the Roman Church the Cathars remained faithful to the customs of the church of Christ, which did not administer baptism without previous instruction. In the West the Cathars thus stressed the uselessness of the *baptism of children*, while in the spirit of the original "Bogomil" attitude to baptism they denied the efficacy of the sacrament of baptism altogether, when it was administered with water, i.e. a material element.[45] In the New Testament texts concerning baptism water was always interpreted in the spiritual sense, as the teachings of Christ. At the most it was allowed that the "ecclesia Dei" had used water for baptism for a time, probably for tactical reasons, in order to use the "symbol of that baptism" to encourage people to accept the baptism of the Holy Spirit.[46] In spite of occasional tendencies to accept a solution which would allow some significance to Church baptism,[47] there was never any doubt that a second baptism, the true spiritual baptism of Christ, was absolutely essential.

We have already touched on the way the sect developed a radically new conception of the Christian idea of sin. The various groups regarded the sin of the angels against God in different ways; there was no unified view of whether they had acted of their own free will or had been forced. There was no room for doubt, however, that the "sin" of the first man and woman was not an act of free will, but the result of violence to the human soul which the evil prince of this world bound in its material prison by means of physical passion. "Original sin", which as we know had not the same character as that attributed to it by accepted Christian doctrine, was only of secondary significance in the dualist view of the world. It was the result of the sinning of the angels, and human will could not change it in any way.[48] No sin committed on earth aggravated that tragic supernatural guilt, nor could any merits gained on earth lessen it. The innocent and the hardened criminal, both were in the power of the Devil. Only the administration of spiritual baptism (consolamentum) could bring about a decisive change. Human beings backslide, however, and it was therefore necessary to guard constantly against mortal sin. The essence of mortal sin was submission to the laws of physical being established by the Devil. From this point of view legalized marriage was as mortal a sin as adultery. Similarly, anyone who tasted meat even in the direst need, to save his life, was committing as grave a sin as the man who took poison to kill himself.[49]

Here the greater importance attributed to dualist metaphysics than to Christian ethics is apparent. We should remember, too, that most of the "believers" (credentes) did not receive spiritual baptism until shortly before death, when they were not likely to commit mortal sin again and endanger their soul's salvation. All the sins committed during the believer's lifetime were now insignificant, for they had been atoned for by the administration of the consolamentum, just as the ancient sin dating from before the creation of our world was now washed away. There only remained the danger that this state of beatitude would be rendered invalid through the sin of the dignitary administering the consolamentum. But even in this case all who had received initiation at his hands would suffer alike, a magnate like Raymond VI, Count of Toulouse, whose life passed in worldly pursuits and who was burdened by many sins, as well as a Cathar saint martyred for his faith.

This whole conception of sin and the efficacy of spiritual baptism gave rise to sharp criticism from the Cathars' enemies. Ecclesiastical writers reproached them for insufficient penitence for the sins committed before the administration of the consolamentum, and for too little effort to reform. They were shocked by the sect's benevolent attitude to usury, which the Cathars did not consider a mortal sin; nor did they feel it necessary to hand back illicit gains.[50]

It was impossible to deny, however, that the select company of the "perfect" led a blameless life of asceticism.[51] It was a very weak point, however, that there existed another code of morals allowing the more numerous "believers" to live without the strict abstinence to which they were enjoined only when the consolamentum was administered to them. In the eyes of Catholics the Cathar faith allowed irresponsible enjoyment of all the joys of life together with the certainty of salvation, guaranteed by the administration of the consolamentum "in articulo mortis".[52]

Looking at it from this point of view some modern scholars have also tended to see the Cathar sect as a passive body to whom salvation was administered by the magic formula of the initiation rite. The Cathar "priest", i.e. the "perfect" member of the sect who had acquired their secret knowledge, maintained this magic power over the mass of "believers" at the price of self-abnegation. This view, of course, denies the sect all the real religious and moral values it possessed and reduces its faith to "cosmogony and magic".[53]

The history of this important religious movement itself shows convincingly the one-sided nature of this view. The vitality of the sect, and its capacity for enduring persecution, sprang primarily from its moral strength. The groups of devoted "believers" were firmly convinced that only their church of "good Christians", "good people", could lead them to salvation. This was expressed in the ritual petition known as the melioramentum, uttered by believers not only at the moment of acceptance among the "perfect" members of the sect, but on many other occasions as well in the body of the faithful, and even when meeting with one of the "good people". The accepted form of words was always repeated there times and accompanied by a deep bow or genuflection. The Catholics who witnessed it wrongly saw in it a form of adoration, and accused the Cathars of allowing themselves to be worshipped like saints. They themselves, however, stressed that this respect was paid not to the man, but to the Holy Spirit in him. The melioramentum was the expression of the believers' most profound longing - that they, sinful people, might receive at last the "good" which God granted to the "good Christians". Through the medium of the latter they asked God to protect them from an evil death and lead them to a "good end", i.e. "into the hands of faithful Christians".[54] The fear of an "evil death", dying without the soul's salvation, was very real in the minds of the "believers", even those whose lives were not exactly a model of piety. It is told of Raymond VI of Toulouse that once when he fell seriously ill he had himself carried home as fast as possible, because there were no "good men" at whose hands he could receive the consolamentum in the region where he happened to be at that moment, on the other side of the Pyrenees.[55] At dangerous times, when "good men" could not be sure of arriving in time at the bedside of the sick, there was a contract which acquired validity at such moments; the believer could rest assured that the consolamentum would not be denied him on his deathbed even if his sickness or battle wounds robbed him of the powers of speech and clear memory. During the last days of Montségur this *convenensa*[56] acquired a very tragic significance.

The life of the "believers" was one of devoted service to the "church of God"; this church had its material needs which had to be met from gifts and legacies made by wealthy members or from collections made among the "believers" and the supporters of the sect. The journeys to far-off sister churches, like the frequent pilgrimages "ultra montes" and "ultra mare" mentioned in the early days of the Cathar churches in Lombardy, cost a great deal of money. We have already seen that the bad times which began for the sect after the introduction of the new system of the Inquisition, inevitably led to a division of labour among the faithful. Some were given the exclusive task of caring for the financial side of the sect's affairs, to find and keep money for the church to care for its "good men" and sometimes to be able to save their lives.[57] Sometimes even the "perfect" members of the sect were forced by evil days to insist on getting what was due to them. They even refused to administer the consolamentum to heirs who had refused to pay the church money left it by their parents. The records of the Inquisition contain references to a case in which the sect refused help to a member who was in prison, until he gave guarantees that he would return money embezzled from their funds.[58] With the ardour of a convert Rainerio Sacchoni sternly criticizes his former fellows for their miserly seeking after money, but does not deny that this reputation for miserliness among the "perfect" was the result of the constant danger threatening them. The people who gave them shelter ran the risk of having their property confiscated and their houses razed to the ground, and they were not usually willing to run this risk without guarantees of substantial financial compensation.[59]

Both Catholic and Waldensian polemists accused the Cathars of profit-seeking;[60] the records of the Inquisition tell not only of Cathar investment in landed property, but also of loans at interest, and mention properly kept account books.[61] All this business activity was in the interests of the church alone, however, and was necessary to defend its very existence in a hostile world. It was certainly not carried on for individual enrichment. We shall see below that only in Bosnia, where the Patarene church grew up in exceptional circumstances, was the relation of the heretics to worldly wealth less simple. In the West the "perfect" were consistent in leading lives of apostolic poverty; the true believers, too, served the church of the "good Christians" without thought of gain and devoted their whole being to it. Even when the remnants of this "ecclesia Dei" were driven from place to place, helpless and unprotected, we see a tiny group clinging to those who were still for them the bearers of "good".

There were of course many believers whose inner bonds with the Cathar church were much looser,[62] although they gave it their good will and also expected it to provide them with salvation for their souls on the threshold of eternity. The main body of the faithful, however, was moved by profound religious feeling.

The religious life of the sect is truthfully reflected in its ritual. The authentic texts which have survived are a noteworthy confirmation of the accuracy of the various descriptions of Cathar rites in polemical treatises, Inquisitors' handbooks and the records of the Inquisition trials themselves.

Today, besides the Occitan version of a Cathar ritual in the Lyons manuscript which has been known for a long time, there is a Latin fragment in a manuscript found in Florence, which also contains the Cathar Book of Two Principles. This precious fragment presents a more highly developed version of the two principal rites of the Cathars, the *traditio orationis* and the *consolamentum*. Analysis of the language points clearly to an Occitan original for this Latin text.[63] There can be no doubt, however, that in the West the principal texts and formulae used in Cathar rites were from the outset spoken in Latin. This can be deduced from the fact that even the Occitan text leaves in Latin those elements which are common to more than one rite. They are given together at the beginning of the Lyons manuscript.[64] They include, besides the Paternoster and the first seventeen verses of the Gospel according to St.John, certain set forms of words used on various occasions. As we shall see below, a Bosnian MS of the fifteenth century has preserved for us a Slavonic parallel to the whole of this introductory Latin section of the Lyons ritual.

The circumstances in which the Latin elements of the western Cathar ritual came into being are hidden in mystery; we do not know whether they were modelled on a Greek or a Slavonic text. It is certain, however, that the ritual forms attested in the West had already crystallised, for the greater part at least, in the eastern Christian world. The practice of frequent prayer by day and by night is borne out by the evidence of the earliest reports from the Bulgarian and Byzantine milieus. In all cases the prayer was the Paternoster taught by Christ, and Zigabenus tells us that it was repeated several times running. Special terms were used in the West to distinguish the number of these repetitions, and in our texts we find the words: sembla, dobla, sezena. They do not refer only to the prayers offered by individuals on various occasions,[65] but also to common prayer at various Cathar gatherings.[66] The "elder" pronounced the formula: "Let us worship the Father, the Son and the Holy Ghost", in Latin, sometimes three times over. The others replied in unison: "It is worthy and just".[67] This adoration formula and the response were sometimes repeated both before and after the prayer itself. Yet long before this *Adoremus* ... and *Dignum et iustum est* were heard in the valleys of the Rhine and the Loire, in Languedoc and Lombardy, the same words could be heard in Greek, far away in Asia Minor, at the secret meetings of the sect. The work of Euthymius of Acmonia, which reflects a very archaic stage of the sect's development, gives us a similar description of their form of common prayer to that we derive from western sources. He quotes both these formulae word for word.[68] The strange pamphlet by a Byzantine monk thus provides invaluable proof of the great age of the Cathar ritual.

In the West the Cathars joined together in prayer on the most varied occasions, from the solemn moment when the consolamentum was being administered, to common mealtimes. There was an exact ritual prescribed even for this everyday occasion, and the sacred act of the breaking of bread was never omitted. Before the meal began the "elder" took the bread in his hands in a white cloth, and after blessing it by reciting the Lord's Prayer over it, distributed it among all present. For the "perfect" it was the "bread of holy prayer". Believers called it "signed bread" (panis signatus), apparently because they still remembered the practice of distributing in church pieces of bread which had been blessed with the sign of the cross. The Cathar *fractio panis* was undoubtedly modelled on this "eulogia", which was a widespread custom in the church both eastern and western.[69] We do not know whether the sect adopted this custom in the West, or whether they had already practised it in the East. In Byzantine sources there are clear references to meals eaten in common.

The actual gatherings of the Cathars for the purposes of worship are described in the records of the Inquisition.[70] They opened with the reading of a passage from the New Testament, followed by its interpretation in the form of a sermon. This always stressed the points on which the Church of Rome had departed from the pure teachings of the Gospels. After melioramentum by the believers the whole gathering recited the Lord's Prayer. The elder in charge of worship recited the Adoremus ... to which the others responded with "May the grace of our Lord Jesus Christ be with us", and the melioramentum was again repeated.

This gathering for worship, held by the Cathars regularly once a month, was often combined with common confession.[71] This practice went back in origin to the earliest days of the sect, for Cosmas the Presbyter already noted that the heretics "confessed to each other".

The gatherings of the Cathar communities drew not only the "perfect" but the believers as well. This was already the case in Byzantium. The epistle of Euthymius of Acmonia and the related abjuration ritual text show that new adherents of the sect, even though they had not yet been initiated by the spiritual baptism, were nevertheless bound to the community through their constant contacts with its true members, and common meals with them. After a time they were allowed to recite the Lord's Prayer with them.[72] This important step in the new adept's progress towards complete acceptance among the "perfect" was consecrated by a special rite, the "delivery of the Lord's Prayer" The form of this rite was probably already fixed in Byzantium, but it is attested only later, in the West. Both versions of the Cathar ritual which have come down to us describe it in detail.[73] The solemn occasion was attended by the "Christians" (i.e. the "perfect") who had decided beforehand that the adept was fit to receive the Prayer. Believers could also be present. The ceremony was conducted either by a member of the Cathar hierarchy (ordinatus) or an "elder" (ancia, ancianus).[74] One of the "perfect" (in the first example the "elder" himself) bowed three times, and then brought out a portable table or lectern for the Gospels, which he covered with a white cloth. He placed the Book on it, repeating the obeisance three times at each stage, and finally recited the customary Latin formula: Benedicite, parcite nobis (Bless and forgive us). After the melioramentum the candidate received the Book at the hands of the "perfect" conducting the ceremony, and listened to a lengthy homily on the meaning of the Lord's Prayer. Both our texts contained models for this homily, but of that in the Latin fragment only the last few sentences remain. The Florentine version, on the other hand, also includes an explanation of the text of the Prayer itself, which the believer then repeated word for word. Finally he was solemnly exhorted to fulfil from then on the duty of constant prayer, repeating the Paternoster by day and by night, alone and in the company of others, and never to take food without first reciting the Prayer. The candidate responded with the words: "I receive it (i.e. the Prayer) from God, from you and from the church"; he then repeated the melioramentum, gave thanks, and the ceremony concluded with communal prayer by all present, "perfect" and "believers".

In the words of the Cathar ritual, the Prayer was delivered to the candidate in a state of abstinence (astenenza), which must have been the period of trial prior to the administration of the *consolamentum*. This *traditio orationis* was certainly a necessary requirement before spiritual baptism could be administered. To those on their deathbed the Prayer was first delivered, followed by the rite of the consolamentum.

The writings of ecclesiastical polemists, beginning with Eckbert of Schönau, as well as the records of the courts of the Inquisition, have given many descriptions of this most solemn rite, which marked the believer's entry into the company of the "good men" filled with the Holy Spirit. These descriptions complement each other and present some significant features which are not to be found in either the Lyons or the Florence text.[75] Basically the formulas given in both versions are in accord, but it is nevertheless possible to see certain differences. (We have already noted that the Latin version represents a more highly developed stage of this Cathar rite.) We shall see below that in one important point the manner of administering the consolamentum in the "Albanensian" church differed from the practice of the other churches. There is also no doubt that the form of spiritual baptism inherited from the East was modified in the West in certain details. It is not easy to trace these changes exactly, since the information about the rite which can be gleaned from eastern Christian sources is scanty. We have only the curiously cryptic description given by Euthymius of Acmonia, the very brief account by Zigabenus, and a few remarks in anti-Bogomil anathemas.

It is beyond doubt that the rich evidence provided in the West contains not a single document attesting the two phases of the spiritual baptism,[76] to which Euthymius Zigabenus refers. Everywhere in the West the rite of *consolamentum* is a single complete act. The basic elements remained the placing of the Book on the head of the candidate, calling upon the Holy Spirit (epiclesis), recital of the Lord's Prayer and the reading of the opening verses of St.John's Gospel. The ritual act of the laying on of hands, which involved the presence of at least two "Christians",[77] was always performed in the West at the same time as the placing of the Book on the candidate's head. The Cathars of Concorezzo insisted that the actual touch of the hands was essential, while for the Albanenses the Paternoster, recited during the rite of the consolamentum,was the decisive moment. There was no need for actual touch of the hands; it was enough for the candidate for initiation to hear the voice of the "perfect" administering it, even if there was a river or a wall between them.[78] The Byzantine texts make no reference to the symbolic kiss with which in the West the "Christians" and after them the believers brought the ceremony to a close; women received the kiss symbolically being touched by the book and by the celebrant's elbow and then kissed each other. We do not know whether the Latin formula Gracia..., repeated three times in succession at the end of the ceremony can be considered the equivalent of the "hymn of thanks" to which Zigabenus refers. On the other hand, the presentation of a black monk's habit is attested in the West as well. The "perfect" were explicitly called *heretici vestiti, induti*, in the West. The dangers which beset members of the sect later forced them to replace this distinctive apparel by a symbolical cord worn under the clothes.[79] The western sources say nothing of the written pledge which the Byzantine Bogomils, according to Euthymius of Acmonia, demanded from new initiates, and refer only to a solemn spoken vow.

The new member of the sect took upon himself a number of very strict ordinances. Social relationships were affected by the injunction to avoid telling a lie or swearing an oath. The great importance of the mission of the "perfect" and the dangers it involved made it necessary to impose even stricter rules of conduct on them; they were to undertake nothing without prayer, never to travel, pass the night or take food without the company of another "brother"; if the two were separated they were to refrain from food for three days. They were never to sleep unclothed; and finally were never, under any threat whatever, to betray their faith. The rules enjoining personal asceticism included those of never touching a woman, never killing an animal, and never tasting foods which came from animals, such as meat, eggs or milk products.[80]

The "perfect" were allowed only a vegetarian diet; the only flesh they could partake of was that which "came of water", i.e. fish and shell-fish. Even this diet, however, was made stricter by the institution of fast-days, the same days of the week as those observed by the Byzantine Bogomils (Monday, Wednesday and Friday). On these days the "Christians" abstained from wine, oil, and fish or shell-fish. In the West there were also three forty-day fasts in the year, one of which coincided with the customary quadragesima before Easter. The other two also reveal a remarkable coincidence of timing with the feasts of the Church from Whitsuntide to St.Peter's day and from Martinmas to Christmas. In fact these two forty-day fasts corresponded to established Church custom in the West, and were not unknown in Byzantium. During these three long fasts the Cathars were supposed to abstain from the flesh of fish and shell-fish; exceptions were allowed in cases of serious illness. The first week in each of these periods was marked by even greater severity, and so was the last week of the fast before Easter; during these "strict" weeks no wine was to be drunk, no oil used, and even vegetables were forbidden.[81]

The rite of initiation not only bound the members of the sect to live a life of extreme asceticism, but also opened to them some mysteries which had been kept hidden from them up to then. It follows from what Euthymius Zigabenus says that the Byzantine Bogomils did not begin to reveal their "simpler faith" to their converts until after their spiritual baptism, while the final step revealed the "pearls" of the doctrine. Euthymius of Acmonia also declared that the gradual revelation of the "diabolic mysteries" did not begin until after the new adept had received spiritual baptism and sworn to preserve the new faith.[82] Nevertheless, at another point the monk of the Peribleptos monastery reveals that this initiation began earlier; his young charge had in fact been taught some of the sect s secrets, although he had not yet entirely fallen into the Devil's power; the infamous "Satanic" formula had not yet been recited over him (i.e. the epiclesis which was customary at the rite of spiritual baptism).[83] This young man had only "believed" the heretics teachings and taken part in their prayers.[84]

The western sources leave us in no doubt that the *believers* were already instructed in the system of dualist doctrine, although not indiscriminately. A number of matters were only revealed to those who were already well known and had proved their constancy in the faith. We have seen that the teachings of John de Lugio were kept secret from the believers of the "Albanensian" church, for tactical reason. The sect also had its special secret doctrines (disciplina arcana) some of which were not even known to all the "perfect".

Anselm's *Tractatus de hereticis* aptly describes the "believers" as people who are within the church only by reason of their faith, but not yet by initiation.[85] The believers took part in various gatherings and were thus often given the opportunity of listening to the explanations which preceded various rites and were informed with dualist metaphysics.

At first sight there is nothing particularly heretical about the "traditio orationis", which has its analogy in the ancient Christian form of worship.[86] Nevertheless the homily given by the "elder" to the new adept begins with an obvious reminiscence of the dualist myth. "God's people" had turned away from Him, deserted the Holy Father, when they believed the lies of "evil spirits" and fell into their power. But the good God had mercy on His people and wanted to receive them in "His peace and faithfulness" through the advent of His son. The proof of God's mercy is the delivery of the Lord's Prayer before the apostles of Jesus Christ – before the "church of God", in the presence of the Father, Son and Holy Ghost. The acceptance of this prayer, which Jesus handed on to His disciples, makes it possible for the human soul's prayers to be heard by God the Heavenly Father.[87] (According to the secret book of the Concorezzenses, when they fell the angels lost the right to praise God in the words of the Paternoster, which previously had risen to the throne of the "invisible Father" as a hymn of praise.)[88]

When the Lord's Prayer was handed on to them every sentence was explained to the believers in the sense of the dualist faith. The Latin ritual in the Florence MS presents such an explanation in connection with the rite of "traditio orationis".[89] Another commentary on the Prayer appears on its own in a manuscript in the library of Trinity College, Dublin. In the Dublin glosses the simple original meaning of this basic Christian prayer is veiled in abstract symbols, so that the seven requests appear in the concept of "seven substances".[90]

The actual words of the Prayer were quite orthodox in every case,[91] although it is a different variant which was used in the West from that customary in the East. In the churches of Byzantium the sect was loudly anathematized for reciting the Paternoster without the final doxology,[92] while the western Church accused them of the opposite.[93] Indeed in all the lands which came into the sphere of the Church of Rome, including Bosnia, the Cathar communities used the Lord's Prayer with the doxology given in St.Matthew s Gospel (Matthew 6,13). Indeed, as we have seen in Byzantium, the sect did not change the wording of Biblical texts. The Latin Bible quoted by the Cathars of Languedoc, as well as the later Occitan version of the New Testament, preserved in the Lyons manuscript, are in absolute agreement with the Vulgate text.[94] Over every sentence in the Scriptures, however, there hung a secret meaning with which the sect invested it. This changed the authentic Biblical tradition, turning it into something else, in the spirit of a doctrine essentially alien to that of Christianity. And yet, like the Valentinians and other Christian Gnostics, the dualist sect of the Middle Ages sincerely believed that they were discovering the hidden "true" meaning of the Christian message. They saw themselves only as "Christians", "true Christians", and they called themselves by the name everywhere in the West as in the East.

Nor can their Christian feeling be denied, although their religious ideas remained within the limits of the dualist myth of the soul as a prisoner in the earthly body. We read in the form of common confession used by the Cathars, preserved in the Lyons manuscript:[95] "We have come before God and before you and before the congregation of the holy church, to make confession and receive absolution and penance for all the sins we have committed in word, thought or deed, from the day of our birth until this moment. We beg for mercy from God and from you, that you may pray for us to our Holy and merciful Father. For our sins are many, by which we wrong God by night and day, in word and deed and thought, of our own will and against it, but most of all through our own will, inspired by the evil spirits in the bodies in which we are clad. Benedicite ..." And again: "Judge and condemn, oh Lord, the wickedness of the body; have no mercy on the body, born of iniquity, but have pity on the soul shut in its prison..."

This gives us a vivid idea of the spiritual force of Catharism and of its true significance in the intellectual history of the Middle Ages. In spite of the change in its attitude to secular learning the sect contributed nothing to the development of philosophical thought even in the West. There, as in the East, the mental world of the sect remained enclosed in a naive mythology. While some aspects of the myth are not without poetic charm, they never found the artistic expression which was so often inspired in late antiquity by the Gnostic drama of the soul. The mediaeval dualist sect left behind no works of art or literature that can compare with the Mandaean hymns or the Manichaean illuminations. Nevertheless the Cathar ritual texts are still powerful in their moving religious feeling.[96] These authentic expressions of the attitude of the sect certainly effectively refute suggestions that they were merely in the thrall of ritual magic.

There remains the objection that the efficacy of the consolamentum could at any time be "mechanically" endangered from without, and was not therefore dependent on the moral qualities of the person to whom this spiritual baptism had been administered. We know very well what uncertainty and confusion arose from this fact. It was for this reason that almost all the Cathar churches administered the consolamentum not once only, but twice or even three times.[97] This solution, while undoubtedly also a "mechanical" one, did give some security against the fateful consequences of a moral lapse on the part of the Cathar dignitary administering the consolamentum. In the last degree, it would seem that the significant element was after all the individual receiving the consolamentum. If he fell into *mortal sin* later, the only way to attain absolution was to receive the consolamentum once more. Before this could be done, however, he had to undergo very strict fasts, which were of longer duration if his sin had been publicly committed. In any case, the sinner was never again allowed to take part in the laying on of hands during the rite of consolamentum administered to new members of the sect, except in cases of extreme necessity.[98]

There were various penances for *minor lapses*, as we learn both from ecclesiastical writers and from the form of confession. The common confession (servitium or apparellamentum) for which this form was intended took place once a month and concerned the minor lapses of everyday life. The confession is proof of the sect's unremitting efforts to achieve the perfect Christian life, which was not the automatic consequence of having received the consolamentum, but depended on the strength of will of each "Christian".[99]

This strength of will was tested frequently in the dangers and deprivations of the life led by the "perfect". At the conclusion of the rite of consolamentum the new member was reminded that from then on he would be "as a sheep in midst of wolves", as Jesus reminded his apostles.[100]

We know that the sect had brought with it from the Byzantine East an ecstatic belief in its apostolic mission and the vivid awareness of continuous apostolic tradition. In the West the Cathar communities had long had a series of proofs prepared, showing that the title of the apostolic church belonged by right to them, and not to the Church of Rome. In contrast to the "blessed" church of the Cathars (ecclesia benigna) they pointed to the *ecclesia maligna*,[101] whose features could already be seen in the prophetic visions of St.John's Apocalypse. It takes the form of a beast rising from the abyss, and of a whore drunken with the blood of the saints – the symbol of that great city "which reigneth over the kings of the earth" – the woman whose name was Babylon.

The image of the Pope as the head of the Roman Church is that of the woman sitting upon a beast, arrayed in purple and decked with gold and precious stones and pearls, having a golden cup in her hand. This interpretation of Chapters 16 to 19 of the Revelation of St. John was common to the Cathars and to the Waldensians of the Lombard branch.

All these attacks on the Church of Rome echo the mood of opposition to the Church and sharp criticism of it, widespread not only among the heretics but in orthodox circles as well. It was partly under the influence of these attitudes that the Cathars developed their antithesis between the Church of God and the Church of Rome, evidence of which was collected in Lombardy by Moneta of Cremona.[102] The tendency towards such a confrontation of the two churches can already be traced among the Byzantine Bogomils, however, in the eleventh and twelfth centuries. In their eyes the Orthodox Church merged with Herod, the Pharisees, the Sadducees and so on, while they applied to themselves the Beatitudes and Christ's references to His apostles as the "salt of the earth" and the "light of the world".[103] This was the spirit in which the Bogomils and before them the Paulicians interpreted the Gospels and such other books of the Bible as they recognized.

This negation of the official church, although it was in all cases conditioned by the times, was fundamentally a continuation of the tradition maintained by the sectarian groups in Byzantine Asia Minor and primarily by the Paulicians. It was certainly from them that the Bogomils took their idea of apostolic succession. In the West this idea was further developed under the influence of earlier native sects. The Italian Cathars (and also the Lombard Waldensians) took over from the Arnaldists (followers of Arnold de Brescia) a curious view of the famous donation of Constantine. According to this theory, which first emerged at the time of the investiture conflict, the apostolic succession was broken in the Roman Church by Pope Silvester. This successor to the throne of St.Peter accepted at the hands of the Emperor Constantine the emblems of secular power. This universal power, however, had in the past been acquired by Julius Caesar through violence (per rapinam) and thereafter remained in the hands of the Roman emperors - the men who persecuted the church of Christ. By accepting this sinful heritage the Christian church was brought to ruin. In this light Silvester appears as the Antichrist, as the "filius perditionis". He and the other popes after him were no longer the successors of the apostle Peter, but only of Emperor Constantine.[104]

From here the Cathars went on to conclude that the apostolic succession broken in Rome continued undisturbed only in their church. It was therefore logical to restore to this church, the true Church of

Christ, a supreme head as successor to St.Péter and the Roman Popes of the pre-Constantine era.

There is abundant testimony to the institution of a Cathar "Pope" in the western sources, of which the most significant is the letter of Conrad da Porto mentioned earlier (1223). This heresiarch referred to by the papal legate as arriving in Languedoc was only the deputy of the "Anti-Pope", whose seat was somewhere in the vicinity of Hungary. The region indicated in the legate's letter can most readily be identified as Bosnia. This is all the more probable as only two decades later a letter of Pope Innocent IV (19. VIII. 1245) refers to a Cathar Anti-Pope in Bosnia. Much later, reports from among the Franciscans in Bosnia explicitly say that the heretics there consider one of their number to be "the Vicar of Christ and successor to St.Peter".[105]

In the Middle Ages Bosnia was in the sphere of interest of the western Church, and the "Patarene" heresy there shows the same specific features as were characteristic of the sectarian movements in the Roman Catholic world. These were partly the legacy of earlier heretical movements in western Europe. They included the Cathar view of infant baptism and the idea discussed above, based on a curious interpretation of the donation of Constantine and leading to the attempt to restore the apostolic succession. Apart from this, the western branch of the sect differed from the eastern in the version of the Paternoster used (complete with doxology) and a different attitude to the drinking of wine.

It cannot be too strongly emphasized that these marked differences nevertheless *created no barriers* between the western and eastern groups. Different variants of the dualist myth were accepted in northern France, Languedoc, and Italy, just as in "Sclavonia", Bulgaria or "Dragovitsa" As we have seen, Rainerio Sacchoni noted that not only the majority of the Cathars of Languedoc, but also "almost all" of the churches "across the sea" (ultra mare) believed in the original form of the radical dualist conception, like that followed by the older generation of the Albanenses under Belesmanza..[106] Rainerio noted a significant change among the western Cathar churches, however, in the manner of securing the succession of their bishops. According to the original custom the "younger son" (filius minor) ordained the "elder son" in episcopal office on the death of the bishop. A new "filius minor" was then elected, who was ordained by the new bishop. In the course of time the Cathars "on this side of the sea" (citra mare) became dissatisfied with this system, finding it a serious fault that the son (filius minor) thus ordained the father (the bishop). They therefore introduced a new manner of ordination; the bishop during his lifetime ordained his "elder son" as his successor. The latter then automatically became bishop on his death, while the "younger son" was elevated to the rank of "elder son" and was ordained bishop the same day, as successor-designate to the new occupant of the bishop's throne. A new filius minor was then elected and ordained in the manner hitherto followed. Thus in the days of Rainerio Sacchoni the Cathar church usually had two bishops, which is why John de Lugio signed his letters "filius maior et ordinatus episcopus".[107]

It follows indirectly from what Sacchoni says that the churches "across the sea" retained the old custom. There is no doubt that the sect established its hierarchical system in the early days in the Eastern Christian world, although it is not attested until later, in the West. The bishop of Concorezzo, Nazarius, as a young man had met the *bishop and "elder son"* of the "Bulgarian" church.[108] It can also be assumed that the institution of deacons, for which there is abundant evidence in western sources, was taken over from the mother-churches in the East. The deacons headed the community (conventiculum) of brothers and sisters in each town; they led the common confession once a month and ran the houses (hospitia) which gave shelter and hospitality to travelling "brothers".[109]

The rank of bishop and his two "sons", as well as that of deacon and sub-deacon, came under the heading of "ordo" and were bestowed by the laying on of hands, like the sacrament of ordination in the official church. The members of the Cathar hierarchy were thus ordained *(ordinati)* and with respect to the other men and women "Christians" they were *praelati*. Besides the "praelatio" the Cathar communities also had a *priorate*, i.e. members in a "foremost" position, granted on account of age and personal qualities.[110] Closely connected with the latter there was the position of "elder", *ancianus*, *ancia*, which is mentioned in both versions of the ritual and in various descriptions of Cathar rites. The "elders" occupied a very important place; not only did they head the common table (with responsibilities of prayer and the breaking of bread), but they like the *ordinati* also enjoyed the privilege of conducting the rite of "delivery of the Prayer" and the administration of the consolamentum. It was an *ordinatus* or an *ancianus* who laid the Book of the Gospels on the head of the candidate, while the other "perfect" present took part in the laying on of hands. As a rule, at least two "Christians" were necessary for the rite to be performed, but they did not need to be members of the hierarchy.[111]

The western sources occasionally refer to "presbyters" and there is an isolated reference to the title *filiolus*.[112] In the fourteenth and fifteenth centuries we no longer find the institution of the two "sons" of the bishop, nor of the deacons, in the "Bosnian church";

instead there is a different hierarchy which seems to correspond to the different path followed by the development of the Patarene sect in this country.

In the time of Rainerio Sacchoni, as in the twelfth century, the western Cathar Churches kept in touch with their mother-churches across the sea, from whom they had originally received their *ordo.*[113] The *Bulgarian* and the *Dragovitsian* churches still survived overseas, and the western Cathars still retained the awareness that all their churches "had come from these two" The *Ecclesia Sclavoniae* is not more precisely defined even by Sacchoni. In his treatise, written somewhat later and complementing that of Sacchoni, Anselm of Alexandria is quite definite in assigning this overseas church to Bosnia. In the days of Rainerio Sacchoni there were two Cathar churches in *Constantinople,* whence "papa Nicetas" had come; one existed among the *Greeks* and one among the *Latins.* It is interesting to see that there was a heretical church in Philadelphia in Asia Minor, one of the cities which the Revelation of St. John gives as the seats of the seven churches. It is not unlikely that the Cathar *ecclesia Philadelphiae* marks the continued existence of one of the original "seven churches of Asia" which carried on the tradition of the seven churches of the Apocalypse.[114]

Rainerio Sacchoni is the only authority to give us information about the numerical strength of the Cathar churches, both in the West and "across the sea". The figures represent the number of "perfect" members only, for the term Cathar or Patarene was reserved for them, and they called themselves "Christians" or "good people". The figures Rainerio gives us in his Summa are very valuable, in that they were provided by the Cathars themselves; the author adds that in early times the census was taken more frequently. We may assume that Rainerio knew the figures yielded by the last census from the time when he was still a member of the sect (up to 1245). He gives the total number of "Cathars of both sexes" as not even 4,000. It is significant that the number of members of the "Latin" church in Constantinople (50) is given separately, while the strength of all the other "overseas" churches is given in the approximate round figure of 500. In Italy itself, there were two churches stronger than the rest; the Concorezzenes numbered over 1,500 and the Albanenses about 500. At this time they were still hostile to each other. Far below them came the church of Bagnolo with about 200 souls. These three most important churches, described by Rainerio as the "three chief sects", were scattered over a very wide area. The Concorezzo church had members throughout Lombardy, as did the "Albanensian" church with its centre in Verona. Besides Mantua, the Bagnolo church had its supporters in Brescia, Bergamo and the contado of Milan, as well as a smaller number in Romagnola. In Sacchoni's day there was a Vicenza church in northern Italy which had already been mentioned in the De heresi Catharorum. This *ecclesia de Marchia,* as Rainerio calls it, was widespread in the Treviso March as far as the Alps, and numbered a hundred "perfect" members. Two churches in Central Italy, also mentioned in De heresi, numbered about the same taken together. The Florentine church (which Rainerio simply calls *de Tuscia*) whose boundaries were set by a document of 1227[115] was in Tuscany. (We know that when the Cathar churches in the West delimited their mutual boundaries exactly they were following the example of the "original" churches [ecclesiae primitivae] in Asia Minor and the Balkans.) The other Central-Italian church was the one called by Rainerio (ecclesia) *de Valle Spoletana.* It seems to have covered the area of Roman Tuscany with such important centres of the Cathar faith as Orvieto and Viterbo.

The *ecclesia Franciae* was settled in Verona and various places in Lombardy in Rainerio's day. This was a relatively numerous exiled church, with 150 "perfect" memebrs.[116]

Unlike the French Cathars the Occitan branch still survived in its own country. Rainerio numbered it at 200 souls, which covered the church of *Toulouse,* the *Albigensians* (Albi) and the church of *Carcassonne,* together with the remnants of the church of *Agen,* then "practically destroyed".[117] The situation of the Cathars in Languedoc was more desperate every day, but the time of their mass flight to Lombardy did not come until some twenty years later.

[1] Guill. Podiol., t. 19, 200.

[2] *Thouzellier,* Languedoc, 381 f.; Cathares en Languedoc, 42 f.

[3] See the Liber contra Manichaeos. For the text of the Cathar tract separately, see *Thouzellier,* Un traité cathare.

[4] Ed. *J.N.Garvin - J.A.Corbett,* The Summa contra haereticos ascribed to Prepositinus of Cremona (Mediaev. Stud., XV, Univ. of Notre Dame), Notre Dame (Indiana) 1958.

[5] On the author, see *Dondaine,* Manuel, 174 f.

[6] See *Guiraud,* Inquisition I, XXI f.; cf. Borst 17 f.

[7] See *Dondaine,* Manuel, 92 f.

[8] On Rainerio Sacchoni see *Dondaine,* Liber, 57 ff.

[9] For these handbooks and "summas" see *Dondaine,* Manuel.

[10] See *Dondaine,* Hiérarchie II, 283 f.

[11] Tract. her., 312,8-11. For further references see *Borst,* 149, n. 21.

[12] Tract. her., 311.

[13] Rain. Sacch., 71 f.

[14] Moneta, 75.

[15] See Chapter XVI, Note 17: referring to the letter of the Patriarch Cosmas and the relevant passage of the Bulgarian Synodicon.

[16] De heresi 311,18-19, 34-35; Tract. her. 311,24 - 312,13; Rain. Sacch., 76.

[17] Rain. Sach., 76.

[18] Manifestatio haeresis Albigensium et Lugdunensium, ed. according to the Rheims MS 495, ff. 135r - 137v by *Dondaine*, Durand de Huesca, 268 - 271. Peter Sarn. I, 9 - 13. Cf. *Thouzellier*, Languedoc, 284 - 291.

[19] See *Dondaine*, Liber, 79 - 147; *Borst*, Anhang III, 254, 318.

[20] Salvo Burci, *Doell.* II, 53 f., *Ilarino*, 309. Rain. Sacch., 77.

[21] Liber contra Manichaeos, 138 - 139; 211,1-2.

[22] Rain. Sacch., 77; cf. ByzSlav XXVIII (1967) 52 f.

[23] Rain. Sacch., 63.

[24] Ibid., 77.

[25] Tract. her., 313,3.

[26] *Dondaine*, Hiérarchie II, 274 f.

[27] Rain. Sacch., 76; cf. already Salvo Burci, *Ilarino*, 309.

[28] De heresi, 311,30-33.

[29] Rain. Sacch., 64 f.

[30] See Chapter IX, p.115.

[31] Peter the Martyr, 330.

[32] In support of their contention that it is not possible for the married to attain salvation, they called upon St. Augustine as well as the Gospels. See Praepositinus of Cremona, cap. IV, ed. Garvin-Corbett, 66. The traditional story of Peter Autier's conversion as told among the last Cathars in the south of France is reminiscent of a similar happening in the life of St. Augustine, and may have been a deliberate echo. See Vidal, Derniers ministres, 64. For the idea of Civitas Dei see Chapter XVIII below, p.288.

[33] The Cathars could have learned of it from Augustine's Contra Faustum.

[34] Peter the Martyr, 331.

[35] Cf. *Borst*, 184, n. 16; cf. also 186 f., n. 26.

[36] See Chapter IX, n. 80.

[37] *Everw. - Steinf.*, col. 678D: "Nuptias damnant, sed causam ab eis investigare non potui..."

[38] See Chapter XI, p.137.

[39] See the report by Heribert on the heretics in the Périgord region: "...vinum non bibunt, nisi permodicum tertia die..." (Herib. Mon., PL 181, col. 1721). Other western sources do not mention the rejection of wine by the Cathars, and we have explicit evidence to the contrary. (Borst, 183, n. 14.) See Moneta, 138 f. (Cap. V. De usu ciborum, carnium, sicut et ovorum. Pars prima, I. Probatio haereticorum, quod esus quorundam est illicitus). The heretics quoted St. Paul, Romans 14,21. Moneta objects: Eadem ergo ratione illicitum est vinum, et deberent ab eo abstinere. Ad hoc autem dicunt, quod secus est de vino, quia illud postea Apostolus concessit Timotheo ... (referring to 1 Timothy 5,23).

[40] Referring to the Commandments of John Chrysostom; see *Solovjev*, Svedočanstva, 33 f. (cf. Chapter XVI above, p.232).

[41] See Atti dei Convegni (Acad. Naz. dei Lincei) XII (1956), discussion: *Hönerbach*, 149; *Puech*, 156.

[42] See Rain. Sacch. (listing communes opiniones), 64: "...quod diabolus fecit hunc mundum et omnia quae in eo sunt."

[43] See Chapter IV, p.56.

[44] Rain. Sacch., 76; Tract. her., 311,14-24; see ibid., 312,11-16, on quotations from the Bible concerning marriage: Nazarius and his followers, like the Bagnolo and Albanensian churches, give them a purely spiritual interpretation. Desiderius takes these passages literally, but with the important reservation that they are intended only for the "believers", i.e. those who have not yet been initiated and are not therefore bound by the vow of celibacy. Moneta records the Cathar view of the body as a mere instrument (organa, instrumenta animarum); eternal remuneration applies only to the soul, not to the body. From the context this view must have been heard in the ambience of Desiderius. Moneta reproduces the words of Desiderius: "quod corpus nullam remunerationem haberet, quia nihil voluntarie fecit, sed coacte" (Moneta, 346 f.).

[45] See Chapter IX, p.114 sq.

[46] Moneta, 292.

[47] The Cathar Latin ritual goes furthest in this respect, in the homily before the administration of the consolamentum. See the Florence Ritual, 163,34.

[48] See Chapter XI, p.140

[49] Rain. Sacch., 67 f.

[50] Ibid., 66.

[51] Valuable objective evidence on this subject is provided by Jacobus a Capellis. See *Dondaine*, Nouvelles sources, 479. On the "Summa" of Jacobus a Capellis see id., Durand de Huesca, 265.

[52] Peter Sarn. I, 15, No 13. Cf. *Griffe*, Languedoc, 56.

[53] *Lucie Varga*, in: Rev. de synthèse 11 (1936) 139.

[54] Melioramentum: *Borst*, 198 f. On the idea of "evil death" and "a good end" see the dèfinition in Bern. Gui, 20: "Per malam mortem dant intelligere heretici mori in fide Ecclesie Romane; per bonum autem finem et per manus fidelium christianorum dant intelligere quod recipiantur in fine suo ad sectam et ordinem ipsorum, secundum ritum eorum; et hoc dicunt esse bonum finem."

[55] Peter Sarn. I, 40 f.

[56] Convenensa (conventia): Bern Gui, 20. For other evidence, see *Borst*, 199, n. 26.

[57] See Chapter XIV, p.200 sq.

[58] *Dossat*, Les cathares d'après les documents de l'inquisition, in: Cathares en Languedoc, 90.

[59] Rain. Sacch., 68. For a criticism of Rainerio's evidence, see *Molinier*, Église et société I, 246.

[60] According to Joachim de Fiore, people who were blinded by money would join the sect because it gave financial support to poor members. For quotations from his "In Apocalipsim" see *Thouzellier*, Languedoc, 116. The Liber Antiheresis accuses the Cathars of making money by trade. Ibid., 70, n. 88.

[61] References to the sources in *Borst*, 126, n. 24.

[62] For criticism of Rainerio's evidence on the immorality of the Cathars, see *Molinier*, Église et société II, 5.

[63] For Occitan linguistic features see *Borst*, 281, n. 7, 8. Linguistic analysis leads the author to conclude that there was an Occitan model for the Florence ritu-

al (see *Borst*, 280) and thus to disagree with *Dondaine* (Liber, 35) who believes the Latin version to be the original and to date perhaps from early Christian times.

64 Complete groups of Latin sentences, as they occur in the forms for the different rites, have been incorporated into this introductory section of the Lyons ritual (pp. VI - VII). Group 1: Benedicite ...; Fiat nobis ...; Pater ... parcat vobis ...; Adhoremus ... III vegadas (i.e. thrice). Group 2 (after the text of the Lord's Prayer): Adhoremus ... III vegadas; Gracia ...

Group 3 (the same as Group 1, but without the triple Adhoremus). - Group 1: see the rite for administration of the consolamentum to the sick (p. XXV). Group 2: see the normal consolamentum rite (p. XXI).

65 Lyons Ritual, XXI f.

66 See Tract. her. 314,40: "...Et statim faciunt duplam" (dupla = the Latin form instead of the Occitan dobla). Ibid., 314, 316, gives the various numbers of prayers recited at common prayer.

67 The Latin formula "Dignum et iustum est" is attested only in Tract. her. 314,9; 316,15,18. (This same formula occurs in the introductory response to the anaphoras of the Church liturgies in East and West.)

68 Euth. Acm., ed. *Ficker*, 77,19-21: Ἵσταται ὁ πρόκριτος τῶν ἀσεβῶν καὶ ἄρχεται λέγων· <u>προσκυνοῦμεν πατέρα καὶ υἱὸν καὶ ἅγ</u>ιον πνεῦμα καὶ ἀποκρίνονται οἱ συνευχόμενοι· <u>ἄξιον καὶ δίκαιον</u> (our underlining).

69 See *Borst*, 201, n. 31.

70 See *Schmidt* II, 115 ff.; *Guiraud*, Inquisition I, 152 f.

71 *Borst*, 199 f., n. 28. The Occitan *apparellamentum* is the equivalent of the Latin *servitium*.

72 For the text of the abjuration rite: *Ficker*, Phundagiagiten, 173: Τὸν φθάσαντα τούτοις συναναστραφῆναι καὶ προσομιλῆσαι καὶ συνεστιαθῆναι καὶ καταδέξασθαι τὴν αἵρεσιν, μήπω δὲ τὴν ἐπῳδὴν παθόντα

Euth. Acm. ed. *Ficker*, 85,10-17; l. 15: συνηύξατο μετ' αὐτῶν τὸ πάτερ ἐκείνων

73 Lyons Ritual pp. XI - XV. Florence Ritual, pp. 37 - 39.

74 Unlike *A.Dondaine* (Liber, 43), we believe that *ancia* and *ancianus*, the Occitan and Latin forms respectively, signify in both the rituals the *elder* of the Cathar community (for this term see below). The *ordinatus*, on the other hand, was a member of the Cathar hierarchy, whose rank was bestowed by ordination. The two rituals thus differ not in terminology, but in the form of the rite itself; in the Occitan rite the leading part is played by an *elder*, but in the Latin rite by a member of the hierarchy. These were two alternatives, as Rainerio Sacchoni suggests, at least for the rite of the consolamentum.

75 For references to the sources for the consolamentum see *Borst*, 193 - 196.

76 Some scholars identify this rite of traditio orationis, attested in the West, with the first phase of the spiritual baptism as described by Euthymius Zigabenus. See *Dondaine*, Liber, 34, and after him *H.-C.Puech*, in: *Puech - Vaillant*, Le traité, 257, n. 1. In fact, however, this first phase in the initiation ceremony of the Byzantine Bogomils includes the laying of the Book on the candidate's head, the reading of the opening verses of St.John's Gospel, the call to the Holy Spirit (epiclesis) and the recital of the Paternoster - all the essential elements of the consolamentum except for the laying on of hands.

77 Rain. Sacch., 63.

78 Tract. her. 313,35 - 314,2. One of the anathema articles in the letter of

Patriarch Cosmas refers to the Bogomil baptismal rite, performed by means of the Lord's Prayer alone. (See the text edited by *Gouillard,* in: Travaux et Mémoires 4 (1970) 371, 1.40. Cf. the Sinodik of Boril, Palauz., No 47.

79 *Schmidt* II, 127. For Byzantine evidence of the handing over of raiment during the rite of spiritual baptism see Chapter V, n. 32.

80 *Schmidt* II, 127.

81 For Cathar fasts see *Borst,* 183 - 185. Cf. *H.-C.Puech,* in: *Puech - Vaillant,* Le traité, 243 f.

82 Euth. Acm., ed. *Ficker,* 30,12-22.

83 See the passage in Euth. Acm. already quoted, ed. *Ficker,* 85,10-17; 1.15-16: καὶ ἐθάρρησαν αὐτῷ μυστήρια ...

84 Ibid., 1.14: ... εἰ μὴ μόνον πεισθεὶς εἰς τὴν αὐτῶν διδασκαλ(ε)ίαν. Cf. the text of the abjuration rite, Ficker, Phundagiagiten, 173,1-2: Τὸν φθάσαντα τούτοις συναναστραφῆναι ... καὶ καταδέξασθαι τὴν αἵρεσιν (our underlining). These two references to believing the heretics' teachings and "accepting" heresy are quite in keeping with the western definition of the class of "believers". (See n. 85 below.)

85 Tract. her., 312,14-16: "...illis qui sunt intra ecclesiam fide tantum sed extra sunt professione, idest credentibus suis."

86 See *L.Eisenhofer,* Handbuch der katholischen Liturgik II, Freiburg i. Br. 1933, 248 f. Cf. Borst, 192, n. 7.

87 Lyons Ritual, XII - XV.

88 Interrog. Ioh., 83.

89 Florence Ritual (after a homily of which only the end is preserved). *Dondaine,* Liber, 47.

90 See *Gonnet,* in: Cathares en Languedoc, 59 - 66.

91 Examples of the reading: panem nostrum supersubstantialem (instead of quotidianum) occurring in the Catholic tradition, are quoted by *Borst,* 191, n. 6.

92 See *Tollius,* art. 7; cf. the anti-Bogomil anathemas incorporated into the recension of the Synodicon for the Sunday of Orthodoxy intended for one of the bishoprics of the Athens Metropole (Synodikon, 11. 336 - 339, p. 67).

93 See Moneta, 445. Cf. *H.-C.Puech,* in: *Puech - Vaillant,* Le traité, 248, n. 1; *Borst,* 191 f., n. 6.

94 See *Thouzellier,* in: Cathares en Languedoc, 42 f. The New Testament text preserved in the Vienna MS 1137 together with the Interrog. Ioh. is also identical with the text of the Vulgate.

95 Lyons Ritual, X.

96 *A.Dondaine* (Liber, 46) stresses "leur tenue sobre et archaïque".

97 Rain. Sacch., 70.

98 Tract. her., 315, rubric 6: De dispositione penitencie catarorum.

99 For penance given for common sins see ibid., 11. 18-20; Rain. Sacch., 69.

100 Et dicunt ei: "A modo eris inter nos et penitus in hoc mundo sicut ovis in medio luporum". Tract. her., 314,39-40.

101 See *Borst,* 215, n. 8.

102 Moneta, lib. V, cap. I, 389 f.

103 Euth. Zig., ed. *Ficker,* 103,13-15; 104,24-26; 105,31-32 = PG 130, col. 1321C; 1324B; 1325A.

104 For the western origin of this legend see *Borst,* 215, n. 9; he also gives references to evidence in the sources to the existence of this tradition among the

Cathars, Waldensians and other heretics.

105 See Chapter XIX below, p. 298.

106 Rain. Sacch., 77.

107 Ibid., 69.

108 Ibid., 76.

109 Moneta, 278.

110 For these concepts see *Mandić*, Bogomilska crkva, 267.

111 Rain. Sacch., 63.

112 Brevis Summula, 121.

113 Rain. Sacch., 70.

114 The name Philadelphia occurs in Anselm of Alexandria, too, in the context of the first beginnings of Catharism (see Tract. her., 308,2-4).

115 *Dondaine*, l. c., 300, n. 65.

116 On the individual Italian Churches and their hierarchy see *Dondaine*, Hiérarchie II, 278 - 305. Cf. *Borst*, Anhang I, 231 - 239, which gives a list of Cathar bishops not only in Italy but also in northern and southern France.

117 Rain. Sacch., 70.

For the Cathar Church of Agen see Chapter X above, n.2. Cf. Chapter IX, n.l.

Chapter XVIII

THE END OF THE CATHAR SECT IN WESTERN EUROPE

After the fall of Montségur the sect in Languedoc, bereft of its most powerful buttress, was thrown into confusion. Nevertheless even round the middle of the thirteenth century "good Christians" there could still rely on a network of "believers" and friends who still successfully resisted the efforts of the Holy Office spies. The records of the Inquisition for these years reflect this situation quite clearly. The Cathar groups were still in frequent contact with each other and the heretics invited and called on each other as they had always done. "Good people" could still move about the country without much difficulty, as long as they observed the utmost caution. They would set out only at night, with experienced guides to take them safely through parts of the country unknown to them. Sometimes a boat had to be prepared, in readiness to take them across a river; or a temporary shelter had to be devised. Local names help us to trace the journeys of these Cathar preachers; their halting places were isolated farms, or even cottages in the village, and everywhere they were taken in by trustworthy people who devoted all their care to them, giving them shelter for the night or even for days at a time. The believers would seek them out in these hiding-places, and gatherings were held in gardens, or divine service in some out-of-the-way farmhouse. Their common faith was nourished by the living word brought by these "good people" and eagerly awaited. One of the Cathar communities kept the Scriptures carefully hidden in a cave, tucked away in a valley, together with three heretical books whose titles we are not told.

In these Inquisition records we still find the names of nobleborn supporters of the Cathars; the sect still exercised a strong influence in aristocratic families, offering material support and a refuge in times of danger.[1] The organs of the Inquisition still had to count with the active dislike of the people of the South, and their records provide clear evidence of this. A manuscript in the Municipal Library of Toulouse records the results of a systematic investigation carried on in 1245-1246 by Bernard de Caux and his companion. Like their predecessors, whose work they took up, these two Inquisitors of Toulouse wanted to get as complete a picture as possible of the extent of heresy in the See. They were not always successful, for people often held back information, or deliberately misled them. In some parts people pledged themselves not to betray anything.[2]

Nor was passive resistance all they had to deal with. Towards the end of 1247 two men in the service of the Inquisition were murdered and the documents they had with them stolen. The records of confessions wrung from their victims by the Inquisitors were a constant threat to large numbers of people, who could find themselves before the tribunal at any moment. It was in the interests of many to destroy the registers of the Inquisition.[3] This incident in Caunes, however, was the prelude to a serious crisis in the work of the Inquisition in southern France. The real cause of trouble was the Inquisitors' objections to the close control exercised over their activities by Pope Innocent IV, who more than once reversed their decisions on his own judgment alone. Finally the Dominicans retired from the work of the Inquisition, which was carried on by the bishops for several years.[4] This temporary crisis could not permanently arrest developments in the South, however, and the course of events was not favourable to the Cathars. After the death of Raymund VII, the last independent ruler of Toulouse (1249), his remaining lands came into the hands of the French Prince Alphonse, as the agreement made in Meaux twenty years earlier had laid down. Besides this, the king of France, Louis IX (Saint Louis), and his brother Charles, married the heiresses to the County of Provence. The political links between the southern provinces and the Kingdom of France speeded up the process of assimilation of the local nobility to that of the north. The next generation grew up in quite a different atmosphere; foreign influences began to predominate and the tradition of southern independence grew fainter.[5]

In these changed conditions Cathar influence weakened, and one by one their safe refuges in the castles of the nobility were closed to them. From 1255 the Inquisition in the County of Toulouse was again in the hands of the Dominicans, who found themselves in a far more advantageous position than they had enjoyed under Innocent IV. Unlike his predecessor, Pope Alexander IV relinquished effective control over the work of the Holy Office, and the Inquisitors no longer came under the authority of the papal legates. They acted absolutely independently, ignoring the local bishops and the decisions of provincial councils; indeed they were almost independent even of their superiors within the Order.[6] By this time Inquisition law had already been codified down to the smallest details. The position of the accused had greatly deteriorated with the introduction of torture as a means of interrogation. This significant change is not attested before 1260 in the south of France, although the obligation to force confessions of heresy from the accused by torture and make them reveal the names of their accomplices was laid on the secular authorities as early as in 1252, by the bull

Ad extirpanda issued by Innocent IV. It only applied to Italian towns, however. The Inquisitors in Languedoc, in the sees of Toulouse and Carcassonne, obtained from Innocent's successor Alexander IV the right to give each other absolution for trespassing against canon law. This privilege removed the legal obstacles in the way of the introduction of torture as regular Inquisition practice, and gave the organs of the Holy Office a free hand.[7]

Thus equipped with vast powers, the renewed Inquisition made great progress against heresy in the South. They were more and more effective in preventing the "perfect" from making contact with their "believers", and it became a very risky undertaking indeed. Life was becoming increasingly difficult in their homeland, and their gaze turned more and more to the "promised land" on the other side of the Alps.[8] Mass emigration to Lombardy began in the early seventies, not only the "perfect", but many of the "believers" leaving their homes, too. For some their new home was not without bitter disappointments, but nevertheless large numbers settled down permanently in the towns of North Italy (Coni, Asti, Alexandria, Pavia, Mantua, Cremona, Piacenza, Milan.)

The "believers" who remained in Languedoc were thrown back on themselves; the "good Christians", hounded "by wolves and curs", did not dare to stay among them any longer.[9] Meanwhile in Italy, despite the Inquisition and the energetic measures of the Popes, the Cathar church still maintained its organization. The very state of Italian affairs, confused as it was, allowed opportunities for protection and it was even possible to escape the notice of the Inquisition for the whole of a lifetime. In Ferrara there were no doubts about the orthodox beliefs of Armanno Pungilupo from the moment he cleared himself of suspicion of heresy before an Inquisition tribunal. After his death (1269) he was almost canonized, and his tomb in the church of Ferrara was surrounded by pious devotion and stories of miracles performed there. Among the Inquisitors, however, suspicion lingered, and finally the whole case was carefully investigated again. The case dragged on for years, and in the records the secret life of Armanno is revealed to us, pieced together from the evidence of witnesses. These facts left no doubt as to his real attitude towards the Church, to its priests and monks, all of whom were "wicked men" *(mali homines)* in his eyes. Only heretics were "good men", and the martyrdom of one whom he himself saw go to his death had moved him profoundly. He looked after the "perfect" while they lived, with deep devotion, providing those who came to Ferrara with all they needed and accompanying them safely through the city to visit their fellows in the faith. He carried the blessed bread from the "perfect" to the "believers" - he himself was a Cathar "believer" who received the *consolamentum* in Verona two years before his death. On this occasion we read the names of the highest dignitaries of the church of Bagnolo, who administered the spiritual baptism to Pungilupo.[10] There were other names, too, in the records, giving us valuable additional information about the Cathar hierarchy in northern Italy, as we know of it at this later period from Anselm's Treatise on heretics. This list shows that in a large number of towns there were still deacons who were in charge of the affairs of the Cathar communities.[11] The ability of the sect to resist was certainly helped by the refusal of the towns to introduce the papal Inquisition, but their most powerful protection was just outside the towns, where those noble families lived who had not been drawn by the course of developments into the municipal body politic, as the other aristocrats had. Not only did these descendants of the old feudal nobility retain their contempt for the citizens; they retained their age-old hostility towards all who represented secular or ecclesiastical power. They were therefore always glad to open their doors to those who were driven from the towns by political conflict or religious persecution. The noble homes in the urban *contado* provided the Cathars with their safest refuge. This was true not only of Florence, but of Orvieto and the towns of Lombardy as well. Heresy also had many supporters among the warlike gentry, who not infrequently took part in the violence directed against the Inquisitors.[12]

The heretical movement was involved not only with the conflicting social forces in the towns, and with the antagonism between town and countryside, but also with the long-drawn-out struggle between the papacy and the Hohenstaufen, which divided the whole of Italian urban society into two camps, the Guelphs and the Ghibellines. After the bloodshed which followed the tragic death of the young Hohenstaufen heir (1268), the tribunal of the Inquisition (composed of Franciscan monks) intensified its activity in Orvieto. Among the Ghibellines who had been behind the disturbances the Inquisition found followers and protectors of heresy, propagated in the Orvieto region by Cathar preachers from Lombardy and elsewhere. Entire families of noble birth were reduced to poverty by the confiscation of their property, driven out of the city and prohibited for two generations from taking any part in public life. The attention of the Inquisitors centred on destroying the "nest of vipers" in the families of Tosti and Ricci, who had already been in trouble after the disturbances of 1239, and who had always been active opponents of papal theocracy. It appears from the records of the trials in Orvieto in 1268-1269 that the Inquisitors only became less vigorous in their activity and pronounced less severe sentences, in the last few months.[13]

In northern Italy the heretical movement had its supporters among the most powerful men of the Ghibelline side. At first it was Ezzelino III da Romano, governor of the Marca of Treviso, who built up a strong personal domain in Verona - the precursor of later signories. The most powerful ally of Frederick II, he was himself ultimately declared a heretic for the protection he had granted the movement, and Rome called down upon him all the sanctions established by the decision of the Lateran Council of 1215. After the fall of Ezzelino (1259) the heretics of northern Italy had a firm defender for the next ten years in the person of Uberto Pellavicino, who stopped all trials of heretics in Milan during his rule.[14]

The fact that a man belonged to the Ghibelline camp, however, by no means meant that he necessarily favoured the heretics. Mastino della Scala, elected podesta of Verona after the death of Ezzelino, was a Ghibelline who took strong measures against heresy, which had certainly gained strength in the region of Verona under Ezzelino. It was in Mastino's time that anti-heretical articles in the spirit of the constitutions of Frederick II and Pope Innocent IV were incorporated into the new version of the city statutes. Under these changed circumstances it was only a matter of time before the decisive blow was struck at Catharism. The Bagnolo church, as well as the Albanensian, were centred in the region of Verona at this time. At the end of the sixties of the thirteenth century the Cathars still maintained a house in Verona itself *(domus Catharorum)*. We learn this from the evidence given concerning the rite of consolamentum administered to Armanno Pungilupo by the Bagnolo bishop and his assistants. However, the real focus of religious life for the Cathars of this part of Italy, apart from Verona, was still on Lake Garda. Instead of Desenzano other names appear. In the village of Lazise on the eastern shore under the jurisdiction of Verona, the local authorities set free a heretic who had been arrested and was waiting to be taken to Verona by an armed escort. The great bastion of the Cathar sect was the *castellum* of Sirmione, at the tip of a narrow peninsula running northward into the lake. This fortified stornghold was overrun with heretics, who seemed to have enjoyed full protection there. We learn, again from the records of the Pungilupo case, that leading figures in the Bagnolo church stayed in Sirmione, and so, too, did important Cathar exiles from Languedoc, among them the former bishop of Toulouse, Bernard Oliba. The Inquisitors in Verona naturally took the destruction of this hotbed of heresy as their primary goal, and in this they had the wholehearted support of the Scaligers of Verona. The brief note in the chronicle for autumn, 1276, reports that the Bishop of Verona, the Inquisitor, and Mastino's brother Alberto della Scala, appeared in Sirmione. They were no doubt accompanied by a strong detachment of soldiers; 166 heretics, men and women, were taken off to Verona. Perhaps other arrests were made elsewhere at the same time as in Sirmione, for the same chronicle of Verona gives the number of victims burned at the stake in the city amphitheatre at about 200. This happened at the beginning of 1278, already under the rule of Alberto della Scala, founder of the Scaliger signory in Verona.[15] This was as exceptional a case in the practice of the Inquisition during this period as was the mass execution of victims after the fall of Montségur. For the Cathars, too, the tragedy of Sirmione was as catastrophic a blow as the fall of Montségur a few decades earlier. The Church's stubborn fight against dualist heresy, which had lasted longer in Italy than anywhere else in the West, was now drawing to a close here, too. After the main force of the Cathar movement had been broken, the Inquisition proceeded in more moderate forms in Italy. Nevertheless the steady infliction of penalties on person and property, with routine persistence, inevitably discouraged those who would have given support or protection to the heretics.[16] The long persecution exhausted the movement and reports of heresy in the various Italian towns gradually die away.[17]

Meanwhile hope had again revived among the exiles from southern France, inspired by exciting news from the homeland.

Towards the end of the thirteenth century tension again became acute between the Inquisition and the authorities of the Languedoc towns. About the middle of the eighties cases were brought in Carcassonne against people accused of plotting to destroy the Inquisition registers, kept in the tower of the *City*. The whole affair was most involved; the record of evidence given before the Inquisition tribunal is full of contradictions and serious doubt arises as to whether there was ever any plot at all. Nevertheless the complaint brought against the Inquisitor John Galand on this occasion, by the consuls of the *Bourg* of Carcassonne, is significant historical evidence of the Inquisitor's arbitrary actions, and especially of the terrible and incredibly inhuman system of the Inquisition prisons. Some time in 1285 or 1286 the consuls themselves sent this letter to the King, the Pope and the Prior of the Parisian Dominicans. Two earlier appeals (1280 and 1284 or 1285) had been presented in collaboration with the authorities of other towns, in particular Albi and Castres.[18]

In Albi things developed in an exceptional way, due to a change in the relations between the town and the bishop. After the fall of the power of the Trencavels Guillaume Peyre, the bishop, took into his own hands practically all the secular powers previously divided between the

Viscount and the bishop. He enjoyed the full support of the community of Albi, which gained enormously by the enlargement of its freedoms. Even under Guillaume's two immediate successors the interests of the town community were fully in accord with the bishop's efforts to defend his gains in the face of claims by the Capet monarchy which would have given it the whole of the Trencavel heritage. From the middle of the sixties, however, the much strengthened royal power did not allow the bishop to extend his gains, and the bishop would not allow further expansion of the municipal power, which could only be at his expense. Relations between the two became hostile, and the town found an ally in the representatives of the monarchy. Under bishop Bernard Castanet the conflict broke out into the open. Between 1297 and 1299 a number of people were tried by the episcopal court for attacking the bishop's property. The secular legal process, however, allowed the accused to manoeuvre skilfully and to prolong the proceedings endlessly, with the willing help of the royal officials. Finally Bernard Castanet seized on the most effective weapon he had. Investigations in connection with the Inquisition trial in 1286-1287 had brought to light a large number of people suspected of supporting the remnants of heresy in the diocese of Albi, and all that was necessary was to choose from this list the names of those who headed the town s opposition to the bishop. And indeed, the thirty-two men condemned to severe penalties by the Inquisition in 1299-1300 were almost without exception members of the town's oligarchy.

The political purpose of this new Inquisition trial is quite clear, yet the accusation of heresy was not a fiction; these were people who were close to the heretics. Sympathy to the Cathar sect which had long been rooted in the diocese of Albi was undoubtedly continuous in the town itself. Some of the names appearing in the case of 1299-1300 are already known to us from the stormy events of 1234. The supporters of heresy in the town were almost without exception from the upper ranks of society - merchants, wealthy craftsmen, lawyers and the lesser nobility. It was also people of these classes who were at the head of opposition to the Inquisition in the years 1302-1308. In Albi the connection between heresy and the political scene thus took an unusual form. During the period under consideration the protection of heresy there did not go hand in hand with the defence of southern independence against French domination. What we see is the very contrary; the town oligarchy, favourable towards heresy, allies itself with the representatives of the monarchy against a southern ecclesiastical lord who is trying to maintain his autonomous position against the centralizing pressures of the monarchy.[19]

Memories of the lost independence of the South had not quite died out in Languedoc, however. The general hatred of the Inquisition was easily turned against the whole Church of Rome. The people of the South had not yet forgotten that "the priests had handed their land over to the French", and the troubadours poured oil on the flames with their songs. At the turn of the thirteenth to fourteenth centuries the growing struggle against the Dominican Inquisition in Carcassonne found an impelling leader in the Franciscan monk Bernard. The greatest success achieved by the movement, albeit a passing one, was the freeing of the prisoners held by the Inquisition in Carcassonne.

Under these conditions the remaining Cathars dared to cross into Lombardy more often, and the Cathar exiles settled there journeyed over the Alps more frequently. A number of Cathar preachers were again active in southern France, but there was of course no question of restoring the bishoprics. The only member of the Cathar hierarchy living in Languedoc now was the *diaconus maior*. The real leader of the movement, however, was the *ancianus* Peter Autier,[20] who came from the region of Foix, where the heretical movement was particularly deeprooted. A notary by profession, Peter was an elderly man before he received the consolamentum in Como in Italy. He then spent four years preparing himself, together with the members of his immediate family, for the difficult task awaiting him in his homeland. It was the end of the century when he went back. From the outset his appearance attracted believers from far and wide, and the old man tirelessly carried out the duties which crowded in on him. By this time the sect was forced to rely on the help of people of the lower and indeed the lowest classes;[21] everything had to be carried out in the deepest secrecy. They could only meet to listen to his sermons, to partake of the "breaking of bread" or to receive the consolamentum, at night. They could only travel under cover of darkness, safely conducted by *nuncii* and *ductores*. Peter spent the winter in his native Ax, journeying tirelessly throughout Languedoc with his companions for the rest of the year. He even dared to enter Toulouse, and his son Jacques assembled a gathering where the Inquisition would never have dreamed of looking - in one of the churches of Toulouse.

From 1307 the organs of the Inquisition in Toulouse and Carcassonne systematically combed the open country and even the inaccessible valleys of Foix, but without success. None of the people questioned betrayed the hiding-place of the Cathar *ancianus*. In 1309 Peter Autier finally fell into the hands of the Toulouse Inquisition, and in a short time almost all those who worked with him on his mission were caught and imprisoned. The minute precision of the proceedings took almost two years, and Peter's condemnation was staged with great pomp and solemnity in Toulouse cathedral (1311). This last of the leading Cathars of Languedoc looked death in the face unflinchingly, and his last words at the stake revealed his unshakeable conviction that his faith would conquer if he were allowed to speak.

After the death of Peter Autier and the imprisonment of the others, there were no more "perfect" for the sect to regroup around in southern France. The last of all, Guillaume Belibasta, escaped across the Pyrenees to Catalonia, followed by a handful of "believers" from the valley of Ariège in Foix. Belibasta had received the consolamentum towards the end of Peter's ministry, and had not been one of his close companions. He was a rough plebeian, often shamelessly miserly in his conduct, and his poor preaching was as little satisfactory to the "believers" as was his moral laxity. But they had no choice; after the death of his sick companion Belibasta remained the only bearer of the "good" to whom they could turn. Three years later this little community of heretics living in safety near Valencia was nosed out by a mercenary spy who treacherously lured Belibasta across the Pyrenees and then handed over to the Inquisition his few remaining followers as well. The burning of Belibasta in 1321 brought to an end this poor epilogue to the story of Catharism in southern France. The hundreds of pages of Inquisition records which have remained, however, from Toulouse, Carcassonne and Pamiers, reflect the life and teachings of these last Cathars in the mirror of their confessions. The register of Jacques Fournier, later to become Pope Benedict XII, is particularly rich in information; as Bishop of Pamiers he spent seven years (1318-1325) systematically investigating heresy in the region of Foix. Some of the interrogations he recorded give complete reproductions of "sermons" by Peter Autier, his companions, and Guillaume Belibasta.[22]

The wealth of material provided by Inquisition records for this later period has greatly enriched our knowledge of Cathar mythology with much picturesque detail. Some aspects had already been noted in earlier Inquisition records; for instance those of the time around the middle of the thirteenth century give various accounts of God and the Devil (Lucifer) collaborating in the creation of man. Stories of this kind belong to the folk cycles built up round the Book of Genesis, and include among others the apple which stuck in Adam's throat.[23]

The myths woven by the fantastic imagination of the Cathars in this later period around the idea of the transmigration of souls also belong to the region of folk lore.

Apocryphal literature remained close to the sect's way of thinking. The story of Christ's life on earth was enriched among the later

Cathars in the south of France by a whole series of new episodes of an apocryphal nature, the source of which is unknown to us.[24] The "Vision of Isaiah" was a constant attraction with its dazzling images of the angelic choirs, the mystery of the Son of God descending through the celestial spheres, and the glowing vision of His later ascent. Individual details, however, seem to have undergone considerable change in the course of oral tradition, just as the form of the dualist myth changed in various ways.[25]

This later tradition preserves the basic outlines of the radical-dualist picture of the world, but the drama of the cosmos shows some unexpected changes. Lucifer waited thirty-two years before the gates of Heaven, hid in Heaven for another year, and finally seduced the angels by showing them the figure of a beautiful woman. According to Belibasta God then declared that no woman should ever enter Heaven.

In one version of the drama the account of the fall of the angels comes close to the original dualist myth as we find it among the Bogomils described by Zigabenus and in the secret book of Nazarius: here, too, Satan builds his own heaven. Then again there is an unexpected change: the glass heaven of Lucifer collapses, and the angels fall into the abyss.[26] There follows their repentance and wish to return to God. The evil prince of this world prevents this by giving them *tunicas*, earthly bodies, in which they will forget their heavenly home.

The Saviour comes to earth to save the imprisoned souls. According to an early fourteenth century version the Saviour is one of the loyal angels of God the Father. This angel, whose name is John, is the only one who resolves to take upon himself the sufferings of mankind and thus becomes worthy to be called the Son of God.[27] We cannot fail to see the remarkable similarity between this and the soteriological myth which formed the secret learning of the ninth century Paulicians of Byzantium.[28] The profound difference between the two, however, lies in the fact that in the Cathar myth the "adoption" of one of the angels as the Son of God depends directly on the fulfilment of the measure of human sufferings (recounted in a book which took God Himself forty years to write). Even during his sufferings on earth the Christ-angel remained a spiritual being, and someone else died in his place on the cross.[29] He simply left the world for three days in order to inform his Heavenly Father that he had fulfilled to the letter all that was written in the book. In return he received the name of Son of God and handed on to his disciples on earth the powers given to him.

The Saviour brought twelve angels with him from Heaven; they had heavenly bodies, and began their sufferings while he was still on earth. Gradually these "heavenly" apostles were replaced by "earthly" ones, and now the Cathar "perfect" held the place of these "earthly" apostles. There were always apostles in the "Church of God", which had also possessed its own "pope", the successor of the Apostle Peter.[30] For the last of the Cathars in the south of France this was nothing but a memory - and yet in Bosnia a century later the Patarene Church still considered their bishop to be the "vicar of Christ and successor of Peter"

The apostolic succession was thus preserved in the church of the "good Christians", and only in that church could salvation be attained through spiritual baptism. Sometimes the soul had a long pilgrimage to make, however, and was incarnate in many people and animals before reaching its goal. The greater the burden of sin in a man's life, the longer the pilgrimage of the soul's reincarnations. An illustration was the story of a horseshoe found by one of the Cathar "perfect"; he had lost it in one of his previous lives, when his soul, originally that of a great sinner, had been imprisoned in the body of a horse.[31]

Baptism by the Holy Spirit, the only means of obtaining absolution, was in principle administered only to adults, but in this later period cases are recorded where it was also received by children. The administration of the consolamentum to the sick was connected with a custom whose name inspires horror in the context of Cathar history; *endura*. Originally it referred to the period of preparatory asceticism *(astenenza)* necessary before the consolamentum could be administered, but a man suddenly stricken with serious illness could not undertake such a trial. If he recovered there was the danger that he might relapse into sin and thus preclude the effect of the spiritual baptism received. This fear led many to deliberately bring death closer after the consolamentum had been administered to them. They would refuse food, and were given only water to drink, which of course only prolonged the death-throes for days. There are records of shocking cases where members of the family refused food to the sick in spite of their desperate pleas. Elsewhere human feelings prevailed and food given in time saved the sick man's life.[32]

The *endura* was a late custom, for which we have no reliable evidence before the last quarter of the thirteenth century.[33] Besides this cruel practice we learn from the Inquisition archives of the more human side of Cathar life. Despite their rejection of marriage on principle, the "perfect" nevertheless found themselves forced to care for the family happiness of their followers still living "in the world". It was undoubtedly in the interests of the sect itself that a "believer" should choose a wife of the same faith; a form of rite was developed for the occasion and one of the "perfect" would solemnize the new relationship somewhat in the role of the priest performing the marriage ceremony.[34]

There was no point in the "believers" using the Lord's Prayer, since they were not yet freed from their sins and could not yet turn to God as to their Father. The last Cathar teachers, however, tried to give them help in this pious custom and prepared special prayers for them; as a rule they formed the simple plea that God would guide them as He had guided the three wise men of the East.[35]

The Inquisition trials staged at the beginning of the fourteenth century, which have provided us with such a detailed picture of the late Cathar sect, put an end to the last wave of the dualist movement in Western Europe. Elsewhere at this time there were only isolated cases of Catharism; this was true of the countries north of the Alps, as well as Italy. At the beginning of the fourteenth century there were still traces of Catharism in the Austrian regions,[36] where the Waldensians were otherwise the most widespread sect. Around the middle of the thirteenth century the wave of Waldensian heresy penetrated into Bohemia from Austria, undoubtedly facilitated by the settlement of Germans in the southern regions of Bohemia and Moravia.[37]

In 1315 the Dominican Inquisitors tried a group of heretics in Prague itself. The trial took an unexpected turn when the Bishop of Prague intervened at the last moment as the condemned were being handed over to the secular arm to be burnt at the stake. John IV of Dražice, Bishop of Prague, rescinded the sentences and recalled the Inquisitors, who had acted without his permission. He himself took over the interrogation of the suspects; some of the sentences were changed considerably, but fourteen victims were nevertheless burned.

We learn of this conflict from a letter of the Pope's in which the text of the complaint against the bishop by one of the Czech prelates is reproduced. The articles of faith attributed to the sentenced heretics are distinctly Cathar in character. The beliefs confessed by a group of heretics condemned in Krems in Austria in the same year also point to a Cathar rather than a Waldensian group.[38]

At the beginning of the fourteenth century the remnants of the Cathars turned to Sicily as to a "good country" *(bona terra)* where they could still live in peace under the tolerant rule of the King of Aragon, Frederico III (1296-1337). It is there that a Cathar *diaconus maior* is still attested for this period.[39]

The remnants of the sect lasted longest of all, however, in the north of Italy, in Piedmont, influencing the Waldensian groups living in the inaccessible Alpine valleys and in the lower regions. At the end of the fourteenth century ruthless action by the Inquisitors annihilated the last focus of Cathar heresy. The record of the interrogation of Giacomo Bech (1388)[40] preserved in the archives of the Inquisition in Turin, reveals significant facts about the life of the Cathars in Chieri. The Cathar community in that little town still kept a great book whose title, Liber Civitatis Dei, cannot be seen otherwise than as a reminiscence of St.Augustine. It contained the names of those who pledged themselves by solemn promise to keep the faith of the sect. This little community of heretics, apparently quite deserted in its Piedmont retreat, was in constant touch with fellow-believers in the Balkans. It is clear from Bech's confession that about ten years previously a heretic from "Sclavonia" had visited the heretics of Chieri. The text of the Inquisition document leaves no doubt that "Sclavonia" here means Bosnia.[41] Giacomo Bech himself was to be sent there as a new adept of the sect, in order to perfect himself in the Cathar teachings under "teachers" living in that country. He was given a certain sum of money by the heretics for his journey, but bad weather prevented him from sailing. Other heretics from Chieri had actually been in Bosnia, however; Bech gave the names of several who had made the pilgrimage there between 1348 and 1382. It is clear that for the Cathars of this North Italian town the Church of Bosnia was the teacher of their faith. This is expressed in the form taken by the sentence passed on Bech: "You listened to and believed in the teachings of the Cathars and of heretics from Sclavonia...".[42]

A quarter of a century after Giacomo Bech and his companion Antonio Galosna were burned at the stake, sentence was passed on some dead heretics in Chieri. In 1412 their bones were publicly burned along with symbolical figures of a number of heretics from several local families. Among the names of the latter we find John Narro who according to Bech had travelled to Bosnia about the year 1360. The words of the Inquisitor at the auto-da-fê in 1412 confirm that Giovanni Narrò had been among the Bosnian heretics in order to learn their doctrines. The heresy which Narro and his son believed and preached is explicitly called "the faith of the heretics of Bosnia".[43]

Like much of the early evidence, the documents of the last period of the sect in western Europe show its close relationship with fellow heretics "across the sea". In the fourteenth century the gaze of the "good Christians" of Piedmont was fixed on Bosnia, the only source of effective help they had.

And indeed, as the last faint gleam of Cathar faith flickered in that Alpine region, extinguished everywhere else in the West, in Bosnia the "Church of God" was not only safe from persecution but enjoyed such moral authority that it was recognized throughout the land and in neighbouring countries.

1 See the record of the Inquisition in Toulouse for the end of 1254 or beginning of 1255 (Archives de la Haute Garonne, Ms 202). *H.Blaquière - Y.Dossat,* Les Cathares au jour le jour..., in: Cathares en Languedoc, 259 - 289.

2 See *C.Douais,* Les hérétiques du comté de Toulouse dans la première moitié du XIIIe siècle d'après l'enquête de 1245, in: Compte-rendu du Congr. intern. des catholiques, Paris 1891, 5 - 19. *Dossat,* Crises, 216 - 244.

3 *Dossat,* Crises, 169; cf. ibid., 34 f.

4 Ibid., Chapter VII, 173 - 188. The Dominican Inquisitors gradually took up their work again throughout France (187 f.). For accounts of the Dominican Inquisition in northern France in and after the second half of the thirteenth century, see *L.Tanon,* Histoire des tribunaux de l'Inquisition en France, Paris 1893, 118 ff.

5 *Schmidt* I, 330 f.

6 For a long time collaboration with the bishops and consultation with other competent authorities remained a pure formality in the conduct of trials by the Inquisition, and had no practical effect. Not until the early seventies of the thirteenth century was there any real change. *Dossat,* Crises, 211.

7 Ibid., 211 ff.

8 See *Dossat,* in: Cathares en Languedoc, 283.

9 Ibid., 75 ff.; *Dupré-Theseider,* in: Cathares en Languedoc, 305 ff.

10 Record of Pungilupo's trial, *Muratori* V, 97 - 150.

11 Tract. her. 324. The list reveals that in some towns both the Concorezzo and the Albanensian churches had a deacon.

12 See *Volpe,* Movimenti, 109 - 112.

13 *Fumi,* Orvieto.

14 *Volpe,* op. cit., 110 f.; *Guiraud,* Inquisition II, 533 ff.

15 For developments in Verona from 1259 onwards see *Cipolla,* Verona, 70 f., 267 - 287. On the Lazise case see id., Nuove notizie sugli eretici veronesi 1273 - 1310, in: Rendiconti della R. Accad. dei Lincei, Cl. di scienze mor. e filol., S.5, t.5 (1896) 336 - 353. On Sirmione see also *Dondaine,* Hiérarchie II, 294.

16 See the case of Florence, *Perrens,* 386 f.

17 For individual cities see *Borst,* 136 f.

18 See *J.-M.Vidal,* Un inquisiteur jugé par ses victimes. Jean Galand et les Carcassonnais, Paris 1903. This work also deals with further developments in Carcassonne. Cf. *Guiraud,* Inquisition II, 303 ff.

19 *J.-L.Biget,* Un procès de l'inquisition à Albi en 1300, in: Le Credo, la morale et l'Inquisition (= Cahiers de Fanjeaux, fasc. 6) Toulouse 1971, 273 - 341.

20 On the whole of this affair see *Vidal,* Derniers ministres, cf. also *Schmidt* I, 347 - 357.

21 See *Molinier,* Église et société III, 273.

22 Toulouse: Liber sententiarum inquisitionis Tolosanae 1305 – 1323 ed. *Ph. van Limborch,* in: Historia Inquisitionis, Amsterdam 1692.

Carcassonne: Reg. 4269 (lat.) Bibl. Nat.

Pamiers: Reg. 4030 (lat.) Vatic. Ed. *Duvernoy.*

23 On four volumes of Inquisition material referring to the years 1240 - 1260 (Doat, vols. XXII, XXIII; ms. 609 Bibl. munic. de Toulouse; ms. 160 Bibl. de Clermont-Ferrand) see *Vidal,* Doctrine I, 362. For these legends of the creation of man (*Doat* XXII, p. 32, and later: XXV, p. 44) see *Guiraud,* Inquisition I, 53. For folk lore tales of this type see *Dähnhardt,* Natursagen I, Chapter 2, 88 f. The apple: ibid., 208 and esp. 209 (the version recorded in a fourteenth - century French MS).

[24] See *Vidal*, Doctrine I, 381; *F.P.Badham - F.C.Conybeare*, Fragments of an ancient (?Egyptian) Gospel used by the Cathars of Albi, in: The Hilbert Journ. 11 (1913) 805 - 818 (not available to the author).

[25] See *Turdeanu*, La vision... (see chap. XI, Note 11), 315 f.

[26] This is really a duplication of the motif of the fall of the angels. See *Vidal*, Doctrine II, 375 f. On the idea of the heaven of crystal see *Linda Sadnik*, Der Balkan und die Hochkulturen des Vorderen Orients, in: Saeculum V (1954) 37. According to the answers given by Arnaldus Cicredi, 122D - 123A (*Duvernoy* II, 34 f.) the repentance of the angels and the idea of the *tunicae* followed immediately upon the angels' fall from their heavenly home. (This version does not mention the fall of the angels after the collapse of Satan's heaven). The image of raindrops used to give an idea of the numbers of the falling angels occurs in Dante (see *Borst*, 146, n. 12). It also appears in apocryphal literature. See the Legend of the Lake of Tiberias, *Ivanov*, Bog. knigi, 291. For the various places where the angels landed after their fall (and where they remained as demons) see ibid.; Palaea 16. cent., in: *I.Ja.Porfir'jev*, Apokrifičeskije skazanija o vetchozavetnych licach i sobytijach po rukopisjam Soloveckoj biblioteki, in: Sbornik otd. rus. jaz. i slov. XVII, 1 (1877) 85. These ideas are already found in patristic literature (see DTC 4, col. 344).

Anselm Alex. noted down the belief of the Albanensian Cathars, attested by their "teacher" Lanfrancinus de Vaure: "...quod non omnes oves sive anime que descenderunt vel ceciderunt de celo incorporantur, sed alique purgantur in aere isto caliginoso sine corpore aliquo, et maiorem penam sustinent quam ille que sunt in corporibus, sed cicius salvantur..." (Tract. her., 312,16-21). According to Art. her. J.Maurini VII (*Duvernoy* II, 508 f.) "...illi spiritus qui crediderunt dyabolo et ei consenserunt et cum eo ceciderunt de celo facti fuerunt demones, et illi qui plus peccaverunt ceciderunt in abissum, et illi qui plus peccaverunt post eos remanserunt in terra, et alii in aqua et alii in aere, et alii iuxta celum, secundum quod magis vel minus peccaverunt. Illi vero spiritus qui expresse non consenserunt, set solum accesserunt ad dictam turbationem, sunt spiritus et anime hominum et mulierum qui postea ponuntur in corporibus humanis." (Cf. *Vidal*, Doctrine II, 375 f.).

[27] Ibid., 385 f. Another version, according to which Christ was bajulus Dei in Heaven was attested during an interrogation in 1243 (Doat XXII, fol. 26v); ibid., 386.

[28] See ByzSlav XVII (1956) 40, 44 f.; XXV (1964) 56.

[29] For examples of this motif in western sources dealing with the Cathar faith, and analogous passages in the Koran, see *Borst* 167, n. 19.

[30] *Vidal*, Doctrine I, 388, 392.

[31] For the folk lore associations of this motif see *Borst*, 169 f., n. 8. He also refers to the motif of a soul inhabiting a lizard during sleep.

[32] *Vidal*, Doctrine II, 36. Cf. *Borst*, 197.

[33] See *Y.Dossat*, in: Cathares en Languedoc, 85. The author rejects the exaggerated conclusions of *R.-L.Riol*, who sees evidence of this practice in earlier sources.

[34] *Vidal*, Doctrine I, 403 f.

[35] Ibid., 16 f. n. 2, p. 17, gives the text of another prayer in the Occitan language. Conf. *A.Cicredi*, 123C (*Duvernoy* II, 37).

[36] See *R.Holinka*, Sektářství v Čechách před revolucí husitskou (Sects in Bohemia before the Hussite Revolution), Bratislava 1929, 174 ff.

[37] *Holinka*, op. cit., 159 f.

[38] Ibid., 167 - 184. The sectarians tried in Prague were accused of believing in the future reign of Lucifer, and the heretics condemned in Krems were also accused of Devil-worship That the teachings of this Prague sect were really of Cathar origin can be seen best from the articles concerning rebaptization, the denial of resurrection, and Christological docetism (see the text of the articles as given by *Holinka,* 182, 183). The Krems case is preserved in four different versions based on the same original, now lost (doubtless an Inquisition record), but nevertheless differing quite considerably (see *Holinka,* op. cit., 174, n. 137). The rejection of baptism, the Eucharist, and of marriage is a Cathar element (see the version published in MGH SS IX, 826; cf. *Holinka,* op. cit., 178). The two contradictory views of the mother of Christ, considered by these heretics to be *angelum et incorruptam, aut mulierem penitus violatam* (ibid.), could reflect the dissension on this point seen among the Italian Cathars of the Concorezzo branch as early as the beginning of the thirteenth century (see Chapter XVII above). Holinka would seem to have been right in seeing remnants of Catharism both in Prague and the Krems group (184 f.), in spite of his belief in the existence of a "Luciferian" fraction among the Cathars. Referring to the anonymous Passau writer (c. 1260) and David of Augsburg (1256 - 1272) (*Holinka,* op. cit., 149, n. 58), he shows that clear Cathar elements had penetrated even into the doctrines of the Waldensians of the Lombardy branch (the Runcarians) in the German regions beyond the Alps; it can thus be characterised there as a "Catharo-Waldensian" sect (op. cit., 184).

[39] See *Borst,* 137.

[40] Text: *G.Amati,* Arch. stor. ital., S.3.,t.II, pars 1 (1865) 18, 45 - 61; *Kniewald,* Vjerodostojnost, 242, published a photo-copy of the MS.

[41] After Kniewald's publication of the photo-copy the reading *Boxena* is indisputable.

[42] Sentence of the Inquisitional tribunal, ed. G.Manuel di S.Giovanni, in: Miscellanea di storia ital. XV (1874) 25 - 26, 75 - 84. Cf. *M.Esposito,* Un, Auto de fê'â Chieri en 1412, in: RHE 42 (1947) 422 - 432 (quotation on 431).

[43] The relevant documents were published by *Esposito,* art. cit.

Note: I have been unable to consult the work of *R.Nelli,* La nature maligne dans le dualisme cathare du XIIIe siècle. De l'inégalité des deux principes, Carcassonne: Éditions de la revue "Folklore" 1969.

Additional note:

See *G.Gonnet,* Casi di sincretismo ereticale in Piemonte nei secoli XIV e XV, in: Bollettino della Società di Studi Valdesi LXXVIII, No 108 (1960) 3 - 36. Cf.*A.Molnár,* Valdenští (The Waldensians), Prague 1973, 104 ff.

At the very beginning of the fifteenth century the Dominican Vincent Ferrer still met some Cathars in the Waldensian valleys. (Cf. *Schmidt* I, 186.)

Chapter XIX

THE "BOSNIAN CHURCH"

How the heretical church in Bosnia developed in the second half of the thirteenth and early fourteenth centuries is wrapped in darkness. Indeed there is little historical light on the history of the country at all during that time. It remains a mystery why the state of Bosnia which was capable of resisting very heavy pressure in the reign of Ninoslav, very soon afterwards found itself in a position of complete dependence and ceased to play any independent part in events for many decades.[1] At first the regions of Bosnia were to be mere links in a chain of *banovinas* along the southern marches of Hungary. The Hungarian state itself, however, was suddenly crippled by disaster. The influence of the Cuman hordes which had remained in the land since the Mongol invasion led to the frustration of the efforts towards a higher level of civilization made during preceding centuries in the Hungarian milieu. In the period of anarchy which followed the Croatian Šubić (Shubitch) family gained control over the whole of Croatia and Dalmatia. Towards the end of the thirteenth century their expansion was directed against Bosnia, except for the northern regions of Usora and Soli which had earlier been attached to the *Mačva* (Matchva) *banovina* and were now ruled by a Serbian king, Dragutin. Bosnia itself came under the Šubić family, but they had to wage a fierce struggle against the local nobles, and to make considerable concessions to them. The native ruling house of Kotromanić maintained a certain position in the country even under Šubić rule. After springing up so rapidly the Šubić domain again disappeared, as soon as the foreign house of Anjou became firmly settled on the Hungarian throne and began ruthlessly building up a centralized state. From 1322[2] Bosnia was once more under the direct domination of the Hungarian throne. In the complicated struggles between hostile groupings of Croatian nobles Ban Stephen II Kotromanić (1314-1353) remained consistently on the side of Charles I, skilfully taking advantage of the situation to extend the territory under his own control. Again the ancient Bosnian territories were annexed - the "Lower regions" in the west, Usora and Soli in the north; Stephen even seized Krajina (Krayina) - the coastal region between Neretva and Cetinje, and its mountainous hinterland of Završje (Zavrshye). Control over much of Serbian Hum gave Bosnia the important valley of the Neretva with its great trade route and access to the sea at its mouth. Thus in a few years the Bosnian state had grown to twice its size. The peak of its power, however, came under Stephen's successor Tvrtko (1353-1391). The first half of his reign was not without violent upheavals. Right at the outset some of the territory gained by Stephen II Kotromanić was seized by the Hungarian king, Louis I of Anjou. In 1363 Tvrtko successfully resisted a Hungarian invasion of Bosnia, but shortly afterwards he was overthrown by a rebellion among his own nobles, and forced to seek refuge and support at the court of Louis. It took several years before he could regain his throne. He then began a series of successful campaigns, first invading the Serbian domains to the east of Bosnia. His advance in this direction fitted in with the policy of Prince Lazar, the most important of the Serbian regional rulers since the disintegration of the great kingdom of Stephen Dušan. The common victory of Lazar and Tvrtko over Nikola Altomanović brought Bosnia considerable Serbian territory, including the seat of the Orthodox Metropolitan in Mileševo (Mileshevo), where the relics of St.Sabbas were housed. Crowned as king in 1377, the Bosnian ban was clearly reviving the tradition of the Serbian state under the Nemanjids. Towards the end of Tvrtko's reign his expansionist campaigns took the direction of Dalmatia and Croatia. Gaining control of practically all the towns and some of the islands of Dalmatia, as well as of much of the territory of Croatia, Tvrtko acquired the right to the title of King of Croatia and Dalmatia.

Looking back at those dark years when Bosnia had lost all independent significance, we are astonished by the fact that Hungary failed to use her position of advantage to renew the Catholic Church in Bosnia. Looking for an explanation of this paradox, we should bear in mind the state of the church in Hungary itself at that time; it, too, had succumbed to the wave of return to barbarism which swallowed the whole of society. All that had been gained in the Cluny and Gregorian reforms was lost, ecclesiastical benefices were again filled by laymen, and the priests themselves fought for them with arms in their hands. Bishops and abbots, ruling as they did over great estates, were caught up in the struggle between secular factions. Neither the fatefully weakened royal power, not the secularized episcopate could be expected to show any interest in stamping out heresy and restoring the Church in Bosnia. Nor was that all. Voices were raised in Rome to suggest that Hungary itself should be the object of a crusade. It was not until the reign of Ladislaus IV (1272-1290) that the papal legate won from the Queen Mother, Elizabeth, and later from the King himself, a promise to persecute the heretics, particularly in Bosnia.[3] This undoubtedly remained no more than a promise, and so the only noteworthy action taken against the Bosnian heretics at this time concerned the northern regions (Usora,

Soli) belonging to Dragutin s domains. At his request Rome sent two Franciscans to northern Bosnia, who were acquainted with the Slavonic language. They were charged with the conduct of the Inquisition against heresy, which according to the papal epistle was widespread in that region.[4] Dragutin's efforts to convert the heretics were appreciated in Serbia itself, and his credit for the campaign is stressed by his biographer, Archbishop Danilo. For the first time, anathema against the names of some of the Bosnian heretics was incorporated into the Serbian Synodicon.[5]

We do not know what effect the work of those two Franciscan Inquisitors had, but it is certain that the papal Curia did not attempt to send the Franciscans to Bosnia again until 1325, when the situation had changed completely. The Papal Curia, then in Avignon, appealed to Stephen II Kotromanić, who ruled the whole of Bosnia and some of the lands around. In fact his ardour for the faith and his good intentions in the matter of combatting heresy had already been appreciated in Avignon when the Ban was still in a subordinate position under the supreme rule of the Šubić family, as we see from an epistle of Pope John XXII, dated July 1319.[6] There is a much more interesting papal epistle, this time addressed to the Šubić ruler, Ban Mladen II,[7] which gives an apposite description of the true state of the church and of religion in Bosnia at that time. While in Western Europe the Church had only to liquidate the last groups of Cathars, as far as any still existed, in Bosnia the picture was one of complete disintegration of church life. The churches were deserted, the clergy wiped out, it was unheard of for honour to be paid to the cross of Christ... This was the situation the papal Curia had to count with when deciding to renew the Inquisition in Bosnia, with the help of the Franciscans. The whole matter was complicated further, however, by a sharp conflict over the right to conduct the Inquisition there, between the Franciscan and the Dominican orders, which ultimately (1330) ended in total victory in Avignon for the Franciscans.[8] On the other hand, the Bosnian Ban was in a difficult position. He was certainly willing to support the papal Curia in its efforts, but he could not ignore the customs of the country, which trammelled him considerably in this direction. The position of the Bosnian ruler was remarkably strengthened, however, by the profound change in the economic life of the country which followed on the opening up of ore mines. This was the work of Stephen II. Ore-mining, especially that of the precious silver mines, gave a new impulse to Bosnian commerce with the maritime towns and at the same time changed the distribution of the population of the country itself. Real urban centres grew up near the ore deposits, inhabited primarily by the "Saxons", who here as in Serbia and Hungary undertook the mining enterprises. Gradually the old centres of the crafts round the castles also became more prosperous. Urban commerce was encouraged by the minting of local coins. The Ban himself profited richly from the yield of the mines and from the coinage and market duties, especially from the famous Drijeva market in the Neretva valley. He entrusted the collection of his rich share in these profits to foreign agents, primarily from Dubrovnik.[9]

The economic policy of Stephen II Kotromanić, clearly modelled on the Hungarian Anjou rulers, certainly strengthened the real influence of the Ban in his own country. Nevertheless the state machinery still preserved many archaic features; the power of the Ban was significantly limited by the nobles (*vlastela*, sing. *vlastelin*), who were bound to their sovereign only by the personal bond of "loyal service". The nobles gradually won from the sovereign legal guarantees of immunity for their persons and their property. This was certainly the line of development in other European countries, too, but the archaic nature of the legal forms was peculiar to Bosnia, for they went back far into the Slavonic past. The general privileges granted to the aristocracy as a whole did not exist there, and instead charters confirmed the Ban's trust *(vjera* – vyera) in each individual vlastelin. The term "trust" implied guarantees of personal security and that of property. These could not be taken away unless the vlastelin was proved to have broken his bond of "loyal service", i.e. the set of duties binding him to his sovereign, but this could only be judged by a body of unbiased persons – often the same men as had borne witness to the sovereign's grant of "trust".[10]

A charter granting the Ban's "trust" to one Vlkoslav, of the important Bosnian family of Hrvatinić (Khrvatinitch),[11] has come down to us from the time of Stephen II Kotromanić. For the first time we find the "Bosnian Church" taking part in negotiations of importance. Moištra (Moishtra), one of the well-known centres of the church, was the scene of a solemn ceremony, in the house of the "great gost" Radoslav. A djed (dyed) (i.e. bishop) of the same name was also present, as well as the *starac* (starats) Radomir and other dignitaries of the Bosnian church; the vlastela present represented "Bosnia". The church is given pride of place in the charter: the granting of the "trust" of the Ban and his brother is explicitly said to have taken place "before the whole Church and before Bosnia". In case of dereliction by Vlkoslav (broken "trust") the case would be judged by a body of "good men", i.e. representatives of the Bosnian nobility, to whom he could present his defence. There is evidence from the same reign, however, that the "Bosnian church" also had the right to express her opinion of a vlastelin's guilt.[12] In the course of the years the weight of church authority in this connection grew even greater.

Unfortunately we know nothing of the developments which led up to this situation, in which the Church had such an important place in public life. The charter of Stephen II's reign only shows the outcome of these developments. The heretical church was firmly rooted in the land and was part of Bosnian life. Even the powerful vlastela could not manage without her help in ensuring their rights with regard to the Bans and later to the kings of Bosnia. The rulers of Bosnia themselves had to respect the Church, however much they may have wanted to maintain good relations with the papal Curia, and in spite of their desire to hasten the return of Bosnia to the Church of Rome.

The policy of the popes towards Stephen II itself changed, however, after the accession of Benedict XII. The latter, perhaps at the instigation of Stephen's enemies among the Croatian nobles, blamed the Ban of Bosnia for the failure of the Inquisition in that country, and tried to get another armed attack on the country under way. The powerful Hungarian nobles were to take part in this crusade, and Stephen II Kotromanić would have found himself in the same position as Ninoslav before him. This time, however, it was the King of Hungary himself, Charles I Robert, who energetically put paid to this hostile action against his most reliable vassal.[13] The situation was very different from that of a hundred years previously.

Relations between the papal Curia and the ruler of Bosnia soon returned to normal. Only three years after he had called for invasion of Bosnia, Benedict XII praised Stephen II Kotromanić for his eagerness to restore divine worship in his country, so that the churches would be restored and the liturgy heard there according to the rite of Rome.[14] The Pope had received a report from the General of the Franciscan order, Gerard Odonis, who had been to Bosnia at the request of Charles Robert of Hungary, and had been received with all honour by the Ban himself. Once again Stephen II Kotromanić displayed the best of intentions with regard to stamping out heresy in his land, but to that end he demanded effective help from the papal Curia and from Hungary, since if driven to extremes the heretics could find allies in neighbouring "schismatic" Serbia. There is no doubt that the Ban was primarily concerned not to run the risk of serious conflict within his own country, where the heretics had powerful protectors among the nobles.[15]

There was still one more occasion on which the Hungarians called for armed intervention against heresy in Bosnia – during the hostility between Louis of Anjou and Ban Tvrtko. Speaking of his invasion of Bosnia in 1363, King Louis I said that he had himself moved at the head of his army, in order "to destroy the vast numbers of heretics and Patarenes" settled in Bosnia.[16] This argument was used no longer, of course, once good relations had been established between the Anjou king and the Ban of Bosnia.

The time had passed for crusades against Bosnia. The Church of Rome had to relinquish the use of force against heresy there, for some time,[17] and there was nothing for it but to rely on the patient work of the Franciscan monks.

The "little brothers" became firmly entrenched in Bosnia at the beginning of the forties of the fourteenth century, when their Vicariate was established there, comprising several other neighbouring territories as well as Bosnia.[18] The Franciscans were a typically urban order, and in Bosnia, too, they found their supporters in the newly-formed urban centres with their primarily Roman Catholic inhabitants - German miners ("Saxons") and merchants from the coast. These urban settlements remained untouched by the heretical church from the outset. Furthermore, under Stephen II Kotromanić and again under Tvrtko, the Bosnian state had annexed whole regions where the people were either Roman Catholic or Orthodox. The heretical church of Bosnia did not penetrate these newly-annexed regions, and only made headway in some of the outlying parts.[19] Close contact with it, however, awoke interest in heretical Bosnia not only in the Catholic world but also among the Orthodox. By his invasion of Hum Stephen II aroused the hostility of Serbia, and the indignation of the Serbian church authorities. More Bosnian names found their way into the Serbian Synodicon and Ban Stephen himself appeared among the "cursed Babuns, lyingly calling themselves Christians".[20] Of all the names written into the Synodicon in connection with the Bosnian heresy from the time of Dragutin onwards, however, the one which interests us most is that which occupies first place: *Rastudije bos'n'skyi*.[21] A person of that name, as we shall see, played no insignificant role in the life of the heretical church of Bosnia.

The references to the Bosnian heresy in the Serbian Synodicon[22] add little to our knowledge of the sect. A far richer source of knowledge is provided by the written documents arising out of the work of the Bosnian Franciscans. Besides clear lists of "Patarene errors",[23] a polemic has also survived, in the form of a dialogue drawn up especially for use in Bosnia. This is clear from the title of the work itself: "A disputation between a Roman Christian and a Bosnian Patarene". The text itself reveals that it concerns sectarians cut off from the church of Christ in their isolated Bosnian mountains.[24]

In the middle of the thirteenth century the Italian Inquisitor Rainerio Sacchoni declared that "almost all the Cathar churches across the sea" follow the teachings of the older generation of Albanenses. We do not know whether this applies to the church in Bosnia, too, for we have no detailed information for this period. A century later the reports of well-informed Franciscan monks show that the dualist doctrines of the Bosnian heretics did indeed correspond to the faith of Belesmanza's trend in the Albanensian church in Lombardy. The only significant difference was that the Bosnians rejected the prophetic books of the Old Testament, accepting only the Psalms.[25] In Bosnia we also find an interesting reason for rejection of worship of the cross of Christ.[26] The heretics quoted Chapter 14, verse 9 of the Revelation of St.John, interpreting the "mark of the beast" as the cross.

We have already mentioned typically "western" features which were characteristic of the dualist heresy in Bosnia as well: in the attitude to infant baptism, in ascetic regulations,[27] no total rejection of wine, the use of the doxology at the end of the Lord's Prayer, and finally the theory that the apostolic succession had been broken by Pope Sylvester of Rome and was continued only in the Cathar "church of God". In the second half of the fourteenth century the Franciscans found among the Bosnian heretics the firm conviction that one of their number was the Vicar of Christ and the successor to the apostle Peter.[28] This agrees with those western sources which referred to Bosnia as the seat of the Cathar "Pope" as early as the thirteenth century. We do not know to what extent this Vicar of Christ was acknowledged throughout the great community of Cathar churches. In the second half of the fourteenth century his influence was certainly limited more or less to Bosnia itself. Reports of Cathar communities in the West had fallen silent dozens of years before, but for the last remnants of the "church of God" Bosnia was the Promised Land. From the little Piedmont town of Chieri people did not hesitate before the trials and dangers of the long journey, to acquire from their Bosnian brethren a thorough knowledge of their faith.

The Disputation mentioned above, drawn up by Franciscan monks in Bosnia, does indeed bear witness to the remarkably high standard the missionaries found among the heretics there. It reveals their profound knowledge of the Scriptures, their ability to find evidence in support of the individual articles of the heretical faith, and their skill in refuting the arguments put forward by their opponents.

The resistance of the sect to external pressure was not the result of their skill in argument alone, but primarily of the strong protection they had enjoyed from ancient times among the upper classes in Bosnia. The Franciscans' missionary work was by no means easy. Converts to the Catholic faith did not feel safe; they would only go to confession in secret and in public continued to show the accustomed respect to the heretics.[29]

The Vicar of the Bosnian Franciscans, Bartholomew of Alverna sent 23 questions to Pope Gregory XI, and the answers throw an extremely interesting light on the state of religion in Bosnia. This document, known as "Dubia" (1372-1373),[30] is of course of a significance which surpasses the boundaries of Bosnia itself. Indeed the "Bosnian" Vicariate of the Minorite order had been extended to cover the Croatian territory to the north-west of Bosnia, by the Una river, and later that of Bulgarian Vidin, ruled for a time by King Louis I; finally it embraced the southern region of Hungary. Some of Gregory's answers undoubtedly concern these regions beyond Bosnia's frontiers. The Franciscans often had to depend on the help of the local Orthodox clergy in these regions, men who were often very primitive, but willing to perform the rites of the Roman Catholic church in return for some reward. The Franciscan monks themselves had the special permission of the Pope to perform the functions of the parochial clergy in Bosnia itself and also in the adjacent Hungarian territory where parish boundaries had not been drawn up and where there was no parochial clergy. In this part of the country the population was both Catholic and Orthodox, with a scattering of heretics from Bosnia. In the eyes of the Catholics Bosnia was a land of "infidels". The people had yet to be baptised, and in addition there were all sorts of deeply rooted customs to be counted with. When the Franciscans first came to Bosnia they found a very curious marriage custom: married life was linked to the condition that the wife must be "good and faithful", so that it was easy to break off one marriage and enter upon another. The Franciscans found they had to deal with this idea of marriage in Bosnia even among people they had succeeded in converting to the Catholic faith.[31]

The only strongholds of Catholicism in Bosnia were the recently founded urban settlements and the few Franciscan monasteries, while whole vast regions which had never yet been incorporated into normal ecclesiastical organization lived in the spirit of traditions and customs surviving from the very distant past. In these rural areas there were here and there "Patarene houses", the real focus of heretical church life.[32] It was here, too, that trade caravans bringing cloth or salt or other necessities from Dubrovnik would break their journey. The ancient roads, long fallen into disrepair, were passable only on horseback, and the Dubrovnik merchants would hire mounts from the nomad shepherds called Vlachs. In records of agreements between the merchants and the "Vlachs" the name of Ljubskovo (Lyubskovo) occurs over and over again. There were "Patarene" or "Christian" houses there, in which the Vlachs could safely leave the goods entrusted to their care. It was here that they usually came, to the "Patarene region" *(ad patarenos, ad patarinorum contratas)*, and here that their journey ended.

Other "Patarene" localities are also mentioned in Dubrovnik trade documents. We have a record of a consignment of salt belonging to a Dubrovnik merchant, dated March 1418;[33] according to contract the "Vlachs" were to keep one half of the salt, while the other half was to be transported to a place called Bradina and stored in the house of a "Patarene", Milorad. Only if conditions were bad was the load to be stored in Konjic (Konyits), on the grounds of manors belonging to the king of Bosnia.

There were sometimes differences and complaints. One of the merchants of Dubrovnik, for instance, accused of stealing a length of precious cloth, blamed the "Patarenes" of Goražde (Gorazhde) for the loss.[34] This place was well known, being a much frequented market especially for grain, brought from Bosnia.

There was an interesting incident near the castle of Borač, belonging to the Radenović-Pavlović family, powerful protectors of the Patarene church. Two merchants from Dubrovnik passed that way on business in 1441; they had to wait before the castle while a safe-conduct was prepared for them, to guarantee their passage through the region *(contrata)* of the Patarene *gost* Gojslav (Goyslav), of Goražde - which was in the territory of Stephen Vukčić (Vukchitch).[35] Meanwhile the document made out in their names had fallen into other hands. The Patarene concerned then held the two merchants up, as they lacked the document necessary for passage, and took their goods off them.

The archives of Dubrovnik are rich in documents illustrating the varied contacts between Catholic Dubrovnik and heretical Bosnia. That such good relations were essential was the result of the geographical situation so aptly described in a letter of instruction sent by the Republic of Dubrovnik to their emissary at the court of Pius II (1459): the city was surrounded on all sides by non-Catholics, heretics, schismatics and already at that time "infidels" (Mohammedans). The Republic had not the necessary military strength to oppose them constantly with force, and therefore it was bound to live in peace with them. The land belonging to the city was not fertile, moreover, and food supplies had to be obtained from the people living in the surrounding country, regardless of their religious opinions. The Republic was left with no choice but to deliver the goods required by her neighbours, and in fact she even courted their favour with gifts of various kinds. The merchants of Dubrovnik therefore asked the Pope for absolution of their sins in trading with the heretics and men of other faiths, to which they were not led by their own will, but by the bare necessities of life.[36]

There is no doubt that in the eyes of the people of Dubrovnik the inhabitants of the Bosnian regions, untouched by Catholicism and susceptible to heretical teachings, were an inferior race and of course a reserve of unpaid labour. The legal code and the custom of the Dubrovnik Republic did not allow Catholics to be made slaves, but they put no obstacles in the way of slave-trading in "Patarene" Bosnia. In 1393 a decision of the court of Dubrovnik set two women free, who had been bought for a certain merchant. This decision was taken as soon as it was found that the women came from a purely Catholic region, that they had both been baptized, and that therefore they had been purchased illegally. The slave-trader protested in vain that he had bought them as "Bosnian and Patarene women".[37]

Bossinenses et patarenae – the words themselves reveal the contempt of the people of Dubrovnik for the *nameless mass* of unbaptized inhabitants of Bosnia, marked with the shameful seal of heresy. Yet this same republic went out of its way to curry favour with *leading personalities of the heretical church of Bosnia.* There are innumerable references in the records of their Assemblies to the voting of gifts to "Patarenes" whose important business brought them to Dubrovnik again and again. Life itself had enforced this apparent paradox; the barriers of faith had to fall before the fundamental political interests of the Republic.

The people of Dubrovnik themselves demanded that representatives of the Bosnian church should be present at important negotiations between them and the Bosnians; the moral authority of the church seemed to be the best guarantee of the reliability of such contracts. It was already known in Dubrovnik that the heretical faith of Bosnia would not allow the Patarenes to take any oath, and therefore their "word" was sufficient, "according to their custom".[38] Their word was well known to be as binding as the oaths of the other witnesses.

The term Patarenes, often used to denote the entire population of "heretical" Bosnia, signified in this context the "perfect" of the sect, known in Bosnia itself as "Christians" (krst'jane) and sometimes as "true" Christians. Even the Franciscan "Dubia" speaks of the Cathar melioramentum as *adoration* bestowed on the "perfect" by their believers. Right into the late period of Bosnia this honour was shown to the *krst'jane* even by the kings of the country.[39]

The Dubrovnik material suggests that the *krst'jane* lived communally in the houses we have already mentioned, which were usually designated by the name of one of the "perfect" who was at the head of the group *(societas)*. It can be assumed that it was from these houses that the *krst'jane* excercised their active influence over the ordinary Bosnian people, among whom they gained their "believers". The urban centres remained closed to heretical propaganda which mainly affected the rural areas, often hidden away in the inaccessible mountains, as we are told in the "Disputation" drawn up by the Franciscans of Bosnia.

Unfortunately neither the Dubrovnik material nor Bosnian sources themselves have anything to say about the part played by these Patarene houses in the ecclesiastical life of Bosnia, referring to them only in their secular context. The document already quoted, dating from the reign of Stephen II Kotromanić, vividly recalls the solemn moment in the house of the Patarene *gost* Radoslav, when "before the whole church and before Bosnia" the Ban and his brother granted their "trust" to one of the foremost nobles. On the other hand, the records of the chancellery of Dubrovnik aptly depict the importance of these Patarene houses as stations on the trade routes between Bosnia and the Republic. No description of these houses has come down to us, but from the evidence of these documents it can be deduced that they had space enough for quite large gatherings, and for the storage of considerable consignments of goods; and that they could house travellers and stable large numbers of horses. It is highly probable that they had their own farmlands, like the Catholic monasteries which they probably also resembled in the basic form of their communal life. We have no reference anywhere in Bosnia to the existence of houses for Patarene women, nor to the education of young people in such institutions. It is certain that the Dubrovnik merchants regarded the Bosnian *krst'jane* as "monks of the Bosnian sect" *regulantes sette Bosne*. This corresponds to the fundamental character of the class of the "perfect" in Byzantium and in the West.

In the hierarchical system of the Bosnian Church, on the other hand, as reflected in fourteenth and fifteenth century sources, we find features unknown elsewhere. We have already noted that the institution of the two "sons" of the bishop, general among the Cathar churches in the East and the West previously, was not found in Bosnia.[40] The designation *djed* (dyed) became accepted for the bishop, while the leading figures supporting him were known as *strojnici* (stroynitsi =stewards). The dignitaries known as *gost* and *starac* (starats) held a special position among them, but their functions are nowhere precisely indicated. The Dubrovnik records and local documents have preserved the names of many *krst'jane* as well as members of the Bosnian hierarchy. The names of important figures in the Bosnian church were preserved on their gravestones after death, and where time has not destroyed the inscriptions cut in the stone they can still be read.

The social standing of the Patarene church and its hierarchy in the state of Bosnia is clearly reflected in a letter sent to their fellow-countryman Cardinal John (1433),[41] by the Republic of Dubrovnik at the time of the Council of Bâle. It contains detailed instructions on how the leading personalities in Bosnia should be addressed. Besides the reigning monarch, Tvrtko II and the Dukes Sandalj Hranić and Radoslav Pavlović, the *Patarenes* are mentioned; in Bosnia itself they were known as monks *(religiosi)* of the Bosnian Kingdom. Four principal ranks are given: *djed, gost, starac* and *strojnik*. Native Bosnian documents make it clear that the bearers of the first three titles *(djed, gost, starac)* were given the attribute *gospodin* (lord) which was otherwise reserved by the Bosnians for the king and the dukes.

We do not know what was the exact position held by the Rastudij whose name is preserved both in the Serbian synodicon and in the late fourteenth century Bosnian manuscript known as the Batalo Gospel. It is very tempting to identify this "Bosnian Rastudij" with the heretic Aristodius, who is reported in the chronicle of Split to have been active in Bosnia.[42] However that may have been, the times of the Rastudij of the Batalo Gospel were not too far removed from the early thirteenth century, the time covered by Thomas of Split in his chronicle.

Only a fragment of the Batalo Gospel has come down to us.[43] The four sheets of parchment contain the conclusion of the Gospel according to St.John, and two notes have been added to the text. One gives the name of the patron commissioning the manuscript, the court dignitary Batalo, and the scribe *(djak)* Stanko Kromirjanin. We learn that the manuscript, illuminated in silver and gold and beautifully bound, was sent as a gift to the *starac* Radin in the year 1393, during the short reign of King Dabiša. The second note directly concerns the Bosnian church. It is a list of twenty-eight names, the first sixteen of which belong to those "who entered the order of the church before our lord Rastudij".[44] These names are followed by that of Rastudij himself, and then by another eleven names. The most interesting thing is that almost at the end of those first sixteen names, that is to say almost immediately before that of Rastudij, we find three of the names attested at the beginning of the thirteenth century in the famous document of the ceremony at Bilino Polje (1203). They are: Dragič (Dragitch) Ljubin and Dražeta (Drazheta). This would make it appear that Rastudij is a person very close to the period in which we first hear of the heretical church in Bosnia. The actual list of twenty-eight names included in the Gospel manuscript of 1393 was most probably copied from an earlier manuscript. We can rightly assume that we have before us a similar list to that found in the book *(liber Civitatis Dei)* where the Cathar community of Chieri wrote the names of those who had embraced the Cathar faith.[45] (It is possible that the Bogomil church in Bulgaria also kept such lists, and that in this way the names quoted in the Bulgarian Synodicon were handed down). *Gospodin* Rastudij was certainly a very important person in the early history of the Bosnian church, since his name forms the turning-point in the list of those "who entered the order of the church" in the Batalo Gospel manuscript.

It is very likely that Rastudij held the rank of Bishop *(djed)*. In the later period of the Bosnian church we have two interesting examples of dating documents by the bishop in office. In 1404 the most powerful of the Bosnian nobles, Hrvoje (Khrvoye) Vukčić Hrvatinić, who had recently attained the position of Duke (Hertseg) of Split, had a magnificent illuminated manuscript prepared. The *krst'janin* Hval who was entrusted with the copying of the text noted at the end that he was writing in the days of the bishop of the Bosnian church, *gospodin djed* Radomjer (Radomyer).[46] Later the *krst'janin* Radoslav dated a manuscript for the *krst'janin* Gojslav (Goyslav) in the time when Bosnia was ruled by King Stephen Thomas (1443-1461) and the head of the Bosnian church was *djed* Ratko.[47]

The manuscript of Radoslav [48] contains not only the text of the Apocalypse, but important elements of the Patarene ritual in Slavonic.[49] It is the exact counterpart of the introductory (Latin) section of the Cathar ritual in the Lyons manuscript. The text of the Lord's Prayer including the final doxology is a sure pointer to a western text on which the Bosnian version is based. In Languedoc and Italy those voices had died out for good, and *Benedicite, parcite nobis, Adoremus patrem et filium et spiritum sanctum, Dignum et iustum est,* were no longer heard; but in Bosnia there sounded: *Blagoslovite, opraščaite nam', Poklanamo se ocu i sinu i svetomu duchu, Dostoino i pravedno jest'.* (Blagoslovite, oprashtchaite nam, Poklanamo se otsu i sinu i svetomu dukhu, Dostoino i pravedno yest). These words hallowed by tradition were carried freely through the places of prayer, sounding at the administration of the *consolamentum*,[50] at the breaking of bread,[51] at common confession - the living testimony to a faith which had lasted since the time of the apostles and would last until the end of the world.

We do not know whether there had ever been a systematic account of the doctrines of the Bosnian church, or a detailed polemic with its enemies; all that has come down to us are the marginal glosses to the text in one of the Bosnian manuscripts of the Gospels.[52] The miracles and the parables of Jesus are freely interpreted in the spirit of dualist doctrine, as abundantly testified by Cosmas, Zigabenus and later by western authors. We see how ingeniously the sect invented constant new interpretations of the stories in the Gospels. This is particularly striking in the case of the parable of the prodigal son, in St.Luke's Gospel; various interpretations are recorded by Cosmas the Presbyter for the Bulgarian Bogomils and by Moneta for the Italian Cathars. The gloss in the Srećković manuscript interprets the younger, prodigal son as the angels seduced by Satan, while the elder son who remained with his father symbolizes the faithful angels. The father of the two sons is of course the good God ("the invisible Father"). A bizarre detail emerges: the fatted calf (killed for the feast) is Christ.

An allegorical interpretation of the story of the good Samaritan can already be found in various collections of "Questions and Answers" *(erotoapokriseis)*, and the sect had its own approach to this apocryphal text.[53]

The author of the glosses in the Vienna manuscript of the Interrogatio Iohannis had also close relations with Bosnia. These marginal glosses give an allegorical interpretation to some passages of the Bogomil book.[54]

The literary remains which originated within the Bosnian church itself confirm the accuracy of Catholic observers who placed it in the context of the sectarian movement with dualist doctrines. Authentic utterances of the heretical church itself are very few indeed. Much has been written about the "Bogomil" character of the Bosnian Gospels. It is true that the literary heritage of mediaeval Bosnia comprises a remarkable number of Biblical manuscripts and especially manuscripts of the Gospels. Their very number testifies to frequent copying and revision of the text. This makes it very difficult to determine their origin. Nevertheless it appears that the Bosnian Gospels are part of a very archaic (and early differentiated) textual tradition. One reason for assuming a very old original is the fact that these Gospels use two ancient ways of dividing the text. This manner of division makes it easier for the reader to find his way about the matter of the Gospel, but does not correspond to the division for liturgical use. This would suggest that the Bosnian Gospels were intended for private reading in the homes of the upper classes, who commissioned the copies. Occasionally, however, we come across traces of liturgical reading left in inadvertently, proving that the manuscript or the original from which it was copied was transcribed from an ancient *aprakos* text intended for use in the church service. The consistent omission of the liturgical division of the text certainly shows that these Gospels did not serve a liturgical function in mediaeval Bosnia. One of the ancient ways of dividing the text, however, could also facilitate the reading of the Gospels to a large audience, and could thus have served the heretics' worship. This very manner of division into "old chapters", however, is of very ancient date and did not originate among the heretics. There is not a single reliable proof of Patarene interference with the texts of the Gospels or other Biblical books. (Later, after the fifteenth century, these old manuscripts were used without difficulty in the church liturgy.) Certain curious lexical features, wrongly supposed to reflect dualist doctrines, have been shown to be merely synonyms with no heretical significance whatsoever.[55]

Equally suspect are efforts to see "Bogomil" elements in the miniatures illustrating the Bosnian Gospel manuscripts.[56] In fact the conception of the individual figures and the whole composition are entirely in accord with the iconography of western religious art. It reflects the Catholic orientation both of the artists themselves and of the patrons commissioning the illuminated manuscripts. The name of Divoš Tihoradić (Divosh Tikhoraditch), for whom a manuscript of the Gospels was completed about 1330,[57] is mentioned in two documents of the time of Stephen II Kotromanić. One of them gives his name among the witnesses from the Završje region, so that he must have been a Catholic nobleman from one of the regions then recently annexed to Bosnia. These were Croatian regions (Duvno, Livno, Glamoč) which tended to look to Split. Indeed the style of the miniatures richly ornamenting the Divoš Gospel is clearly in the tradition of the art then flourishing in the coastal regions and reaching from there into Serbia. The initials of the Bosnian manuscript show the same fantastic animal motifs as the relief carvings on the marble facing of the outer walls of the cathedral of Dečani (Detchani) - ancient motifs of oriental origin, taken over by Romanesque art in western Europe. This archaic style of illumination became popular in Serbia at the time of Miroslav's Gospel, towards the end of the twelfth century. Bosnian illuminators remained faithful to it up to the fourteenth century, when Serbian miniatures were already imitating Byzantine models and becoming markedly illustrative in character. The symbolical and decorative aspects of illumination were still uppermost in the work of the illustrator of Divoš Tihoradić's Gospel. This conservative style was still followed in the illustration of later Bosnian manuscripts which were far behind those of the Divoš Gospel.

The links between Bosnian illuminated manuscripts and the art of the coastal towns was not restricted to Romanesque art; the illuminations in Hval's Cyrillic manuscript (1404)[58] are purely Gothic in style, except for the Cyrillic initials where traditional Bosnian models had to be followed. The first of the two limners concerned reveals bonds with Dalmatia and specifically with Split, in some of the thematic details of his work. There is no doubt that this was where Hval's manuscript was illuminated; it was intended for Hrvoje Vukčić, the most powerful of the Bosnian noblemen, who had attained the position of "hertseg of Split" shortly before. This same limner was also responsible for most of the illuminations in Hrvoje's Glagolitic missal, copied about the same time (probably 1403-1408). The iconography of all these illuminations is purely Catholic in character, although the latter MS was transcribed by the priest Butko, while the scribe copying the other manuscript was a member of the Patarene church, the *krst'janin* Hval.

Analysis of the style in the illuminations of another fifteenth century Bosnian manuscript (in the Bibliotheca Marciana, Venice) presents a more complex picture. Two artists can again be distinguished. Typically Romanesque motifs long current in Bosnia appear in the title and initials by the first artist, while his miniatures and illustrations show a synthesis of Gothic art and oriental influences. The large number of illustrations by this first painter in the Venice manuscript tends to obscure the remarkable work of the second artist, whose delicately drawn portraits of the evangelists already herald the early renaissance art of Dalmatia.[59]

There is a clear link with the traditions of the art of the western world even in so specific a form as the sculptural ornament of Bosnia tombstones.

Several tens of thousands of mediaeval tombstones have survived in what is now Bosnia, Hertsegovina, Imotska Krajina and some neighbouring areas.[60] Standing in smaller and larger groups on the mountainsides and in the valleys, and on the banks of rivers and lakes, they create an atmosphere of mystery which has sometimes led to the growth of legends about their origins. Systematic research into this abundant material has gradually revealed the significance of these *stećci* (stetchtsi) in the social life of Bosnia.[61]

The greatest period of these tombstones came in the fourteenth and fifteenth centuries; they continued into the Turkish period and did not die until the sixteenth century. The tombstones themselves and some features of their ornament are certainly rooted in ancient traditional funeral rites, in cult symbols. Sometimes they are found on tumuli left behind by the original Illyrian inhabitants. The nearest model for one type of stone monument in mediaeval Bosnia and surrounding regions was certainly the Roman sarcophagus. In ornament the ancient motifs such as rosette and crescent were abundantly used, their original significance in the funeral cult forgotten; indeed Bosnian craftsmen could have taken them from folk embroidery or wood carvings.[62] The artist's invention often attributed different significance to the same plant motif of geometrical pattern, according to its place on the walls.[63]

The ornament on the tombstones, however, was not inspired by native sources alone; it drew much on the art of the western world. The nearest point of contact was in the coastal towns of Dalmatia, where architectonic motifs like the arcade were observed. The decorative arts were highly developed in these towns, and offered plenty of models for imitation. The craftsmen of Dubrovnik used the same motifs in their decorative canvasses as we find on the tombstones of Bosnia; Bosnian craftsmen could find inspiration in the work of silversmiths and engravers, ivory carvers, sculptors and book illustrators. In the fifteenth century the nobility of Bosnia was in lively contact not only with the towns on the coast of Dalmatia, but with the countries of the West. Artists from France and the Netherlands worked in Hertsegovina. Indeed there are some striking parallels between the decoration of the tombstones and some examples of western art.[64] There is a magnificent tombstone from Donja Zgošća with relief carvings of the same trees with stylized branches above the figures of hunters, dogs and lions, as we see in the mosaics decorating the Stanza Normana in Palermo, dating from the twelfth century. In each case the motifs are of ancient provenance, and form part of the normal repertoire of secular iconography in the West. The scene depicting the "siege of the castle of love" on a tombstone in Radimlja (Radimlya) on the other hand, has been based on a theme from contemporary western literature of chivalry. Scenes of chivalrous games, hunting and dancing are repeated over and over again in the art of the late Middle Ages; in the dance scenes on Bosnian tombstones, however, native folk-lore could also have contributed to the inspiration. This combination of heterogeneous motifs in the same scene was quite in keeping with the tendency in late mediaeval secular art to allow aesthetic criteria to prevail over the logic of the subject matter. The effect of the ornament on these tombstones was certainly heightened by the vivid colours so popular during this period, and long ago washed away by the rain.

On the tombstones in Donja Zgošća (Donya Zgoshtcha) the motifs are faithfully treated in carved relief, whereas elsewhere the scenes are reduced and remarkably simplified in content and form, in a manner reminiscent of modern art.[65] The dance scenes are drawn in simple outlines and the harmonious composition creates the illusion of rhythmic movement. The geometrical patterns and formal plant motifs on tombstones in Eastern Bosnia reveal subtle artistic feeling. A great deal of the figural ornament, however, is unskilful in treatment; the folk craftsmen had not mastered the art of proportion when depicting the human figure and the scenes have the naive charm of children's drawings. Yet it cannot be denied that many of these primitive monuments are fresh and effective in expression.

In spite of the different taste in the different regions, in spite of the varying quality of the craftsmanship, the ornament on such tombstones as has survived presents a single whole such as in seen nowhere else. The imagination of their creators made free with the motifs adopted from elsewhere, putting them together in new combinations towards a certain aesthetic effect and sometimes to serve a new symbolism whose meaning escapes us.[66]

Nowhere, however, can we find any trace of the dualist doctrines of the "Bosnian church".[67] Nor do the tombstones under which the dignitaries of this church are laid to rest differ in general character from the others and their ornament. The figure of an old man with an open book, on the tombstone of a *djed* in Hočevlje (Khotchevlye), probably represents one of the evangelists. It has a surprising parallel in the figure of an evangelist in one of the Bosnian Gospels. In the West the tombstones of bishops were adorned with carved portraits of the evangelists as early as the eleventh century.[68]

Thus we see that not even in Bosnia did the dualist doctrine of the "Patarene" church leave its mark on the development of the nation's culture. Nevertheless the heretical church remained an important factor in the life of the Bosnian *state*, so long as it maintained a united existence. The great state of Tvrtko I, made up of heterogeneous elements, could not preserve its unity once the personality of the ruler had disappeared. As in neighbouring Serbia earlier, the country inevitably moved towards splitting up into several independent units. Right at the beginning of the fifteenth century King Ostoja (Ostoya) (1398-1404), who still controlled important castles and the centres of economic life, could still attempt to strengthen the royal power once more. But the resistance of the nobles brought about his fall. At this time the authority of the Bosnian church was a mainstay of aristocratic opposition to the king. In the summer of 1403 "Patarenes" brought Paul Klešić (Kleshitch) to Dubrovnik, where he was given asylum.[69] By the beginning of the following year the growing discontent of the nobles forced Ostoja to return Paul's confiscated lands and castles to him. The king then guaranteed his undisturbed possession of them in the future and entrusted him to the Bosnian *djed*, his *strojnici* and "the whole church of Bosnia".[70] This new form of guarantee first appeared under Tvrtko I, who entrusted one of his *vlastela* to "the trust" of the *djed* and of the whole Church of Bosnia and all the *krst'jane*.[71]

The unquestioned authority with which the representatives of the Bosnian church negotiated with the king over Paul Klešić is reflected in an authentic letter from the bishop to the Republic of Dubrovnik: "...We have compelled the king to return his property to him (i.e. Paul Klešić), because it was confiscated without his guilt." The Duke could now safely return to Bosnia; the bishop *(djed)* sent his *strojnici* and *krst'jane* to Dubrovnik to fetch him.[72] Their actual names have been preserved for us in the records of the Dubrovnik scribe Rusko.[73]

During this period the Patarene church also gave its protection to one of the Radišić brothers, in conflict with the king. One other brother was imprisoned while a third (Paul) took refuge in Dubrovnik along with another Bosnian noble. In spite of Ostoja's protests the Republic energetically defended its right to afford asylum. The king could hardly deny this right to the fortified city *(città murada)* especially since in his own country he did not interfere with the freedoms *(franchigie)* of the Patarenes, under whose protection Paul's brother was living a free life, "eating the king's bread" in one of the Patarene houses *(in caxa de' Patareni)*.[74]

The asylum afforded to Paul Radišić and his fellow *vlastelin* increased the tension which had already existed between Ostoja and Dubrovnik for some time. In the summer of 1403 war broke out. At the beginning of the following year the Dubrovnik mission sent to negotiate peace with Bosnia was given precise instructions on the formal procedure to be followed. All was to be guaranteed by the oath of the king, the queen and the *vlastela* and the *"word" of the Patarenes, according to their custom.*

When the situation changed with the fall of Ostoja, the Republic ordered its emissaries to continue negotiations with the new king; if none had yet been chosen, the treaty was to be sworn to by the greatest man in Bosnia, the Hertseg Hrvoje Vukčić Hrvatinić (Khrvoye Vukchitch Khrvatinitch), and together with him by the leading nobles and the Bosnian assembly. In addition the emissaries were again to demand the "word" of the Patarenes. Similar instructions were again sent from Dubrovnik in the early autumn of 1404, after the new king Tvrtko II (1404-1409) had ascended the throne.[75]

We see that at this time (1404) the government of Dubrovnik appealed to the Patarene church, as to the sovereign and the nobility, in a matter concerning the *whole of Bosnia*. Yet in a few years the unified state of Bosnia had fallen apart into separate regional domains. This was the consequence of rapacious struggles between powerful houses each trying to grab the largest portion. A few able individuals emerged victorious, having managed to repress the others. They took upon themselves all the rights of the ruler in the territory they had seized, and in this situation the King of Bosnia became no more than one of the regional rulers. The unity of Bosnia was still symbolized by the "Crown of Bosnia", reserved for the person of the ruler, and the institution of the Bosnian assembly which continued to exist. The territory of the country was divided up, however, and the major part of it ruled by three powerful nobles: Hrvoje Vukčić of the Hrvatinić family, Sandalj Hranić of the Kosača family, and Paul Radenović, founder of the Pavlović family.

The decisive struggle for power carried on by the nobles took place in the years following the fall of King Ostoja (1404), at a time when Bosnia was also threatened by the Hungarian king Sigmund, who aimed at getting control of the country. Finally in 1414 Hrvoje Vukčić called upon the Turks for military help against Ostoja, who was again on the throne (1409-1418) and had allied himself with Sigmund. A great Hungarian army sent into Bosnia at the beginning of 1415 suffered a decisive defeat. The moment of victory served to settle political differences within Bosnia, and for a very short time the whole country was united against the Hungarians. Pressure from that quarter ceased, and King Sigmund, who had other political interests to pursue, never again took an army on to Bosnian soil.

The bringing in of Turkish troops in 1414, however, was the first step in what was to be a fateful development for Bosnia. From then onwards the Turks frequently interfered in the ceaseless internal disturbances in the country. This meant not only that Turkish troops devastated the Bosnian countryside, but also that the land became politically dependent upon the Porta. The King of Bosnia, like his regional rulers, all became vassals of the Turks.

Murderous struggles broke out inside Bosnia again with full force, soon after the victory over Hungary. The death of Hrvoje Vukčić in 1416 caused a significant shift in the balance of forces. Most of his vast domain was in the hands of King Ostoja, whose goal was to break the resistance of the regional rulers and restore the full authority of the royal power. In 1418 Ostoja died, however, and three years later his son was forced to vacate the throne to Tvrtko II, this time helped to the throne by Turkish intervention. During his second reign (1421-1443) this son of the great Tvrtko increased the weight of the royal authority considerably, but his method was one of skilful fostering of the rivalry between the houses of Kosača and Pavlović. The king's own domain, *contrada del re*, formed the firm foundation of his power, for it was on his territory that the most important ore mines were situated.

After the death of Sandalj Hranić (1435) his nephew *Stephen Vukčić Kosača* (Vukchitch Kosatcha) became a significant factor in the gamble for political power. Inheriting the great Sandalj domains he very soon strengthened his position on all sides and in particular won Turkish support. The Turkish invasion of Serbia (1439) gave him the chance to extend his power to the territory of Serbian Zeta without the slightest scruples. In Bosnia itself he already governed the southern lands of Duke Radoslav Pavlović. There was usually intense hostility between the Pavlović family and the powerful house of Kosača, but from the middle of the fifteenth century the former ceased to play an independent role in the life of Bosnia.

There was a short period of growing prestige for the power of Bosnia as a whole, when Stephen Vukčić and Tvrtko's successor made peace, strengthened still further by the marriage of King Thomas (1443-1461) with Stephen's daughter (1446). The peace did not last long, however, and in 1448 we find Stephen Vukčić assuming the title of "Hertseg (i.e. Duke) of St.Sabbas", intended to show to the world at large what an important position he held in Bosnia. Open hostilities broke out between him and the king, and lasted almost three years. The heavy expense of the war and of diplomatic negotiations, as well as the cost of maintaining a court and administering the country, forced both the king and the Hertseg to look for ways of increasing their revenues. Stephen Vukčić made far-reaching changes in the existing organization of the salt trade and built a new urban centre (Novi) where he introduced the weaver's craft. These "novelties" aroused sharp opposition from Dubrovnik, and the growing tension between the Hertseg and the Republic turned to open war in 1451, which ended in 1454 with the restoration of the status quo. By this time, however, Bosnia began to be urgently menaced by the Turks.[76]

The Patarene church could no longer maintain its authority over the whole of Bosnia, in this ceaseless clash of contrary political interests. Indeed some of the leading figures in the church found themselves in dangerous situations in the course of hostilities between the Bosnian rulers. After the murder of Paul Radenović (1415) the *krst'janin* Vlatko Tumarlić had to seek refuge from his master's murderers in the Franciscan monastery of Sutjeska (Sutyeska).[77] In the years that followed Vlatko did good service to Paul's sons in the diplomatic field. This is only one of the examples of the influential role played by outstanding individuals in the Bosnian church after the regional domains had been established. The archives of Dubrovnik have preserved much historical evidence of important negotiations carried on in Dubrovnik by *krst'jane* or even members of the Patarene hierarchy, in the services of the Kosača or Pavlović families. These *krst'jane* did not live in the communities in the "Patarene houses", but spent all their time at the courts of these aristocratic families. An inscription on the tombstone of *gost* Milutin speaks of the respect in which he was held at the court of the "Bosnian lords" and of the gifts he received "from the lords and the *vlástela* and from the Greek lords."[78]

Close contact with the feudal world, and high social standing in that world could not fail to leave their mark on the attitudes of these foremost members of the church of Bosnia. Their view of worldly wealth, at any rate, was not as uncompromisingly negative as that found elsewhere.[79] They inevitably adapted their way of life to that of the high society in which they lived, in the manner of the late Middle Ages. Just as in France, the Netherlands and elsewhere, we find the same tendency to luxury in Bosnia, the same love of showy ceremonial, and also a very highly developed sense of social differences. Precious cloth, valuable vessels and other works of art commissioned from the craftsmen of Dubrovnik and Venice were an essential feature of the luxurious life of the Bosnian rulers and nobles. As in Serbia, the regional courts of Bosnia offered favourable conditions for the development of artistic traditions.

Hrvoje's missal and Hval's manuscript give us some idea of the nature of the art which was developing in the circles round the most powerful of the Bosnian nobles, at the beginning of the fifteenth century.[80] It was a reflection of conditions peculiar to Bosnia that the transcribing of the manuscript was entrusted to a *krst'janin* of the Bosnian church, while the illuminations were the work of artists from the maritime region. We know that one of them also painted the miniatures in Hrvoje's Glagolitic missal. This liturgical book was drawn up "according to the order of the Holy Roman Church". Even the choice of biblical texts included in Hval's manuscript does not entirely agree with the doctrines of the Bosnian *krst'jane*.[81] It is understandable that the Bosnian church was forced to make such concessions to powerful individuals on whose protection they could safely rely. In 1403 Hrvoje Vukčić was at the height of his power; for the unconditional support he had given Ladislaus of Naples in his fight for the Hungarian throne, he received from the Anjou king not only the title of Hertseg (Duke) of Split, but actual government over Split and three of the Dalmatian islands.[82] The political context of his new position inevitably led him to turn towards Catholicism. His letter to the Hungarian queen explicitly states his wish to turn away from "heathen" and "infidel" ways to the Catholic faith.[83]

The younger generation of the Hrvatinić family, who played a very small part in the political life of Bosnia, were already Catholic. When the nephew of Hrvoje granted his "trust" to the Radivojević brothers he no longer took *krst'jane* of the Bosnian church as guarantors, but Franciscan monks.[84]

Right up to the end, however, the Patarene church had the support of the Pavlović family and of Stephen Vukčić. Soon the *krst'janin Radin*[85] appears alongside Vlatko Tumarlić in the services of Radoslav Pavlović.

His name appears in connection with important negotiations at the beginning of the twenties of the fifteenth century, and in 1432 he was one of the emissaries of Radoslav sent to make peace with the Republic of Dubrovnik.[86]

After 1437 however, Radin appears in the services of Stephen Vukčić Kosača; by this time he already bore the title of *starac* (starats) and was thus a member of the hierarchy of the Bosnian church. The business which brought him to Dubrovnik in the services of his new master was purely secular, however. In the first place there was the property left by Sandalj Hranić; on Stephen's orders *starac* Radin first took charge of part of the treasures of the dead Duke, and then of almost everything; up to that time it had been in the keeping of the Republic of Dubrovnik. Under Radin's care quantities of gold ducats and silver coins were brought from Dubrovnik to Bosnia, along with precious vessels of wrought gold and silver. Duke Sandalj had also had money invested in Dubrovnik at interest, and Stephen Vukčić sent Radin once more to collect the interest.[87]

From the end of 1450 Radin is given in the Dubrovnik records as *gost*, that is to say as holding even higher rank in the Bosnian ecclesiastical hierarchy. His diplomatic experience became even more important both for Hertseg Stephen and for the Dubrovnik Republic at the time when conflict between the two was imminent. The affair is illustrated in a most interesting fashion by the instructions given to the emissaries of the Republic negotiating in Bosnia. At the end of 1450 they were told to begin secret negotiations with *gost* Radin as the "good and principal counsellor to Hertseg Stephen"; they were to appeal to his friendly feelings towards the Republic and at the same time to hint that the efforts he made towards settling the conflict would not remain unrewarded. The Republic *(signoria)* was always grateful to those who expended efforts on her behalf.....[88] In the summer of 1451 the sum of 200 ducats is already mentioned "not as a reward" for Radin but as a sign of goodwill towards him *(in segno di amor)*.[89] Two years later (1453) two emissaries in Bosnia again received instructions to negotiate with *gost* Radin. If he were to further the conclusion of a treaty with the conditions and guarantees of the time of Duke Sandalj, his efforts would be appreciated by the sum of 400 gold ducats.[90]

In 1455 the Republic granted *gost* Radin the right to settle in Dubrovnik for good, whenever he pleased. A house was put at his disposal in the town, for himself, his familiars and his servants. At the same time he received the significant guarantee that nobody would force him to desert the faith he confessed.[91] Several years later, at the very end of his life, *gost* Radin wrote his famous last will and testament,[92] in Dubrovnik. The instructions he gave to the executors of his will resemble those in the will of Stephen Vukčić himself. In Radin's will, too, the first bequest (600 ducats) is "for the soul" of the dead man. *Gost* Radin left 300 ducats to be distributed among the poor of the town, who should pray for him every Sunday and Friday and on certain feast days specifically laid down. Candles were to be lit in the churches of Dubrovnik on each occasion (as he said, "for my soul - the soul of *gost* Radin"). A certain sum was laid aside for the chapel and tomb where the body would be laid to rest. These are certainly concessions to the form of worship of Catholic Dubrovnik. There is no doubt that Radin kept a number of church feast-days and also celebrated his *krstno ime*. We know, however, that the festivals of the Catholic church year retained a certain significance among the Cathars in the West, too. There is nothing to suggest that his concessions to his Catholic environment in any way affected Radin's fundamental religious beliefs,[93] to which he openly confesses in this testament. Even in his bequest for care "of his soul" it is primarily the people of his own church (*od našega zakona* - od nashega zakona) that are concerned, the *krst'jane* and *krst'janice*, men and women, who hold "the true apostolic faith". These are simple people of the peasantry, *kmeti* and *kmetice*, who are to receive money from another sum of 300 ducats; Radin entrusts this distribution to his nephew-namesake, also *gost* of the Patarene church, charging him "by the faith he confesses and the fast he keeps", to perform his task justly. In the first place the poorest of these villagers are to receive money (but men of noble birth come down in the world are also included), and those who are afflicted by age, sickness or great infirmity. This concerns only the *krst'jane* men and women, i.e. the "perfect" members of the church of Bosnia who "live without sin". In addition, the money should be used to help unfortunates among the "people living in the world".[94]

Even in the distribution of bequests to his relatives and servants Radin mentions first those who are *krst'jane* and only then the others.

This will says nothing of landed property belonging to Radin, nor do we know of the existence of any. It accounts for his movable property, in particular 5,640 gold ducats, and precious vessels and valuable clothing, among these treasures a velvet fur coat presented to him by King Matthias of Hungary, similar to the one the King gave to Stephen Vukčić.

Gost Radin, in the words of the Venetians: *principalis baronus* of the Hertseg Stephen, appears in his will as a secular noble surrounded by his retinue and possessed of great wealth. Yet his position in the Bosnian church is not pushed into the background; the latter was still a significant force in the entourage of Stephen Vukčić. Indeed the Hertseg himself relied on the authority of the church, of its *djed* and *strojnici* (stroynitsi) - and primarily of *gost* Radin - when settling conflicts within his own family.[95]

The broad domains of Hertseg Stephen, however, also included regions where the people were Catholic or Orthodox, and above all the Serbian metropole of Mileševo. These facts were bound to determine Stephen's ecclesiastical policy. It was in his time that an anathema against the leading Bosnian heretic found its way into the Serbian synodicon: "Radin, the Hertseg's *goet*, be anathematized!"[96] The name of Hertseg Stephen himself, however, was much more favourably pronounced in Orthodox circles, for he had an Orthodox church built in Goražde (Gorazhde), on his territory, and sent gifts to the monastery of Sinai.[97] In the period following the fall of Constantinople the Byzantine patriarch Gennadios Scholarios (1453-1456) wrote of the hertseg as an Orthodox ruler who was only deterred from openly confessing the true faith and breaking away from the heretics by his fear of the noblemen of his country.[98]

In Rome, however, a much less indulgent view was taken of Stephen's benevolent attitude towards the Patarene church. In 1449 he was threatened with excommunication if he did not lead his subjects to the Church of Rome. It is true that during the pontificate of Eugene IV (1431-1447) Stephen Vukčić expressed his wish to be baptized and to become a good Catholic, for which the Pope rewarded him with his favour and considerable privileges. A bull issued by Nicholas V (1447-1455), however, took it all back because the hertseg persisted in infidelity *(in la infidelita)*. Nevertheless Thomas, king of Bosnia, fearing that Kosača would turn to the Porta, persuaded the Bishop of Hvar, Thomas, who was papal legate in Bosnia, to hold back in his own hands the bull which rescinded the privileges granted to Stephen by Nicholas's predecessor. Our knowledge of the whole affair comes from an instruction sent from Dubrovnik, dated 18th August 1451.[99] A little earlier (at the end of June) armed conflict had broken out between the Republic of Dubrovnik and hertseg Stephen, and in this situation the former were naturally eminently interested in having the contents of the papal bull made public.

In spite of the disfavour of Nicholas V, who called upon Stephen to stop the fighting against Dubrovnik, the Hertseg again showed that he was anxious for good relations with Rome.[100] In fact he never broke off relations with the Papal Curia at any time during his reign, and in 1454 he even asked King Alphonse of Naples to send some Franciscans to work as missionaries in his country.[101] In spite of all this, he never withdrew his support from the Patarene church.[102] Hertseg Stephen, who was bound in internal affairs to respect the attitudes of the powerful Orthodox Church, and who in foreign affairs relied on the Turkish alliance, never developed so thoroughly a pro-Rome policy, with such close co-operation with the Curia, as the last kings of Bosnia did.

This line of development, which was to seal the fate of the heretical Church of Bosnia, already emerged in "royal" Bosnia during the second reign of Tvrtko the younger. The situation of the Franciscan monks in Bosnia at the beginning of the fifteenth century seemed desperate indeed, and some of them even left the country.[103] The effective help they were given by Tvrtko II, however, strengthened their position considerably, and the order already began to take up positions in close proximity to the Bosnian king. At this time (1433) a proposal was made to the Council of Bâle that this oecumenical gathering which was concerned among other things with reconciling the Czech Hussites with the Church, should also make efforts to convert the "Manichaean" inhabitants of Bosnia. The Council used the awful example of the Bosnian heresy as an argument against the Hussites. It is noteworthy that the Vicar of the Bosnian Franciscans, Jacobus of Marchia,[104] whose task it was to combat heresy in Bosnia, was also charged with the conduct of the Inquisition against the Hussites in Hungary and the neighbouring countries. (The two anti-ecclesiastical movements - the Patarene and the Hussite - could have come into contact in the region of Srěm, but we have no knowledge of this.)[105]

None of the plans of the Council of Bâle for the conversion of heretical Bosnia were put into effect. Nor was there anything gained by the gesture of Tvrtko II in 1442, when although himself a Catholic he solemnly abjured heresy in the name of his country.[106]

Unlike his predecessor, the new king of Bosnia Stephen Thomas (1443-1461) was not originally of the Catholic faith. His first marriage was "in the Bosnian manner" and he himself seems to have been the son of such a union.[107] An epistle of Eugene IV written in the summer of 1446 credits Thomas of Hvar with the conversion of the King[108] at the outset of his reign or perhaps even earlier.[109] Two papal bulls of 1445 made the king's birth legitimate and annulled his illegal marriage[110] The second wife of King Thomas (the daughter of Stephen Vukčić Kosača) became a Catholic before her marriage, and the ceremony took place with all magnificence according to the Catholic rite.

During the great Crusade which the combined forces of the Catholic world waged against the Turks, not even the new Bosnian king could fail to show his Christian ardour.[111] This explains his marked orientation towards Catholicism at the very outset of his reign. He nevertheless had to count with the strong position of the heretical church in his own country and with the respect accorded it both abroad and in Bosnia itself. During the reign of Tvrtko II (1428) the Republic of Dubrovnik gave its emissaries letters of credit to the Bosnian *djed*. They were to request him to use his influence over the Bosnian king in order to remove the "novelties" recently introduced, which adversely affected Dubrovnik's trade with Bosnia.[112] The noblemen of Bosnia still regarded the authority of the Patarene church as the essential guarantee of their rights with regard to the sovereign. In the summer of 1446 King Thomas presented the castle of Ključ (Klyutch) and a number of villages to the three sons of Duke Ivan Dragisić (Dragishitch), on which occasion he commended them according to the established custom, "into the hands of the church", specifically, of the bishop *djed* Miloje (Miloye) and of his successor after him.[113]

During the first years of his reign Thomas had to make the usual "adoration" to the Bosnian *krst'jane*, something which disturbed his relations with the Franciscans, who refused him the sacraments. They regarded him as a protector of heretics and in accordance with valid ecclesiastical law as excommunicated from the Church. Pope Eugene IV, however, appreciated the tactical reasons which led the king to treat the Patarenes with caution, protected as they were by the Bosnian noblemen.[114] The time for energetic action against the Bosnian church had not yet come. For the moment all the king could do was to give assiduous support to the work of the Franciscan mission; the results soon made themselves felt among the Bosnian upper class.

The correspondence around the sojourn in Bosnia of Thomas, Bishop of Hvar, who was papal legate 1439-1461,[115] throws penetrating light on developments in the country from the very beginning of Thomas' reign. The letters written to him by Pope Nicholas V about the middle of the fifteenth century[116] speak of the conversion of a large number of highborn Bosnians to the true faith. The atmosphere was changing fast. The Franciscans, until recently limited to a very narrow field of activity in Bosnia, began to take charge of the situation in the second half of the century. Their influence at court was decisive and the king provided them with all their needs.[117] They now had plenty of horses in their monastery stables, and rode all over the country to perform their duties.[118] In the words of Thomas of Hvar, the heretics melted away before them like wax before the fire.[119] We know, on the other hand, that in some regions their missionary activity came up against obstacles placed in its way by the Orthodox clergy.[120] We have no factual information on the results of their efforts to convert the heretics. It is certain, however, that not even the most earnest missionary endeavour sufficed in Bosnia, as elsewhere, to give the Roman Church decisive victory over heresy; the secular power had to intervene with violence. The moment when the final blow fell was hastened by external factors determining the fate of the land. Earlier, some two hundred years before, the danger of Hungarian invasion with the support of Rome, which threatened the very existence of Bosnia, had indirectly saved the existence of the heretical church. Now, on the contrary, the threat of Turkish invasion, driving the Bosnian king to an open pro-Rome policy, led to the disappearance of that church.

The fall of Constantinople in 1453 heralded the approach of catastrophe to the remaining semi-independent domains in the Balkan peninsula. This was not immediately realised in Bosnia; the King and the Hertseg carried on their petty quarrels and Stephen Vukčić relied more than ever before on his alliance with the victorious Ottoman Empire. In 1456, when the defeat of the Turkish army before the walls of Belgrade led the Porta to make unheard of demands on their Bosnian vassals, they at last realized the full extent of the danger threatening them as the others. King Thomas and Hertseg Stephen joined forces to defend the land against the Turks and began to look for outside help. The king even considered a great crusade, which he himself would lead as John Hunyadi had in the recent past. At the given moment, however, such a campaign was quite impracticable, and finally the king made peace with the Sultan. He now concentrated on a far less ambitious goal - that of gaining control over neighbouring Serbia.

The country of the Serbian despots - by this time considerably reduced in size and torn by dissensions - was still tempting for both Bosnian and Hungarian political purposes. King Thomas had gained the consent of King Matthias of Hungary to unite Bosnia and Serbia by marriage. In the early spring of 1459 the marriage of his son to the daughter of the deceased despot Lazar was celebrated in the Serbian capital Smederevo, and the young Stephen Tomašević was declared despot. By June of the same year Mehmed II was besieging the fortress with a vast army, and the son of the Bosnian king surrendered the city to the Turks to buy his own safety. The grave accusation of treachery was now levelled at the "infidel" Bosnians, whose rule had in any case roused discontent in Serbia. It was essential to clear the reputation of Bosnia in the eyes of the Christian world, and to do so quickly. King Thomas now turned his attention to the Patarene church.[121]

And so at last the "church of God" was subject to persecution even in Bosnia. An echo of those days, when the "accursed king" took their lands away from the *krst'jane* and distributed them among the villagers, is found in a tax register dating from the early years of Turkish domination.[122]

In addition to confiscating the property of the heretics, the Bosnian king endeavoured to convert leading members of the Patarene church. Three of them were carried off to Rome, where the task of converting them to the true faith fell to Cardinal Torquemada. The Cardinal had a list of Bosnian "Manichaean" errors, prepared by the Franciscans, working in Bosnia. On the basis of this material he wrote his polemic refuting the fifty Patarene articles one by one.[123] The three Bosnian heretics were converted to the Catholic faith, under pressure, but one of them managed to escape on the way back home.[124] He took refuge in the territory of Stephen Vukčić, where many *krst'jane* had fled from royal Bosnia. According to the figures recorded by Pope Pius II (1458-1464) in his Commentaries, there were forty of them, while two thousand had been baptised in Bosnia. Pius II, the learned humanist Aeneas Silvius, also mentioned the "Manichaeans" in Bosnia in his "Cosmography".[125]

In 1490 the Pope assured his legate that he had already given instruction to Hertseg Stephen, not to shelter the Patarenes in his country.[126] The heretical church of Bosnia had indeed found its last refuge on his lands, while in royal Bosnia it had been broken up at the last moment before the country fell into Turkish hands.

In the spring of 1463 the great Turkish army invaded Bosnia, and after the castle of Ključ was lost the head of the last king of Bosnia fell. The Turks ruled the whole of his territories and those of the Pavlović family; part of the Hertseg's land was also overrun, but Stephen Vukčić managed to get it practically all back. Following on this success, however, Stephen had to deal with his son's rebellion and by 1465 the Turks were in control of most of his domain.[127] Hertseg Stephen had already prepared himself a place of refuge on Venetian territory. In March 1466 the Republic of Venice accepted his offer of the cession of two coastal towns, giving him in return the island of Brač (Bratch) and a house in Split or some other Dalmatian town under Venetian rule. To this intent the Republic sent letters to the Dalmatian cities, instructing them to receive hertseg Stephen (together with his relatives and servants) should another Turkish attack force them to leave their homeland.[128] The same day (10.III.1466) the Venetians gave the same guarantees to *gost* Radin, who had asked for refuge for himself and fifty to sixty people "of his law and sect" *(legis et secte sue)*. We see that here, too, *gost* Radin speaks for his church. The people concerned were undoubtedly upper class Bosnians like the relatives Radin mentions in his will. The Venetian Republic took them under its protection, together with their property, guaranteeing them safety in their territory and freedom to depart. The Venetians regarded their presence as an advantage, and in their situation the Bosnians had most confidence in the Venetian Republic.[129]

Neither the Hertseg nor his gost were given the chance to make use of the privileges granted them by Venice. By the end of May 1466 Stephen Vukčić was dying in his town of Novi, having dictated his will to the Metropolitan of Mileševo a few days previously; the witnesses to this document were three "intimate people of his house". Besides the secular noble "prince" Pribisav Vukotić there were Stephen's two *redovnici* (redovnitsi = monks): the Metropolitan of Mileševo, David, and Radin, *gost* of the Bosnian church.[130]

Gost Radin had written his own will earlier, at the very beginning of 1466, in Dubrovnik, where he died in the following year.[131]

The remnants of Stephen Vukčić's domain, for which the name Hertsegovina became accepted, held out until 1481. We have no news of the fate of the leading figures in the Bosnian church, who belonged to the entourage of *gost* Radin, nor of those who had fled to the Hertseg's territory when their church was overwhelmed by disaster in royal Bosnia.

We do know, however, that the Bosnian *krst'jane* were of various social standing; besides those who moved in aristocratic Bosnian circles or lived in the privileged "Patarene houses" there were many simple country people who worked on the land.[132] These were the *kmeti* and *kmetice* (kmetitse) mentioned in Radin's will. Most of them remained on the land even after the Bosnian church was broken up. Their presence in various parts of Bosnia and Hertsegovina is attested by the *defters*, Turkish tax registers. The name "Kristiyan" clearly marks these people off from the rest of the non-Muslim population. Whole villages inhabited solely by *krst'jane*, and individual farms belonging to them, are recorded. In other places all that is left of the *krst'jane* are traces in toponyms.[133] The names in Turkish documents significantly fill in the picture of the distribution of the Patarenes before the break-up of the mediaeval Bosnian state;[134] the material is still being studied, however.

The latest evidence of this church is provided by references in Byzantine sources. Apart from the epistle of Gennadios Scholarios there is the historical work by Laonikos Chalkokondylas, which refers to the "heretics in Sandalj's country".[135]

Both these late Byzantine authors use the word *kudugeroi* for the Bosnian heretics; it was also used for the Bogomils in Byzantium itself during this period. We must now turn to the scattered later reports on this dualist sect, preserved in the environment from whence it originated, the Byzantine Empire and Bulgaria.

[1] *Ćirković*, Istorija, Chapter 3.

[2] Ibid., Chapter 4.

[3] *Theiner*, Mon. Hung. I, No. 566, 348 (= *Smičiklas*, CD VI, No 320, 378).

[4] *Wadding* V, 261 (291 f.) Cf. *Mandić*, Bogomilska crkva, 29.

[5] See *Solovjev*, Svedočanstva, 77 f.; *Mošin*, Sinodik, 378.

[6] *Theiner*, Mon. Slav. I, No. 192, 135 (= *Smičiklas*, CD VIII, No 411, 508). Cf. *Mandić*, Bogomilska crkva, 30, n. 18.

[7] *Smičiklas*, CD VIII, No 436, 535 f. Cf. *Mandić*, Bogomilska crkva, 30, n. 19.

[8] *Ćirković*, Istorija, 109 f.

[9] Ibid., 96 f.

[10] *Ćirković*, "Verna služba".

[11] Text: *Thallóczy*, Studien, No. 7, 11. Cf. *Ćirković*, art. cit., 102.

[12] *Ćirković*, Istorija, 104.

[13] *Ćirković*, Istorija, 110.

[14] "...ut omnes ecclesie, ... in pristinum statum resurgant et restaurentur, ut in eis divina officia secundum ritum et morem ecclesiasticum ac fidelium celebrentur..." *Theiner*, Mon. Hung. I, 633 (= *Smičiklas*, CD X, No 372, 526 f.)

[15] *Ćirković*, Istorija, 110 f.

[16] Ibid., 128. In a letter to the Venetian Republic Tvrtko himself said of his nobles: "...nobiles nostri regni sicut primo deo et nobis aliqualiter infideles..." *Ljubić*, Listine IV, No 146, 84. Cf. *Mandić*, Bogomilska crkva, 67, n. 212. Somewhat later Tvrtko's brother Vlk complained to the Pope that the *ban* was giving refuge to the heretics who came streaming into Bosnia from all sides (epistle of Urban V, 14.XII.1369: *Theiner*, Mon. Hung. II, No 175, 91. Cf. *Mandić*, op. cit., 67, n. 213).

[17] Bosnia's immediate neighbours were constantly on guard against the Patarene heresy. See the decisions of the assemblies of Split and Trogir as early as c.1335: *Farlati* IV, 377 (cf. *Mandić*, Bogomilska crkva, 66, n. 208). In 1369 Pope Urban V reminded the Archbishops of Split and Dubrovnik of the church rules against Catholics trading with heretics (*Theiner*, Mon. Slav. I, No 366, 265; cf. *Rački*, Bogomili i patareni, 438). There is an interesting letter of Innocent VI, dated 28.VI.1357, in which he mentions that Dionizije Babonić, Prince of Krupa in what is now Bosenska Krajina, was afraid of the people of his country becoming infected with the Patarene heresy from neighbouring Bosnia (*Smičiklas*, CD XII, No 318, 419), cf. *Mandić*, Bogomilska crkva, 67.

[18] On the establishment and extent of the Bosnian Vicariate of the Franciscan order see *Šidak*, Dubia, 211 f.

[19] *Ćirković*, Istorija, 111 f.

[20] Serb. Sinodik, 345; cf. *Mošin*, Sinodik, 384.

[21] Serb. Sinodik, 302. In the Zagreb recension the name Beliz'men'c appears before that of Rastudij (ibid., 345). The identification of this person with Belesmanza, bishop of the "Albanensian" Cathars (see *Solovjev*, Autour des Bogomiles, 91-96, and in another form *Mandić*, Bogomilska crkva, 147), remains pure hypothesis.

[22] On the prosopographic details in the Dečani version of the Serbian synodicon (Vlach Dobrojević and his possible identity with the Serbian heretic Vlach, said by the Life of the Patriarch Jefrem [1375 - 1378] to be a follower of the Messalian heresy) see *Šidak*, Današnje stanje, 133, n. 20; *Dj.Sp.Radojičić*, Jedan bogomil kod Srba iz osme decenije XIV veka, in: *Dj.Sp.Radojičić*, Književna zbivanja i stvaranja kod Srba u srednjem veku i u tursko doba, Novi Sad 1967, 113 - 115; *Mošin*, Sinodik, 387, n. 195.

[23] Isti sunt errores hereticorum bosnensium (quoted below as: Errores), ed. *F. Račkі*, Starine I (1896) 138 - 140 according to the Marciana MS, Venice (14th - 15th century); Msc lat Cl II Clod. LXIV, fol. 146, taking into account the variants in the MS Jugosl. Akad., Miscell. I a 57, fol. 48v - 79 (c. 1375). Cf. *Kniewald*, Vjerodostojnost, 163 f. Text: ibid., 168 - 169.

[24] Hic sunt omnia puncta principalia et auctoritates extracte de disputatione inter christianum romanum et patarenum bosnensem (quoted below as: Disp. bosn.), ed. *F.Račkі*, Starine I (1896) 109 - 138. See pp. 112, 113, 127.

[25] Errores, art. 6, ed. *Kniewald*, Vjerodostojnost, 168.

That there were close contacts between the Bosnian heretics and their fellow-believers in Italy is indirectly shown in the third scholion to the Interrogatio Iohannis: "...locus ... sicut Bossina et Lombardia et Tuscia." Interrog. Ioh., 75, n. 1.

[26] Disp. bosn., 130.

[27] Infant baptism: Errores, l.c., 171; Torquemada, Symbolum, art. 31, ed. *Kniewald*, Vjerodostojnost, 180; Disp. bosn., 130. Ascetic regulations: Errores, art. 30, l.c., 168; Torquemada, Symbolum, art. 43, l.c., 180. (Prohibition referred only to *carnes et lacticinia.)*

[28] Errores, art.11, ed. *Kniewald*, Vjerodostojnost, 168. "Item dicunt se esse ecclesiam Christi et successores apostolorum habentes unum de seipsis qui dicit se vicarium iesu christi et successorem sancti petri..." Cf. Torquemada, Symbolum, art. 17, ibid., 179. On Pope Silvester, ibid., art. 19: "Quod omnes pape a beato Petro usque ad beatum Silvestrum fuerunt de fide eorum et quod beatus Silvester fuit primus qui apostatavit ab eis." Cf. Disp. bosn., 112.

[29] See "Dubia", art. 21, ed. *Kniewald*, Vjerodostojnost, 162.

[30] The text is preserved in the MS Jugosl. Akad., Miscell. I a 57, where it is followed by the text of the Errores. The tenth article, incomplete in Kniewald's edition, was published by *Šidak*, "Dubia", 208, n. 8.

[31] See the penetrating analysis by *J.Šidak*, "Dubia", 221 - 230. For the epistle of Gregory XI to the Bosnian Franciscans, empowering them to perform the functions of the parochial clergy, see *Smičiklas*, CD XIV, No. 370, 494.

[32] Information in the sources, concerning the Patarene houses, was collected by *Mandić*, Bogomilska crkva, 308 ff.

[33] *Dinić*, Patareni, No. 18, 187 f.

[34] Ibid., No. 41, 196.

[35] In the document Gojslav is said to have issued the safe conduct together *cum valioso vayvode Stiepani*. *Dinić*, Patareni, No. 38, 194 f.

[36] Ibid., No. 77, 213.

[37] *Dinić*, Trgovina robljem, No. 162, 64.

[38] Examples are given by *Dinić*, Prilog.

[39] "Dubia", art. 21. Cf. *Torquemada*, Symbolum, art. 26, ed. *Kniewald*, Vjerodostojnost, 179.

[40] The attempt by *Mandić* to bring the Bosnian ecclesiastical ranks into line with the hierarchical system of the western Cathars of the twelfth to thirteenth centuries is pure fabrication (*Mandić*, Bogomilska crkva, 219).

[41] *Dinić*, Patareni, No. 31, 193.

[42] See *Solovjev*, Doctrine, 519.

[43] Ibid., 516 f.

[44] Professor Šidak restored the important word *prije* (before), omitted by Lj.

Stojanović and after him by Solovjev. (See *V.Mošin,* Rukopis pljevaljskog Sinodika pravoslavlja, Slovo 6-8 (1957) 168, n. 22. A new edition of the list in Batalo's Gospel, by *Dj.Sp.Radojičić,* Odlomak bogomilskog jevandjelja bosanskog tepačije Batala iz 1393 godine, in: Izv. na Inst. za Ist. BAN 14-15 (1964) 495 - 509.

[45] See *Kniewald,* Vjerodostojnost, 143.

[46] *Vrana,* Književna nastojanja, 812, n. 96.

[47] *Mandić,* Bogomilska crkva, 87.

[48] Ed. *Rački,* Starine XIV (1882) 21 - 29. Cf. *Vrana,* Književna nastojanja, 818.

[49] *Solovjev,* Doctrine, 524, edited the corrected text of Radoslav's MS.

[50] In the Disp. bosn. (109 f.) spiritual baptism is called *baptismus libri* (as distinct from *baptismus aquae).*

[51] "...nos etiam frangimus panem aliis". Ibid., 125.

[52] The glosses to the text of Srećković's Gospel, itself now lost, were edited by *M.N.Speranskij,* Ein bosnisches Evangelium in der Handschriftensammlung Srećković's, in: AfSlPh XXIV (1902) 172 - 182. They are analysed by *Solovjev,* Doctrine, 499 - 512.

[53] See our study, Byzantinobulgarica III, 29, n. 20.

[54] *Ivanov,* Bog. knigi, 67.

[55] On the Bosnian Gospels see *Irena Grickat,* Divoševo jevandjelje, filološka analiza, in: Južnoslovenski filolog XXV (1961 - 1962) 227 - 293. The other later studies are listed by *Ćirković,* Bosnische Kirche, 559, n. 27. On the supposed "Bogomil" elements in the text of the Bosnian Gospels see also *A.Vaillant,* in: RESl 28 (1951) 273.

[56] See *S.Radojčić,* Stare srpske miniature, Belgrade 1950.

[57] *J.Djurić - P.Ivanišević,* Jevandjelje Divoša Tihoradića, in: Zbornik rad. Viz. Inst. 7 (1961) 153 - 160 and VII.

[58] *V.J.Djurić,* Minijature Hvalova rukopisa, in: Ist. Glasnik (1957) No 1-2, 39 - 52 and XI.

[59] *Jovanka Maksimović,* Ilustracije Mletačkog zbornika i problemi minijatura u srednjovjekovnoj Bosni, in: Ist. Glasnik (1958) No. 1-2, 117 - 130 and XII. A review of both these studies by *J.Šidak,* Dva priloga o minijaturama u rukopisama "crkve bosanske" in: Slovo 9-10 (1960) 201 - 206.

[60] Most (about three-quarters) of this material is contained in *Marian Wenzel,* Ukrasni motivi na stećcima - Ornamental Motifs on Tombstones from Medieval Bosnia and Surrounding Regions, Sarajevo 1965. The ornament on the tombstones is classified according to specific motifs, accompanied by the author's drawings. This great quantity of documentary material, some of it previously unpublished, makes this a very valuable work. Her interpretation of the individual motifs, however, is often open to question. See the review by *M.S.Filipović,* in: Godišnjak društva istoričara B. i. H. XVI (1965) 290 - 292.

[61] *Benac,* Olovo, 59 f., 72.

[62] *Benac,* Stećci, XXXIV.

[63] *Nada Miletić,* Náhrobní kameny středověké Bosny. Prague 1968, 7.

[64] Dealt with in a very important study by *S.Radojčić,* Reljefi bosanskih i hercegovackih stećaka, in: Letopis Matice srpske 137 (1961) januar, 381, sv. 1, 1 - 15.

[65] Ibid., 7.

[66] *Nada Miletić,* art. cit., 10.

[67] It was primarily *A.Solovjev* who sought for Bogomil symbols in the ornamental motifs on the *stećci* (stetchtsi); see his study: Simbolika srednjevekovnih grobnih

spomenika u Bosni i Hercegovini, in: Godišnjak Ist. društva B. i H. VIII (1956) 5 - 66; Les Bogomiles vénéraient-ils la Croix? in: Bull. Acad. Belg., Cl. des Lettres, 5e série XXXV (1949) 47 - 62. See the critical note on this article, by *A. Frolow* (A.F.), ByzSlav X (1949) 358 f. The conclusions reached by Maja Miletić are also unconvincing; she interprets certain ornamental motifs and even the forms of letters engraved on the *stećci* in the spirit of early Christian symbolism (*Miletić*, "Krstjani"). Cf. also *S.Radojčić*, art.cit. and *Ćirković*, Istorija, 234. The latest to accept Solovjev's view is *G.Wild*, Symbol und Dogma in Bogomilentum, in: Saeculum 21 (1970) No. 4, 383 - 392. I have not as yet been able to consult this author's book: Bogomilen und Katharer in ihrer Symbolik, Teil I 1, Wiesbaden 1970.

68 *S.Radojčić*, art. cit., 10.

69 *Dinić*, Odluke veća, Nos. 1, 2, 221 f.

70 *Stojanović*, Povelje i pisma I 1, No. 439, 433 f.

71 *Ćirković*, "Verna služba", 105 f. For Tvrtko's letter itself see *J.Šidak*, O vjerodostojnosti isprave bosanskog bana Tvrtka Stepanu Rajkoviću, in: Zbornik rad. Filoz. fak. II (Zagreb 1954) 37 - 48.

72 *Stojanović*, Povelje i pisma I 1, No. 440, 434.

73 Ibid., No. 441, 435.

74 *Pucić*, Spomenici I, primj. V. Cf. *Mandić*, Bogomilska crkva, 69 f., n.233; *Ćirković*, "Verna služba", 107.

75 *Dinić*, Prilog, 34 f.

76 See *Ćirković*, Istorija, Chapter 5,6.

77 *Dinić*, Patareni, No. 14 a, 186. See *Ćirković*, Istorija, 285.

78 *V.Skarić*, Grob i grobni spomenik gosta Milutina na Humskom u fočanskom srezu, in: Glasnik Zem. Muz. XLVI (Sarajevo 1934) 79 - 82.

79 In this connection authors frequently quote the treatise on trade by Benedetto Cotrugli of Dubrovnik, who reproached his fellow-countrymen with overdue respect for wealth. He thought this was due to the influence of the Bosnians, "li quali seguono l'costume de Manichei, honorando li ricchi, et ricevendoli volontieri nelli loro sospity, et discacciando li poveri, allegando sequir in cio l'ordine di Dio et della fortuna." See *Ćirković*, Bosnische Kirche, 567, n. 60. This passage may refer to the Bosnians in general, who were often indiscriminately described as heretics or "infidels". The mention of the hospices, however, would seem to suggest that the author was thinking of the Patarenes who lived in a privileged manner in their communal houses.

80 See *V.J.Djurić*, art. cit., 51.

81 Traces of Patarene beliefs may perhaps be seen in Hval's inscriptions over the figures; John the Baptist is the only one without the word "Saint", while the Crucifixion and the Assumption have been left without inscription. Caution should be exercised in drawing conclusions, however, in view of the lack of comparative material, as *Djurić* rightly points out, ibid., 47.

82 See *F.Šišić*, Vojvoda Hrvoje Vukčić Hrvatinić i njegovo doba, Zagreb 1902, 159 f.

83 Acta Bosnae No. 526, 95. Cf. *Mandić*, Bogomilska crkva, 90. For his adherence to the Patarene faith see the testimony of Mateo de San Miniato: "Homo ipse patarenus est..." Ibid., 90; cf. *Šišić*, op. cit., 154.

84 See *Ćirković*, "Verna služba", 106 f.

85 On the "career" of Radin Butković see *Truhelka*, Testamenat; *Solovjev*, Gost

Radin; *Mandić*, Bogomilska crkva, 79.

[86] 1423: *Stojanović*, Povelje i pisma I 1, No. 598, 583; *Miklošić*, Mon. Serb., 322. 1432: *Stojanović*, op. cit., No. 609, 627 f.; *Miklošić*, op. cit., No. 314, 366 f.

[87] *Stojanović*, Povelje i pisma I 2 Nos. 651, 652, 653, 659. pp. 46, 47 f., 59 f.

[88] *Dinic*, Patareni, No. 51, 200 f.

[89] Ibid., No. 57, 203.

[90] Ibid., No. 64, 208 f.

[91] *Stojanović*, Povelje i pisma I 2, No. 729, 151 f.

[92] Text: *Truhelka*, Testamenat, 371 - 375.

[93] Most of the contradictions which earlier scholars found between the text of Radin's will and the teachings of the Patarene church have been resolved, partly by careful study of the text itself, and partly by the publication of a new document dating from 1493 and referring to part of Radin's legacy. *M.Dinić*, Nekoliko ćirilićkih spomenika iz Dubrovnika, in: Prilozi za knjiž., jezik, ist. i folklor 24 (1958) 104. See *Ćirković*, Bosnische Kirche, 561, n. 39; *J.Šidak* (review of Solovjev's study, Le testament du gost Radin, in: Mandićev zbornik, Rome 1965), Slovo 17 (1967) 196, 197 (on the meaning of the word *kćerša*). Apart from the celebration of church festivals doubts may be raised by Radin's belief in the Last Judgment. According to the will the *krst'jane* should also pray for his soul on feast days, as well as every Sunday and Friday, "the holy prayer of God that He may free us from our sins and have mercy on us at the Day of Judgment." (Text: *Truhelka*, Testamenat 372). This plea has a parallel in the Lyons Ritual, in the form of common confession: "...que nos no siam jujat, ni condampnat al dia del judici cum li felo." (Lyons Ritual, XI.)

[94] In our opinion the view put forward by *M.Barada*, that the original text read *mirskiem ljudem* and not *mr'sniem ljudem* as it stands in the only extant copy, is highly probable; this copy contains a number of scribe's errors. See *M.Barada*, *Šidakov* problem "bosanske crkve", in: Nastavni vjesnik XLIX (1940 - 1941) 405, n. 1. Cf. Lyons Ritual, X: *gent mondana*.

[95] *Stojanović*, Povelje i pisma I 2, No. 68. Cf. *Ćirković*, "Verna služba", 108. In the West there is testimony that the "perfect" were called upon to settle disputes in aristocratic families. See *Schmidt* I, 313.

[96] Serb. Sinodik, 302, art. 34.

[97] See *Ćirković*, Bosnische Kirche, 572.

[98] Quotation from the letter of Gennadios (in Old Serbian translation): *Lj.Kovačević*, Nekoliko priloga za crkvenu i političku istoriju Slovena, in: Glas srp.uč. društva 63 (1885) 12; cf. *Miletić*, "Krstjani", 83.

[99] *Dinić*, Patareni, No. 59, 204.

[100] See the epistle of Pope Nicholas V to his legate Thomas of Hvar (1.VII.1452): *Theiner*, Mon. Hung. II, No . 427, 263 ff. (= Farlati IV, 260).

[101] *Ćirković*, Bosnische Kirche, 572.

[102] The same applied to the Pavlović family. A document issued by the sons of Radoslav Pavlović, 15.VII.1454 (*Stojanović*, Povelje i pisma I 2, No. 728, 148 f.) testifies to the very important position still held by the Patarene hierarchy at their court at this time. Before sending emissaries to conduct solemn negotiations in Dubrovnik, Peter and Nicholas Pavlović took counsel "as is the custom" with the *strojnici* (stroynitsi) of the Bosnian church. Two of the latter (one *gost* and one *starac* - starats) figure in the document as guarantors "for both sides".

[103] Acta Bosnae. No. 348, 65. On the number of Franciscan monasteries and monks given in the sources as existing in Bosnia at the end of the XIVth century, see

J.Šidak, O autentičnosti i značenju jedne isprave bosanskog djeda (1427), in: Slovo 15-16 (Zagreb 1965) 282 - 297.

104 *Ćirković*, Istorija, 285 f. On Tvrtko's request to Pope Martin V to have his orthodoxy investigated before his marriage to a relative of King Sigmund: ibid., 260 f. On the Council of Bâle, with quotation of the relevant text: *Mandić*, Bogomilska crkva, 33, n. 33; *Šidak*, Heretički pokret, 14. Jacob of Marchia wrote a polemic against the Bosnian heretics which was still known in 1697, but since then cannot be found (ibid., 33, n. 32b). We only know the contents: Summarium "Dialogi contra Manichaeos in Bosna" S.Jacobi de Marchia compilatum a censoribus ecclesiasticis J.B.Lucini et J.B.Barberi an. 1697. (Printed by *Mandić*, Bogomilska crkva, 442 - 444).

105 On this point see *D.Prohaska*, Husitství a bogomilstvo, in: Čas. pro mod. filol. V (1916) 1 - 13; *J.Macůrek*, Husitství v rumunských zemích, Čas. Matice Mor. 51 (1927) 33 ff.; *Šidak*, Heretički pokret.

106 See *Mandić*, Bogomilska crkva, 408, quoting a letter from Benedict Ovetarius to the bishop of Padua, Peter Donatus.

107 Ibid., 411.

108 30.VII.1446: *Farlati* IV, 70.

109 On the evidence of the letter of Eugene IV, dated 17.IV.1444, *Mandić* concludes that this happened before Thomas came to the throne (op. cit., 407 f.).

110 29.V.1445: *Theiner*, Mon. Slav. I, No. 555, 388; Acta Bosnae No. 845, 198. Cf. *Mandić*, op. cit., 411.

111 *Ćirković*, Istorija, 286.

112 *Dinić*, Patareni, N°s 23, 24, p. 290. Cf. *Dinić*, Državni sabor, n. 22. A charter issued by the Bosnian djed Mirohna (Mirokhna), the authenticity of which is proved by *J.Šidak*, O autentičnosti ... jedne isprave..., loc. cit. (see above, n. 103), dates from the year 1427. Cf. also *P.Anđelić*, Originalni dijelovi dviju bosanskih povelja u falzifikatima Ivana Tomke Marnavića, in: Glasnik Zem. Muz. BiH (Arheologija) XXVI (1971) 347 - 360.

113 Charter of 22.VIII.1446: *Miklošić*, Mon. Serb., 438 - 440. Cf. *Rački*, Bogomili i patareni, 464; *Ćirković*, "Verna služba", 106.

114 Letter from Eugene IV to his legate Thomas of Hvar, 3.XI.1445: Farlati IV, 257 - 258. In this letter the Pope lists the Patarene errors, on the basis of what the legate has reported; cf. *Mandić*, Bogomilska crkva, 34, n. 35.

115 See Acta Bosnae N°s 790, 856, 858, 870, 872, 1012, 1014. Cf. *Mandić*, op. cit., 68, n. 216, 217.

116 1.II.1449: *Raynaldi*, Annal. eccl. ad. an. 1449, N° 9, vol. IX, 538. Cf. *Mandić*, op. cit., 412, 413. According to this letter there were still "nonnulli ejusdem haeresis labe infecti tam nobiles et barones, quam alii qui religiosi nuncupantur". The *religiosi* were undoubtedly the *krst'jane* of the Bosnian church. The principal heretic among the aristocracy of Bosnia is given as Stephen Vukčić, followed by Duke Ivaniš Pavlović, who had once adopted the Catholic faith but again relapsed into heresy. 1.VII.1452: see n. 100 above.

117 *Thallóczy*, Studien, 411.

118 Cf. *Ćirković*, Istorija, 320.

See the document printed by *Thallóczy*, Studien, 411. Professor Ćirković has kindly informed me that this letter from one "frater Helyas" is in the Capestrano archives. Thallóczy wrongly dates it in the pontificate of Nicholas V (1447 - 1455); in fact it was written some time between the death of this Pope (24.III.1455) and the death of John of Capistrano (23.X.1456).

[119] Acta Bosnae, No 901, 211.

[120] Acta Bosnae, No 954, 225. Cf. *Ćirković*, Bosnische Kirche, 572 f.

[121] *Ćirković*, Istorija, 313 - 319.

[122] *T.Okiç*, Les Kristians (Bogomils Parfaits) de Bosnie d'après les monuments turcs inédits, in: Südostforsch. 19 (1960) 128, cf. 129.

[123] Ed. *Rački*, Starine XIV, 5 - 20; *D.Kamber*, Kardinal Torquemada i tri bosanska Bogomila (1461), in: Croatia sacra 2 (1932) 39 - 40; Juan de Torquemada, Symbolum pro informatione manichaeorum (El Bogomilismo en Bosnia). Edición critica, introducción y notas por *N.L.Martinez* y *V.Proaño*, Gil. Burgos 1958 (we have not had the opportunity to consult this edition). The text of the 50 articles was published by *Kniewald*, Vjerodostojnost, 177 - 181 (the edition referred to in this study). Cf. *Kniewald's* commentary, 174 f.

[124] Letter of Pius II, 2.VIII.1461: *Theiner*, Mon. Hung. II, No 546, 363 f. Aeneas Silvius, Commentarii, ed. 1584, lib.V, 227, quoted by *J.Matasović*, Tri humanista o patarenima, in: Godišnjak Skop. Filoz. fak. I (1930) 247.

[125] Aeneas Silvius, Historia de Europa, Basel 1571, 407.

[126] *Theiner*, Mon. Hung. II, No 540, 358 f.; Acta Bosnae No 1016, 241. See *Mandić*, Bogomilska crkva, 35.

[127] *Ćirković*, Istorija, 329 f.

[128] *Ljubić*, Listine X, No 364, 356 f.

[129] Archivio di stato, Venezia, Senato - Terra V, 151. Ed. *M.Šunjić*, Jedan novi podatak o gostu Radinu i njegovoj sekti, in: Godišnjak Društva istoričara B. i H. XI (1960) 267.

[130] *Pucić*, Spomenici II, 126 f.

[131] *Truhelka*, Testamenat, 365.

[132] There were analogous cases attested in the West, too. *Dossat*, Les cathares d'après les documents de l'inquisition, in: Cathares en Languedoc, 72: "Certains hérétiques travaillaient la terre et versaient les redevances habituelles" (ref. *Doat* XXIII, fol. 117 - 117v; 257v - 258).

[133] *Okiç*, art. cit. The relation of the rapid Islamization of Bosnia to Patarenism is analysed by *A.Solovjev*, Nestanak bogomilstva i islamizacija Bosne, in: Godišnjak Ist. društva B. i H. I (1949) 42 - 79; Le témoignage de Paul Ricaut sur les restes du Bogomilisme en Bosnie, in: Byzantion 23 (1953) 73 - 86. But see *Šidak*, Današnje stanje, 140 f.; *Ćirković*, Bosnische Kirche, 574 f. This problem cannot be clarified until Prof. Okiç has published the complete results of his study of Turkish tax registers.

See also: *G.Stanojević*, Jedan pomen o kristjanima u Dalmaciji iz 1692 godine, in: Godišnjak Društva istoričara B. i H. XI (1960) 273 - 274; *J.Lučić*, Prilog pitanju nestanka bosanskih bogumila (Patarena), in: Hist. zbornik 14 (1961) 239 - 242.

[134] See the highly informative sketch map drawn by *S.Ćirković* reconstructing the topography of the Patarene church on the basis of data in Bosnian, Dubrovnik and Turkish documents. *Ćirković*, Istorija, 284.

[135] Laonikos Chalkokondylas, Ἀποδείξεις ἱστοριῶν, ed. Bonn, 249.

Chapter XX

THE LAST TIDINGS OF THE BOGOMILS IN BYZANTIUM AND BULGARIA

From the middle of the thirteenth century, when Rainerio Sacchoni wrote his Summa, we learn nothing from western sources of the dualist churches of Constantinople or anywhere else in the Byzantine Empire or in Bulgaria. Since the times of Germanos II we have no document from Byzantium which could compare in extent with the anti-Bogomil encyclical sent by that Patriarch to Constantinople. After the recapture of their capital by the Byzantines (1261) all that we have on the subject from ecclesiastical sources is an allusion to the Bogomil heresy in the polemic by Georgios Moschabar.[1] The subject of his work is *processio Sancti Spiritus*, for long a point of conflict between Byzantine and western Theology. The author was the chartophylact of the Patriarch of Constantinople, and he knew the Bogomils as a living, contemporary sect. He mentions their doctrine that the bodies of the saints are in their essence "no different from the bodies of animals". He is undoubtedly thinking of the belief in the inevitable destruction of all material bodies, held without exception by all branches of the sect in the East and in the West. In this connection he adds that these heretics do not believe in the Last Judgment.

This reference to the Bogomils in the work of Moschabar, made in passing in the course of a polemic addressed to a western monk or a Byzantine of "Latin" inclinations, tells us nothing about the strength of the sect in the capital or in the provinces. However, we have clear evidence in the two following centuries that this heresy was still alive in the Balkan regions of the Byzantine Empire. We owe them to chance, and this paucity of information is all the more striking in contrast to the situation in the Catholic world. Wherever it cropped up in the countries of the West we find the Inquisition recording the activities of the sect down to the very last. In Byzantium there was no such institution systematically concerned with seeking out and persecuting heretics however inaccessible their retreat, and preserving a careful record of their replies before the judges.

There are two documents in the records of the Permanent Synod in Constantinople for the first half of the fourteenth century, which testify to the existence of the Bogomils. Both concern the same village in Thrace, bearing the Slavonic name of Bukovik. The accusation was brought (1316) against a rural priest Garianos, who was accused of being converted by the heretics of that village when he settled there for a time, and of joining their faith. The authorities went on to accuse him of stealing the church vessels and hiding a heretical symbol under his priestly headgear. All these accusations, however, proved unfounded.

Fourteen years later (1330) the name of Bukovik again appeared in a court case, when two brothers were contesting the ownership of the estate of Irini near the town of Vira (Fere, north of Enos). Their conflict led to bloodshed. John Laskaris went secretly to Bukovik to "the godless living in that place" and brought back with him a number of armed men who helped him to attack the estate which his brother had inherited from their father.[2]

The first of these documents is clear testimony to the heretical beliefs of the people of this Thracian village. It is worthy of note that the name *paterini* is used for the heretics, which as we have seen was the usual designation in the West but rare in Byzantium. From the other synodical document we can safely assume that in the year 1330 the people of this village still had the reputation of heretics.

The biography of Gregorios Palamas gives us another clue. According to the account of his life by Philotheos, on his way to Athos the young Gregorios arrived in a village of heretical "Messalians", i.e. Bogomils, and devoted a great deal of endeavour to convert them.[3] This village lay somewhere halfway between Constantinople and Salonika, on the borders of Thrace and Macedonia, close to the mountain of Papikion which was famous for its many monasteries. The "teachers" of these heretics accepted only the Lord's Prayer, given by Christ to his disciples, and rejected all other prayers and hymns as added by the Church. They were also hostile to the symbol of the Cross. We learn no more about their beliefs from this Life of Palamas, but the points referred to are so concrete that there is no need to doubt the authenticity of the episode, even though the endeavour to deflect from Gregorios Palamas himself any accusation of "Messalian" heresy is apparent.

In 1325 Palamas left the Holy Mountain, menaced by Turkish raids. Others accompanied him, men like himself who had been leading a life of meditation on Athos, but not as members of the great monasteries. Gregorios Palamas and his companion Isidore took their way to Salonika, the second city of the Byzantine Empire. Here they did not shut themselves away in ascetic seclusion but moved in the society of the spiritual élite, monks and laymen, men and women of high social standing. Isidore collected a number of disciples around himself, instructing them in the ascetic life and the virtues of monasticism. He thus fulfilled the mission entrusted to him on Athos by his "spiritual father",

Gregorios Sinaites. Among the enemies of Palamas, however, Isidore's work was regarded most unfavourably, and he was severely criticized for not following the rules of monastic life.[4] In addition, some of the writers of the anti-Palamas school accused both Isidore and Gregorios Palamas of succumbing to the influence of heretics while they were in Salonika. A letter written by one of the enemies of Palamas some time in 1345 mentions one Porini, a nun whom Palamas and Isidore honoured as a prophet. This woman, according to the writer of the letter, was a disciple of one Georgios who had lived in Salonika, in fact in the same house as Isidore, until he was finally driven out for his "Messalian" heretical beliefs. According to Akindynos this heretic was also driven out of Athos branded on the forehead.[5]

The report of Akindynos that the "Messalian" heresy had been detected in Salonika, on Athos and even in Constantinople itself agrees in part with the testimony of another representative of the anti-Palamas school. The historian Nicephorus Gregoras, writing practically at the same time as Akindynos, deals with the occurrence of heresy on Mount Athos and lists several heretics whose views were condemned by the Synod of Constantinople. In his work of 1347 he links the name of Gregorios Palamas with these heretics, declaring that he left Athos before they were condemned, in order to hide in the crowd "of many tongues" in the great city of Salonika.[6] In his History, however, Nicephorus Gregoras did not repeat this accusation that Palamas was a Bogomil heretic.[7] That there was no foundation for the contention can best be seen in the writings of Palamas himself. His spiritual system is based purely on Biblical tradition, in which the human being is seen as a whole and the body is not placed in opposition to the soul. Christ united his divinity with humanity, became "one of our body", and this alone made possible any contact between God and men. The union of each individual Christian with the living Christ is realized in the Church through the sacraments of baptism and the Eucharist.[8] Here we can see that Palamas' mystical system was fundamentally incompatible with the dualist teachings of a sect which regarded the body as the prison of the soul, and denied (or at least depreciated) Christ's incarnation and thereby the meaning of both the fundamental sacraments. In spite of this, the chief opponent of Palamas, Barlaam of Calabria, called him and his followers "Messalians" (i.e. Bogomils) in the very title of his polemical work against them.[9]

The possibility cannot be excluded that Palamas and Isidore came into contact with heretical circles in Salonika accidentally. Unfortunately the very nature of the heresy detected in Salonika, according to the testimony of Akindynos and Gregoras, remains obscure. The reports mention only the rejection of holy pictures.

The accusation of Bogomil heresy in Salonika and on Athos appears again in another form, in the work of the Byzantine Patriarch Kallistos, who belonged to the Hesychast school of Palamas. In his account, too, we find a heretical nun in Salonika, this time under the name of Irene. According to Kallistos, some monks of Athos who were then in Salonika came under her influence and brought the heresy back with them to the Holy Mountain. These facts are given in the life of St.Theodosius of Trnovo, the leading figure in the Hesychastic school in Bulgaria, written by Kallistos, who goes on to describe the work of two heretics who deserted Athos and took refuge in Bulgaria.[10] The Greek original of this hagiographical work by Kallistos is unknown to us, and the late Slavonic version[11] which has survived is clearly corrupt, so that it is very difficult to determine the true sense.

It would appear from this hagiographical source that in some of the teachers of this heresy the endeavour to attain absolute poverty in the evangelical sense was carried to the lengths of Adamite nudity.[12] Two of a group of monks driven out of Athos for "Messalian" (i.e. Bogomil) beliefs, Lazar and Cyril "Bosota", arrived in Bulgaria where they converted a priest named Stephen to their heresy. Kallistos gives Theodosius the merit of calling an assembly at which these heretics were refuted. Some of the views imputed to them are obviously those of the Bogomils, but there are others which cannot be reconciled with the teachings of the sect. Since we have no text that can be relied on it is dificult to draw any conclusions about the nature of the heretical tendencies then current in Bulgaria,or of their contemporary context.[13] The account refers to the reign of Tsar Ivan Alexander (1331-1371), when Bulgarian learning reached its last peak in the Middle Ages, and when signs of the fateful disintegration and catastrophic end of the Bulgarian state were already apparent.[14]

There are several passages in the Life of Theodosius which are a clear indication that Bogomilism was still felt as a very real danger, in spite of the fact that ecclesiastical circles had other things to worry about. In Bulgaria, as in Byzantium, the Hesychastic trend was victorious; opponents were declared heretics and placed on a level with the Bogomils. At yet another Church assembly, called - according to the Life - against the Jews (1360), the Bogomil heresy was again condemned and along with it the "newly appeared" erroneous teachings of Barlaam and Akindynos. And shortly before his death the Bulgarian saint exhorted his disciples to preserve the purity of the true faith and beware of both these errors.[15]

Within the Byzantine Church the pro- and anti- Palamas factions had no scruples about using the accusation of Bogomil heresy as a weapon against their adversaries. Nicephorus Gregoras describes how a group of bishops opposed to the Patriarch Kallistos, after his second enthronement in 1355, accused the head of the Byzantine Church of "Messalianism". The Patriarch accused them of various transgressions in his turn, among them that of accepting the Bogomil heresy.[16]

In the late forties of the fourteenth century a monk Niphon was accused of Bogomilism; as *prôtos* he was entrusted with supreme supervision of all the monasteries on Mount Athos. The accusation was brought against him by Serbian monks, and was clearly a political move. The Tsar Stephen Dušan, on whose territory Mount Athos was at that time, was anxious to get the office of *prôtos* into Serbian hands. He actually achieved this (1348) after he managed to get Gregorios Palamas away from Athos, for the latter defended Niphon. Niphon, falsely accused of heresy, was given complete satisfaction both on Mount Athos and in Constantinople (1350).[17]

We have evidence of the *real appearance* of the Bogomil sect on the territory still left to the Byzantine Empire, as late as the first half of the following century. The Metropolitan of Salonika, Symeon (+1429) refers to it in his anti-heretical "Dialogue".[18] He seems to have been in possession of precise information about these sectarians, living quite close to Salonika, and known as Bogomils or as Kudugeri. The name *kudugeros*, which we find in other late Byzantine authors too, derives without doubt from corruption of the word *kalogeros* (= good old man), commonly used in the Byzantine world to mean "monk".[19] In Symeon's account of the Bogomil teachings we are struck by the archaic feature of total rejection of the books of the Old Testament, including the prophetic books and the Psalms of David. At the end of his treatise there is an interesting reference to the fact that many orthodox are misled by the heretics shortly before death, so that they are torn "away from Christ" at the very end of their lives.[20] It can hardly be questioned that what our Byzantine ecclesiastical writer is referring to is the administration of spiritual baptism to the seriously ill and dying. This is the only reference to the subject in Byzantine sources, a mere hint in fact, while in western sources we have innumerable cases of this phenomenon, as of many other details revealing various aspects of the life of the sect. In the East, where everything around the sect had long fallen into oblivion, we must be grateful for the palest ray of historical light.

We have interesting testimony to the existence of the Bogomil sect in Bulgaria less than two decades before the Second Bulgarian Empire disintegrated, from the Bosnian Franciscans to whom we owe so much of our knowledge of the Patarene church in Bosnia. The Franciscan mission was working in the north-west regions of the land at the time when King Louis I of Hungary was in temporary possession of the town of Vidin (1365- late 1369 or early 1370).[21] Reports of the work of the Bosnian Franciscans in this area refer – certainly with exaggeration – to the conversion of 200,000 people to the Catholic faith. Besides "schismatics" (i.e. Orthodox) there is explicit mention of "heretics" (Patarenes, Manichaeans) in the Vidin region. This handful of Franciscans was not able to cope with a missionary task of these dimensions alone, and asked for reinforcements. The Hungarians did not rule over Vidin for long, however, and soon the whole of Bulgaria was under Turkish domination (1393, 1396). It was two centuries before a Catholic mission again took root in the region,and by then nothing was heard of Bogomils or Patarenes in Bulgaria. From the end of the sixteenth century missionary work in the country - again mainly in the hands of the Franciscans - concentrated on the remnants of the Bulgarian Paulicians.

[1] This work was written between 1.IX.1277 and 31.VIII.1278. See *I.Dujčev*, Edno prenebregnato vizantijsko izvestie za bogomilite, in: Izv. na Inst. za Bălg. Lit. 6 (1958) 247 - 250.

[2] 1316: MM I, No. 34, 59 - 61. 1330: MM I, No. 70, 151 - 154. The second document was first pointed out by *D.Angelov*, Zur Geschichte des Bogomilismus in Thrakien in der I. Hälfte des 14. Jahrhunderts, in: BZ 51 (1958) 374 - 378.

[3] Philotheos, Gregorii Palamae Encomium, PG 151, col. 563A - 565D. Cf. *Meyendorff*, Introduction, 50 f.

[4] *Meyendorff*, op. cit., 53 f.

[5] For the quotation from Akindynos' letter: ibid., 55. The existence of Bogomils in Salonika is mentioned in a synodical document of 1325 (MM I, 140 - 143: a sick monk of the monastery τοῦ Φιλοκάλου in Salonika was in contact with a man who "appeared" to belong to the Bogomil sect of heretics.)

[6] This was the treatise directed against Palamas, λόγοι στηλιτευτικοὶ καὶ ἀντιρρητικοί, in MS Gen. gr. 35. See *Meyendorff*, Introduction, 56, n. 64. The reference to heretics on Athos and to Palamas' flight from Athos to Salonika is on fol.7v and 8. (Ibid. n. 65).

[7] Nicephorus Gregoras, 'Ρωμαϊκὴ ἱστορία, lib. XIV, 7, ed. Bonn II, col. 718 - 719. The passage taken from his work attacking Palamas has been changed in this context and no longer refers to the accusation against him. See *Meyendorff*, 56, n. 66; 57.

[8] See *Meyendorff*, Introduction, 199 f., 213 f., 226 f.

[9] Ibid., 83. He also deals with Barlaam's criticism of the Hesychasts for their "Jesus prayer", which Barlaam compares to the Bogomil practice of using only one prayer, the Paternoster. On the accusation that Palamas accepted the heresy of the priest Theodore of Blachernae (of the time of Alexius I) see 76, 351 f.

[10] ЖИТЇЄ И ЖИЗНЬ ПРЕПОДОБНАГО ОТЦА НАШЕГО ѲЕѠДОСІА ed. *V.N.Zlatarski*, Sofia 1904, 19 f.

[11] See ibid., Introduction by *V.N.Zlatarski; V.S.Kiselkov*, Žitieto na sv. Teodosij Tărnovski kato istoričeski pametnik, Sofia 1926. Cf. *Obolensky*, Bogomils, 258.

[12] For parallel phenomena in the West see *E.Werner*, Adamitische Praktiken im spätmittelalterlichen Bulgarien, in: ByzSlav XX (1959) 2o - 27.

[13] *H.-C.Puech* rightly comments: "...mieux vaut n'en pas tenir grand compte" (*Puech - Vaillant*, Le traité, 135).

[14] *I.Dujčev* gives an apt characterization of the time, Miniatjurite na Manasievata letopis, Sofia 1962, 9 - 16.

[15] ЖИТЇЄ ... ѲЕѠДОСІА ed. cit., 25, 33.

[16] Nicephorus Gregoras, 'Ρωμαϊκὴ ἱστορία lib. XVIII, ed. Bonn III, 532 - 546.

[17] MM I, No. 133, 296 - 300. For the political background to the removal of Niphon from office see *Meyendorff*, Introduction, 136 f.

[18] Dialogus contra haereses, PG 155, col. 65D - 73B.

[19] For a survey of the different explanations of the word *kudugeros* see *Miletič*, "Krstjani", 82 f. The author pointed out the earliest Byzantine use of the word κουτουγέρος in the polemic of Michael Rhetor against the Franciscans of Constantinople, c. 1250. She refers to *Du Cange*, Glossarium ad scriptores mediae et infimae graecitatis II, Appendix, col. 113.

[20] ... οὐ μόνον ἑαυτούς, ἀλλὰ καὶ ἑτέρους πλείστους καταμολύνοντες, ὡς μανθάνομεν, καὶ μᾶλλον ἐν τῷ τέλει πολλοὺς πλανῶντες τῶν εὐσεβῶν, καὶ τοῦ Χριστοῦ ἀποκόπτοντες. Κατὰ γὰρ τὸν καιρὸν τοῦ τέλους πρὸς ἄρνησιν ἐκκαλοῦνται ...

(col. 73B). The expression these heretics are said by Symeon to use for the Devil remains obscure:... ὃν καὶ Τόπακα γῆς βαρβάρῳ φωνῇ ... καλοῦσι (ibid.).

[21] *Dujčev*, Franciscanesimo, 413 - 424.

Chapter XXI

THE ULTIMATE FATE OF THE BULGARIAN PAULICIANS

The first report for centuries which drew attention to the Balkan Paulicians came not from Catholic missionary circles, but from heretical Bohemia. It was at the end of the fifteenth century, when the young Unity of Brethren was sending its members out to journey through Bohemia and Moravia, and even to travel to far-off lands, in search of groups of Christians who shared similar beliefs with them. We have a record of one Brother Tůma, who travelled with this intent "at home and in foreign countries, as to Hungary and beyond". Finally he came across "some brethern who were known as Pavlikans, who declared that their teachings came from St.Paul".[1] The Unity of Brethren was an ascetic community, originally small in numbers, sternly closed to the world and to higher education, but towards the end of the fifteenth century the influence of educated men became predominant. Tůma "the scribe", who knew Latin and had some knowledge of medicine as well, was one of these, which explains why he was referred to in a historical work on the Brethren written in the first half of the sixteenth century, the author of which commented on this significant change. The report on the sectarian group Tůma met with on his travels mentions that baptism was administered to its members on the deathbed, that they might "die baptised, pure and holy, and without all sin," for "baptism is performed for the forgiveness of sins." The next sentence, concerning the possibility of the dying man recovering and relapsing into sin again, is not clear. Nor do we learn anything of other rites or religious beliefs held by these "Pavlikans", or details of Tůma's travels. Nevertheless the words "to Hungary and beyond," indicating the direction he took, allow us to assume that he may have reached the Balkans and found the remnants of the Paulician sect still lingering on in Bulgaria.

Another century passed before this small heretical enclave in the Ottoman Empire attracted the attention of Rome. In 1580 news of them came to Constantinople, to Pietro Cedolini, the bishop of Dalmatian Nona, who had been sent by Pope Gregory XIII to the European regions of the Turkish Empire as "apostolic visitor". The news brought by Dubrovnik merchants spoke of twelve villages on the Danube, between Nikopol and Ruse, where Christians lived who called themselves *Paulini* and refused to worship the Cross because the Saviour had died on it. The Orthodox Church regarded them as excommunicated, but the Paulini themselves declared that they were subject to the Pope of Rome, and wished to be able to send representatives there.

Pietro Cedolini charged the Franciscan monk Girolamo Arzengo to find out more about the "Paulians" in this part of the Danube region, when on a visitation to Silistra and Provadia. The two letters sent to the "Paulians" by the bishop of Nona, through the mediary of Arzenzo, marked the beginning of systematic efforts to gain these sectarians for the Catholic faith; one of the letters was in Latin script and one in Cyrillic, but both in the "Bosnian" tongue. There is a great deal of documentary evidence of this missionary activity in the archives of the Congregatio de Propaganda Fide and here and there in the Vatican archives.[2]

The Catholic missions could rely on the firm support of the colonies of Dubrovnik merchants in the towns of Bulgaria, but the centre of their activity was in Čiprovec (Tchiprovets), an old mining settlement where the German ("Saxon") miners had long been assimilated with the native population, but had handed on their Catholic faith. Čiprovec was the seat of the Catholic bishopric of Sofia, later raised to the Archbishopric. At the beginning of the seventeenth century the bishop of Sofia was Peter of Soli (Pietro Salinate), a Bosnian Franciscan who did a great deal to convert the Bulgarian "Paulians". In 1611 a report from the Archbishop of Bar spoke of the many dangers and deprivations suffered by Pietro Salinate at the hands of the Turks and also of the Orthodox hierarchy. Nevertheless he successfully converted large numbers of these heretics both round Trnovo and in the vicinity of Plovdiv. At this time the "Paulians", some of whom still persisted in their errors, lived in seventeen villages or hamlets stretching from Nikopol on the Danube up to Plovdiv.

From the time of Pietro Salinate sons of the sectarians were sent to Italy, largely through the charity of wealthy Dubrovnik merchants in Bulgaria. They would study at the Collegium Clementinum in Rome, later at the Collegium Illyricum in Loreto; then they would return home as Catholic priests. In 1635 Elia Marinov, Bishop of Sofia, was anxiously awaiting the return of Philip Stanislavov, who came from Oreše (Oreshe) on the Danube and was then studying in Rome. He looked forward to the young convert's certain success in bringing all the inhabitants of his birthplace to the Catholic faith. The bishop sent news of the effective Catholic propaganda among the "Paulians", some one of whom was converted to the true faith every day.

In Thrace the heretical elements were reinforced by many families who left their homes on the Danube because of the unbearably oppressive taxation, and settled in villages round Plovdiv. In the middle of the seventeenth century Peter Bogdan, then already Archbishop of Sofia, began earnest missionary work among them (1649, 1650). Some of these new arrivals were obstinate in clinging to their heretical beliefs, and to make his arguments more convincing the Archbishop did not use the Latin Bible but took his proofs from the Slavonic text of the Gospels which the heretics themselves possessed.

Like Bogdan, Philip Stanislavov, now Bishop of Nikopol, attests (1659) the poverty of the inhabitants of these "Paulian" villages. The taxes imposed on them by the Turks were exorbitant, and drove them to flee to Thrace with their families, or to move in the opposite direction, over the Danube to Rumanian Wallachia.

In 1640 Bogdan wrote a very full report on his visitation of the Bulgarian countryside. Besides much interesting comment on the country and the people, this account is the richest source of information on the Bulgarian "Paulians". In some villages near Trnovo Peter Bogdan found considerable numbers of unbaptised heretics - mostly very primitive people, living in abject poverty. He also encountered most obstinate sectarians who greeted the preacher's words with ridicule. On the whole, however, this heretical minority which had been surrounded by the hostility of the Orthodox for generations, was more ready to seek the support of Rome. Indeed the declaration of the "Paulians" themselves, that their faith is that of Rome, is recorded. All the reports sent from Bulgaria to Rome are agreed on that point. Reading them,[3] we cannot fail to see that what they say about the "Paulian" heretics is always factual and definite, and that the information is never contradictory, but complementary. The fundamental feature of their heretical faith, stressed in all reports, is the violent hatred of the Cross as the instrument of the death of Christ. Together with the cross they rejected the worship of holy pictures as "idolatry". Instead of the usual church baptism they had a ceremony of their own, during which their "priest" touched the candidate's head with a candelstick, symbolising "baptism by fire". Peter Bogdan described this rite, which was connected with Epiphany Day, the only day on which they went to church. Philip Stanislavov also spoke of "Paulians baptised by fire". Bogdan also mentions a simple marriage ceremony. Apart from this, the heretics received no sacraments and made no sacrifices, but celebrated some sort of sacrificial feast with singing and dancing.There were neither altars nor pictures in their churches, but they kept fast-days and also observed Sundays and other holidays, including some which were peculiar to them.

While it was unknown for the faithful of the Orthodox Church in Bulgaria to read the Bible, the Bulgarian heretics had preserved the

Gospels, the Epistles of St.Paul, the Acts of the Apostles and Apocalypse, handing them down for generations. They read these New Testament texts from ancient manuscripts written in Cyrillic. These books were of Bosnian origin, and for this reason the well-informed Peter Bogdan believed that the heretics had come from Bosnia. He did not however accept the explanation put forward by the historian Orbini of Dubrovnik, who derived them from the Patarenes driven out of "the whole of Italy", who had passed through Friuli into Bosnia, whence some went even further and reached Nikopol. In the spirit of the Byzantine writers attacking the Paulicians, the Catholic priests also connected contemporary "Paulians" with the teachings of Paul of Samosata. According to one of the early reports, glosses deriving from this ancient heresy were written in the margins of their New Testament manuscripts. It was the visible signs of heresy that held the attention of the missionaries, however, and so far as we know none of them attempted to define the substance of the heresy, perhaps hidden in those marginal notes.

The heretics themselves, whose language did not differ from that of the rest of the population of Bulgaria, knew absolutely nothing about the Paulician sect under Basil I, John Tzimisces or Alexius I. Nevertheless their particular reverence for the Apostle Paul connects them to the sect. The name Paulini which they themselves used seems to suggest this link. (In their reports the missionaries used the forms Paulini, Pauliani, Paulianistae, at will.) As late as 1730 there is evidence of the tradition, still alive among them, that they had been converted to Christianity by the Apostle Paul himself. This is attested in a letter from some "Paulians" already converted to Catholicism, who had been moved to Rumanian Banat.[4] There were still remnants of the sect in Plovdiv at the beginning of the eighteenth century, however. In 1717 Lady Mary Wortley Montagu passed this way, accompanying her husband who had been appointed ambassador to Constantinople. This learned Englishwoman, whose mind was in some respects ahead of her time, wrote some of her remarkable letters on her journey through the Balkan peninsula. In a letter sent from Drinopole in the spring of 1717 we read penetrating comments on the state of religion under Turkish domination. There is also a reference to a Christian sect she found in Plovdiv. They called themselves Paulines, and shewed an ancient church where they said St.Paul had preached. She added that the Apostle was their "favourite" saint, as St.Peter was that of Rome, and that they gave him preference over the rest of the Apostles.[5] This is surely proof of the remarkable strength of this tradition, still persisting among the remnants of the sect a thousand years later.

[1] This note comes from the work by Vavřinec Krasonický, O učených (Library of the National Museum, Prague, MS F 41, fol. 11b). I have compared the text as published by *F.Hýbl*, Dějiny národa bulharského, Prague 1930, 274 f., n. 1, with the MS. In the opening remarks the words "a trochu lékař" are followed in the MS by: "jakož já jeho knihy mám Pandekt v jeho ruku psány"; at the end of the text Hýbl's transcription omits the word *opět* ("...došel psích hodův opět").

[2] The source material is collected in: Acta Bulg. Eccl.; further documents were published by *I.Dujčev* and *C.Giannelli*, in: *Dujčev*, Cattolicesimo. See also *L.Miletič*, Novi dokumenti po minaloto na našite pavlikjani, in: Sbornik za nar.umotvor., nauka i knižnina 21 (1905). The literature is given by *Obolensky*, Bogomils, 266, n. 2, to which we should add *L.Miletič*, Našite pavlikjani, in: Sbornik za nar. umotvor., nauka i knižnina 19 (1903); *N.I.Milev*, Katoliškata propaganda v Bălgarija prez XVII vek, Sofia 1914; *I.E.Anastasiou*, Οἱ Παυλικιάνοι Athens 1959, 125 - 134.

[3] See the relevant reports: Acta Bulg. Eccl. Nos.6, 15, 17, 18, 23, 33, 37, 51, 55 (esp. p. 79 ff.), 111, 116, 153.

[4] Ibid., No. 242.

[5] The Letters and Works of Lady Mary Wortley Montagu, ed. Lord *Wharncliffe* I. London 1893, 290.

[6] The Bulgarian folk legend of the origin of the Paulicians was published by *J.Ivanov*, Proizchod na pavlikjanite spored dva bălgarski răkopisa, in: Spisanie na Bălg. Akad. na naukite XXIV (1922) 20 - 31; a later version was published by *J.Javorskij*, Legenda o proischoždenii pavlikian, in: Sbornik Otd. rus. jaz. i slov. AN SSSR 103 (1928) No. 3 (not available to the author). The work of *R.M.Bartikian*, Legendary origin of the Paulicians according to an Old Bulgarian MS (in Armenian), in: Izv. AN Arm SSR, ser. obščestv. nauk (1957) No. 1, remained inaccessible to me for reasons of language.

In conclusion we should mention the dualist folk legends which earlier and more recent authors have sometimes derived from the Bogomils. Dragomanov's theory that these legends were of Iranian origin was further elaborated by Dähnhardt. See *M. Dragomanov*, Zabeležki vărchu slavjanskite religiozno-etičeski legendi, in: Sbornik za nar. umotvor., nauka i knižnina 7, 8 (1892); 10, 11 (1894); *Dähnhardt*, Natursagen I, 13 ff. Cf. also *Linda Sadnik*, Der Balkan und die Hochkulturen des Vorderen Orients, in: Saeculum V (1954) 34 - 40. Ethnological comparative research into these legends has since shown irrefutably that fundamentally they reflect the religious ideas of the first two stages in the civilization process, and that the influence of mature religions can be seen only in the later versions developed in the different environments. See *W.Schmidt*, Der Ursprung der Gottesidee, vol. VI, XII, Münster i. W. 1935, 1955; *J.Henninger*, L'adversaire du Dieu bon chez les primitifs, in: "Satan" (= Études carmelitaines, 1948) 107 - 121; *U.Bianchi*, Il dualismo religioso. Saggio storico ed etnologico, Rome 1958.

Legends of this type cannot therefore be derived from the dualist teachings of the Bogomils, since their own dualist foundation is of ancient origin. The form they take in the Bulgarian and East Slavonic milieu arose through the accretion of motifs from the apocryphal tradition, on which, as we know, the Bogomils too drew for their myth.

Epilogue

When we come to consider the ultimate fate of mediaeval dualist heresy, we find the last period of the movement in the Byzantine sphere anything but clear; indeed its whole history there has been known to us in much less detail than that of the movement in the West. To a large extent these lacunae in our knowledge are due to the very fact that the struggle against this heresy in the Byzantine East was not carried on so systematically and consistently there as in Western Europe, so that far less evidence has come down to us.

Let us briefly recapitulate some of the basic facts. The energetic measures taken by Alexius Comnenus were directed against the Plovdiv Paulicians and the nucleus of Bogomils in Constantinople itself. Later in the twelfth century, isolated cases of Bogomilism were cruelly punished, but in the rural areas the heresy often spread with little to hinder it. From the early thirteenth century the troubled times brought with them an instability in political and ecclesiastical affairs that undoubtedly offered fertile ground for the growth of heresy. The Bogomil sect again grew stronger in Constantinople, where around the middle of the thirteenth century Rainerio Sacchoni knew of a dualist church organized among the Greeks as well as among the Latins. Reports from native sources continued to be scarce, but even in the first half of the fifteenth century the sect was still alive on what remained of Byzantine territory; at that time Symeon of Thessalonica gave a very factual account of the Bogomils.

In neighbouring Bulgaria, which the Papal Curia believed to be "full of heretics", the ancient "ecclesia Bulgariae" still existed in the time of Rainerio Sacchoni. The strong measures taken by Tsar Boril (1211) seem to have been of limited range and ephemeral effect. Boril did not have the whole land under his control, and it is probable that his fall put the heretics and their patrons in a stronger position. Even in the last decades of the independent existence of the Bulgarian Empire the Church there regarded Bogomilism as a very real danger, and yet the only factual report comes from the Franciscan mission, and refers to the north-western regions of the country. Much later surprising light is thrown on the fate of the remnants of Paulicians in Bulgaria by the systematic accounts of Catholic missionaries, but we know nothing of Bogomilism in Byzantium or Bulgaria under Turkish rule.

In the West the Inquisition kept exact records of the campaign to exterminate tha last groups of Cathars, right up to the burning of the bones of dead heretics in the Piedmont town of Chieri (1412) which brought to a close this chapter of history.

In Western Europe the Cathar movement can rightly be said to have been "exterminated". The historian cannot avoid asking whether it was only external forces that brought this about, or whether the chief factor should not be sought in the nature of the sect itself; this is the view of *A. Borst,* whose conception[1] cannot be passed over in silence in this summing-up.

Borst's great work of synthesis dealing with Catharism in Western Europe appeared (1953) after Dondaine had published his discoveries and finally dispelled any doubts about the eastern origin of the movement. It could be said that Borst's reconstruction of the development of the sect in the West is based on the author's personal view of its eastern, "Bogomil" origins. As Borst sees it, Bogomil's teachings stemmed from the fundamental belief that the world was evil; the solution lay in a life of ascetic humility following in the footsteps of the Apostles. Bogomil's negation of earthly life derived from emotional experience rather than from dualist metaphysics; while the Bulgarian heretical preacher may have met the Paulicians or some other heterodox group, this fundamental feeling (Gefühl) oscillating between Christianity and dualism was entirely his own spiritual contribution. The dualist doctrine was only in its initial stage: Satan, already identified with the Old Testament Creator, was still regarded, as in Christian doctrine, as a good angel who had fallen from grace. At that time the rites of the sect were very simple and no hierarchy had yet been formed.

A radical change took place, however, in the Byzantine environment; under the influence of the higher and more educated strata of society the Bogomils' simple dualistic ideas developed into a complete cosmogony. The fateful question arose: did evil come into existence at the fall from grace, or has it existed from all eternity? Difference of opinion on this fundamental question led to a split: the followers of a "radical" dualistic doctrine formed the "Dragovitsian" Church while those who continued to believe the earlier doctrine of the origin of evil kept the name "Bulgarian" Church. Other churches emerged and the sect inevitably distinguished between the "perfect" and the "believers". Byzantine monastic life exercised a strong influence on the sect in the system of ritual and ascetic practice now forming. Some time in the early twelfth century there thus was formed a "sect of monkish rites and speculative doctrine". The conflict between Christianity and dualism within the sect grew sharper and sharper, and was transmitted to the sect in the West, too.

The persecution of the Bogomils, renewed by Manuel I after he came to the throne, drove many of them from their Byzantine homeland and thus encouraged missionary expansion. It was an excellent moment to appear in the West. The remarkable consistency of life and beliefs to be seen in members of the eastern sect made an immense impact on the people of Western Europe in the atmosphere of great religious fervour flowing wild and untrammelled outside the confines of church life.

The first wave of missionary activity was informed by the spirit of the original Bogomils; the ancient "Bulgarian" trend predominated, with a belief in the origin of evil which could still be accomodated to the Christian doctrine. Dogma soon began to assume greater importance in the West, too, however, and as early as 1163 Eckbert of Schönau, a well-informed authority, recorded serious differences over dogma among the Cathars of the German Rhine valley. At the gathering in Lombers (1165) there might still be doubts about the basic character of the new movement, whether it lay in their conception of the Christian life or in doctrines whose anti-Christian character was already beginning to show through here and there. Another mission from the East was soon on the spot to give an unequivocal answer. The arrival from Constantinople of Bishop Nicetas and the words he addressed to the Cathar assembly at St.Félix (1167) marked a turning-point. The western Cathar groups now acquired a firm ecclesiastical structure. The ideal of the apostolic life was not lost from view, but dualist metaphysics became their chief preoccupation. Thanks to the influence of Nicetas, the radical-dualist cosmology, incompatible with the Christian view of the world, took firm root in the West.

In the period of approximately 1170 to 1215 the Cathar movement was still growing, in spite of all the dissensions within the body in Italy during this time. People were powerfully attracted not only by the example of the perfect life as lived by the elect, but also by the possibility of salvation at the end of their earthly journey offered to the body of believers by this new sect, though they might find the thorny way of consistent self-denial too difficult to travel. Then, too, the fact that the movement for reform became an organized church meant that its leaders were burdened with economic problems; in what is now southern France they were also emmeshed in the political interests of the nobles of Languedoc. A striking change also affected the sect's rejection of secular education. Indeed the Cathars' sense for earthly things seems to have been one reason for their notable success in the western world. It was far from a stable success, however, for the essential conflict between the dualist doctrines of the Cathars and Christianity could not long be hidden from the eyes of contemporaries. Before long the Cathars lost the leading role they had played for a time in the religious movement outside the Church, and it passed to others. Waldensianism, a purely western trend which emerged in the seventies of the twelfth century, was sharply opposed to Catharism from the outset.

From the end of the seventies the Roman Catholic Church, too, took pains to work out ways of dealing with this sect, whose doctrines and close-knit organization made it the most dangerous enemy of all. In the end Innocent III mobilized the military strength of the Christian West against the heretics abd their powerful protectors in what is now southern France. This "holy war", however, ultimately turned into a struggle to gain the southern lands for the French crown. Attention was distracted for a time to this political goal, allowing the Cathar sect a breathing-space, but meanwhile even more dangerous forces were gathering against it. The Fourth Lateran Council (1215) summed up the decrees so far issued against the heretics, and made their terms more general; the Church's attitude was now clearly formulated in opposition to the Cathar heresy. The enemy was now well known, and the mendicant orders formed at this time provided a most effective weapon. The sect which the horrors of the Albigensian War had failed to destroy was now defeated by the theological learning of Dominic and the moral sincerity of Francis od Assisi.

The period from 1215 onwards was already marked by stagnation in the Cathar movement, which had assumed regular ecclesiastical forms. The Cathar ritual, now more complex, betrayed a Catholic model in many instances - including the use of Latin as the liturgical tongue. The process of ecclesiasticalization was felt, too, in the ethics of the sect. The demand that the "perfect" should be of unblemished integrity was not relaxed, but they could be absolved of grave sin if they again received the Consolamentum. Thus the sacramental office, together with a precise set system of penance, acquired the force of absolution. Meanwhile there was a visible desire to come closer to Catholic dogma in the teachings of John de Lugio, and even more so in those of Desiderius. However closely these efforts corresponded to the needs of the time, the new ideas were not accepted at all widely, most of the Cathars holding on to out-dated forms of the dualist myth.

During this period the Cathar faith was becoming a religion of the "middle classes", a fact which resulted in more intense utilization of the wealth accumulating in the hands of the sect. The "perfect" still obey the rule of asceticism and poverty, as individuals, but the church itself is rich and material interests begin to assume greater importance than the demands of conscience. The original "revolutionary" fervour is giving way to quietist tendencies. The Cathars are willing to come to terms with the evil world, if that world will leave them alone.

They were not to be left in peace, however. The secular power took a hand in the fight against heresy, and here their interests were identical with those of the Catholic Church. In 1224 the Emperor Frederick II promulgated his terrible law against heresy, which concerned Lombardy. The Italian cities tried to prevent the application of this law, as later they resisted the establishment of the Inquisition. They were principally concerned to preserve their independence. When their political interests were served by so doing, they were ready to turn against the heretics as did the city of Milan in 1233. It was the Church itself, however, which dealt the decisive blow to the Cathar sect. The Dominicans and Franciscans worked not only with learned argument, but also with living example; the Cathars could find nothing to do but imitate them, and it was inevitable that they should be defeated. Here, too, we see that Borst lays primary stress on the spiritual weapons - assigning them greater significance than the merciless proceedings of the courts of the Inquisition. It was the pitiless system of the Inquisition which carried out and speeded up what was in effect already decided. Power alone cannot suppress ideas, and therefore a number of factors were involved in the final destruction of the Cathar heresy. Fundamentally the author sees the "lively West" (das lebendige Abendland) as finally emerging victorious over the stagnating doctrines and pure but quietist ethics of the Cathars.

This is A.Borst's view of the development of Catharism in western Europe, a conception which is based on careful study of the material from western sources but not on an equally thorough knowledge of the eastern documents, which he seems to know only superficially and not always at first hand.

However, a more penetrating analysis of the early testimony to the way in which the sect developed in the Eastern Christian sphere does not confirm Borst's interpretation. There is certainly no reason to doubt the historical existence of the priest Bogomil, designated by Presbyter Cosmas as the first to teach the heresy in Bulgaria under Tsar Peter. That is all we know of Bogomil, and indeed it was the mystery in which his figure was shrouded which led earlier scholars to exaggerate his importance.[2] The idea which Borst propagates, of the originality of Bogomil's teachings, is close to this way of thinking. In fact, in the earliest period from which evidence has come down to us (10th-11th centuries), the teachings of the "Bogomil" sect (which may have come into existence in Byzantium even before the priest Bogomil himself appeared) are so close to those of the Paulicians that there can be no doubt that the Bogomils were linked very closely indeed to this earlier dualistic heresy.

Both movements were inspired by a consciousness of the apostolic mission and of direct continuity of the apostolic tradition, although the Bogomils did not adopt the Paulicians' cult of the Apostle Paul. For both sects the name "Christians" expressed their conviction that they alone preserved the true Christian faith which the official Church had long deserted. In fact, the interpretation of this faith was determined by the dualistic view of the world inherited from the Gnostic thought of late antiquity. The earliest evidence we have of the Bogomil doctrine already gives clear form to the fundamental idea that this world is cut off from God; in the second half of the tenth century Presbyter Cosmas leaves no doubt in our minds that those who belonged to this sect believed absolutely that the visible world around us has nothing whatever to do with God, but is the work of the force of evil. Although the Bulgarian heretics already called the creator of this world the Devil, they undoubtedly identified him with the God of the Old Testament, all books of which they rejected as completely as did the Paulicians.

In the teachings of the Paulician sect the contrast between the God of the Gospels and the Creator of the Old Testament retained the sharp outlines imprinted by Marcion in the second century, following in the footsteps of the Gnostic teacher Cerdon. In the Paulician view, too, the God of the Old Testament was of a lower order, and his reign was limited by the duration of the visible world. The Orthodox Church acknowledged this creator and ruler of things ephemeral, while the faith and hopes of the true "Christians" strained towards the unknown God of the world to come. Thus dualistic doctrines provided the Paulicians with a metaphysical explanation of their inexorable rejection of the ruling Church, which was backed by the power of the whole Byzantine state.

From the outset, however, the dualist view of the world had more profound consequences for the Bogomils, determining not only their attitude to the Church, but to the whole of their society, and deeply affecting their way of life. The evidence of Presbyter Cosmas clearly reflects the essential connection between the asceticism of the Bulgarian heretics and their doctrine of the origin of the world. All things connected with earthly life are the commands of Mammon, i.e. the Devil, the evil creator and ruler of this world of ours. Evil is regarded as an external force which has man in its power from the very beginning. Tiny children are "the children of the Devil" and the baptism given them by the Church is powerless to help them. It is difficult to refrain from the conclusion that as early as the tenth century the sect undoubtedly already professed its own "spiritual baptism", although the earliest proof is not found until a hundred years later, in the writings of the Byzantine monk, Euthymius of Acmonia. Those who had undergone this purifying rite were no longer in the power of the forces of evil, and by their pure life earned the Greek name of "Cathars". In the earliest times of the sect this congregation of the elect had its ritual customs of which mutual confession and the frequent repetition of the Lord's Prayer were noted by Cosmas in his Sermon.

It would appear that there was an insurpassable gap between the few "whose converse is in Heaven"[3] and the rest, whose life "in the world" furthered the designs of Mammon. Yet from the very outset the sect needed the support of those who lived in the world - the *gent mondana* mentioned later in the Occitan Cathar ritual, apparently known in the Bosnian milieu as the "mirski ljudi". In Bulgaria, indeed, suffering from severe internal crisis in the second half of the tenth century, at the end of the Preslav empire, the rejection of the world by the Bogomils could not fail to stir broad sections of the people. It is not difficult to see how the heretical preachers' attacks on the powerful and wealthy would appeal to those who suffered most from cruel poverty and pitiless oppression. At the same time the life of self-denial led by the heretics, and the sufferings they were willing to bear for their faith, earned them the sympathy of many who were not strong enough to follow their example. Nor can we doubt that the select company of the pure gathered round them even then a number of followers who were given an insight into the teachings of the sect. The term "believers" does not appear before western sources, but Euthymius of Acmonia already speaks of people who had not yet undergone the sect's rite of purufication, but already "believed" their teachings.

The doctrines the Bogomils gradually revealed to their converts were constantly being developed and varied by free speculation and the play of the imagination. In the world of ideas inhabited by the Bogomils of Cosmas we can already discern the basic motifs which went to form the myth as recorded later, in the early twelfth century, by Euthymius Zigabenus. This cosmological myth elaborated by the Bogomils in Constantinople drew both on Biblical sources and on apocryphal legends, but in the context of the dualistic view of the world these ideas took on another significance. In Byzantium the Bogomil sect remained at the same primitive level of thought as that recorded by Cosmas; the sect spread in monastic circles, and deliberately turned its back on secular learning. There is no trace whatsoever of philosophical influences on the Bogomil cosmology. Not even in the Byzantine world did the Bogomils turn into the "philosophical sect" suggested by *D.Obolensky*.[4]

The developed form of the Bogomil myth, recorded by Zigabenus, seems to have been current among the sect for several decades, and was undoubtedly the outcome of longer development. The curious thing is that in the eleventh century there existed side by side with it a much more primitive variant, found in the diocese of Acmonia, on what had once been Phrygian territory. We do not know how far back into the past the history of the sect in Asia Minor went; in any case the myth current among the "Phundagiagites" of Asia Minor, as fragmentarily revealed by Euthymius of Acmonia, shows some very archaic features. The Marcionite-Paulician basis of their dualistic view of the world stands out particularly clearly; there is no more than the first hint of the idea of the "prince of this world" assimilating to that of the Christian idea of Satan as the angel fallen from grace.

The idea of the "fallen angel" which was already part of the beliefs of Cosmas's Bogomils, became the basis of the highly developed myth recorded by Zigabenus and occurring in a somewhat different form in the Interrogatio Iohannis, a Bogomil work which may also have come from a Byzantine milieu. There is no Byzantine text to prove the existence of a "radical dualistic" version of the Bogomil myth, based on the idea of two eternal opposing principles, as we found later on in the West. Nor is there any evidence to justify the assumption that the "Dragovitsian Church" came into being when the Bogomil sect split over the question of the origin of evil.

There is nothing to be found in eastern Byzantine sources which would testify to this distinction between an original "Bulgarian" and a later "Byzantine" form of Bogomilism - the distinction Borst throws into such a glaring light. This fact, however, is decisive for our attitude towards the early history of Catharism on Western soil; there is no reason to distinguish two stages, with the year 1167 as a fateful turning-point between them.

A wave of dualistic heresy from the East flooded Western Europe from the forties of the twelfth century, and it is only from that time onwards that we can speak of a "Cathar" movement in the West. Today this can no longer be a matter for doubt. There still remains, however, the question of how far the native factor was involved in the emergence of the movement. *H.-C.Puech* holds the view that this new doctrine from the East took root among "heretical" groups which had long been in existence and which adapted it to their own needs and ideals. He admits that this impetus from the East evoked a fundamental change.[5] It must, however, be said our sources provide no documentary evidence that the Cathar communities in the West grew out of earlier local groups. The Cathars recorded in Cologne at the beginning of the forties of the twelfth century came into contact with a local group of "heretics", but they did not merge with them; on the contrary, conflicts arose.

Undoubtedly the sect adopted some of the attitudes of the native anti-ecclesiastical movement, once it began to develop in the western environment. Like others, Cathar preachers, too, attacked the custom of child baptism and declared the sacrament of the Eucharist invalid if it was administered by an unworthy priest. These and similar arguments were intended for the broader public; the true Cathar doctrine remained the affair of a small number of initiates, and in the same way the sect hid its secret rites from those around it. It was impossible to remain hidden for long, however. Ewervin of Steinfeld (1143 - 1144) knew the Cathar "baptism by fire and the Holy Ghost", the administration of which created the special company of the "elect". Eckbert of Schönau came up against the dualistic view of the world which lay behind the sect's abhorrence of certain foods - for they believed they had been created by the Devil. This German Benedictine monk had also learned the sect's view of the nature of Christ and glimpsed their myth of the origin of the soul. What he discovered amply sufficed for him to classify Catharism unequivocally as a branch of Manichaean heresy.

Shortly before Borst, *A.Dondaine* had suggested that Catharism in the true sense of the word dated from the appearance on the Western scene of Bishop Nicetas, for it was then that the doctrine of two eternal principles also appeared - "the fundamental error" which, according to Dondaine, upset Christian feeling in the West more than anything else.[6]

In reality, the significance of Nicetas' mission lay elsewhere. The doubts it gave rise to did not centre on the dualistic cosmology but struck at the very heart of the dualistic faith. To receive spiritual baptism was indeed the only way to escape from the power of the "ruler of this earthly world", and the soul's salvation therefore depended absolutely on the validity of the consolamentum received at the hands of a Cathar bishop.

This was the fundamental conviction which united all groups of the sect. The western Cathar communities remained linked to their mother churches "across the sea", whence their bishops received ordination; it was thus inevitable that doubts arising among eastern communities of the sect, concerning this question, should find their way to the West. It was this which brought about a split among the Cathars of northern Italy The newly formed churches turned to the three "overseas" churches whence they took their *"ordo"* and with it the variants of the dualistic myth which then went on developing in the West.

The different forms of the cosmological myth which can be distinguished did not emerge in isolation; they should be regarded merely as various expressions of the flow of ideas discernible in the history of the sect from the outset. In the history of the movement as a whole, the difference between "moderate" and "radical" dualism did not play such a significant role as is generally supposed. The different explanations of the origin of evil aroused many a conflict, but could not in themselves lead to an open break between different Cathar groups. The split between the two most important Cathar churches in Lombardy went deeper; yet although it lasted well into the thirteenth century, it was not extended to the other branches of Catharism, sharing the doctrinal views of the "Albanensian" church or that of Concorezzo.

From the standpoint of the Catholic Church, the doctrines of the "Bulgari" and "Sclavini" had the same monotheistic foundation as those of orthodox Christianity, but were inacceptable because of their rejection of the visible world as the creation of the Devil. This basic tenet, with all it involved in the Christological and anthropological spheres, was common to all groups of the sect in both West and East. These were the *communes opiniones* which Rainerio Sacchoni so aptly defined, and which excluded all possibility of reconciliation between the Cathars and the official Church. Not even the bold system of ideas put forward by John de Lugio, or the Christological doctrine of Desiderius, made any inroads into this nucleus of Cathar belief, although both appeared to come closer to Catholic dogma.

The dualistic interpretation of the Cathar doctrines came up against objections even among members of groups themselves in conflict with the Church and placed without its bounds. The Cologne case of 1143 showed that the Waldensians were not the first Western heretics to find themselves in conflict with the Cathars.

Nevertheless, the Cathar mission moved victorious through the Western world, for it offered a lively example of the apostolic life as well as inexorable anti-clericalism. Both these trends were in complete harmony with the local movement inspired by the goal of consistent church reforms but inevitably finding itself in anti-ecclesiastical positions.

From the last years of the second decade of the twelfth century, the eyes of leading Roman Catholics were turned to what is now southern France, where the situation was growing particularly dangerous. The Cathar mission which must have penetrated from the north perhaps before the middle of the century was following here in the footsteps of the monk Henry, whose work had set the whole region of Toulouse at odds. The rural nobility willingly opened the gates of their fortified mansions *(castra)* to Cathar preachers whose eloquence resembled that of the wandering monk who called upon bishops and priests to live a life of modest poverty, and who everywhere instigated the people not to pay tithes.

In the neighbourhood of Toulouse, in the southern Albigensian region, the *castrum* of Lombers offered the Cathars the safest refuge. This seems by all accounts to have been the centre of the Albigensian church, the seat of their bishops. This was the first Cathar church to be established in the south, and in the minds of the northern French the word "Albigensian" came to mean any southern Cathar. In a short time three more sees were established in the south, and the *castrum* of St.Félix in the Lauragais region was the scene of the gathering we have already mentioned, in 1167: representatives came hither from the southern Cathar churches, from that of Northern France and from the transalpine community of Lombardy; they all subordinated themselves to a Greek - to the Constantinople bishop Nicetas, who had come from so far away across the sea.

There is no reason why the early history of Catharism in the West should be seen as a development "from the reformation movement towards the Church". The Cathar sect had brought with it from the East, from the very outset, certain fixed ecclesiastical forms - this was, in fact, the guarantee of continuity and, too, of a lasting influence of the sect where conditions were favourable for an organizational structure to be created. In Languedoc Catharism was able to develop in security, protected by the nobles who found themselves sharply opposed to the Church and therefore indulgent towards all heretics. The Cathars, the "good people", had a profound influence on the life of these noble families; they were assured of entry there, and brought a distinctive feature to the cultural ambiance of this sector of society. The lyrics of the troubadours continued to echo within the *castra* of the nobility, presenting the fundamental theme of chivalrous love in the most varied forms evolved by mastery of the poetic *métier*. But there were many noble ladies of the day who lent their ear not to the words of earthly poets, but to those of the Cathar preachers[7] opening a very different scene before their gaze. There were some who received the consolamentum and thus dedicated themselves entirely to the new faith. Here all attempts to turn the noble families of Languedoc away from the heretics came to grief. The "good people" were even linked to them by family ties; they were everywhere regarded with respect and devotion, and their authority was immense. Here they could no longer be pushed aside by the Waldensians, who came upon the scene much later; in some noble families, in fact, the two heresies existed side by side.

A very difficult situation now faced the Roman Curia. Some of the Waldensians could still be won over to the side of the Church and used for work among their former fellows. Catharism, however, was a far more serious threat; it was heresy *par excellence*, as contemporary documents eloquently confirm, distinguishing as they do between *heresis* and *Valdesia*, between *heretici* and *Valdenses*. In this region Rome could not fall back upon the support of the secular authorities. Raymond V, Count of Toulouse, openly admitted that he was helpless in the matter, and his son Raymond VI was even favourably disposed towards the Cathars. The young Raymond-Roger de Trencavel was powerless against his vassals who protected the heretics. It was not until later that the Cathars put down roots in the County of Foix,[8] where they found support in the ruling family itself.

Only interventiom by force could bring about a change here, and Innocent III took upon himself the terrible responsibility for unleashing a disastrous war which soon got out of hand and ultimately took a quite unexpected direction. The war certainly did not destroy the Cathar sect, but the destruction of the power of the house of Trencavel, and the subjection of Raymond the Sixth's son to the Church completely changed the political situation in the South. In those lands annexed by the French crown immediately, as well as in those domains still left to the Count of Toulouse, the sect found itself in very difficult conditions, especially after the institution of the Dominican Inquisition; nevertheless it was not quite without protection against its persecutors. Although many of the noble families of Languedoc fell victim to the catastrophic upheavals which agression from the North brought in its wake, that social stratum which had always provided the Cathars with their surest support was far from being wiped out.[9]

The ecclesiastical organization of the Cathars survived, under the protection of Montségur, and even after this strong refuge had fallen the "good people" did not lose contact with their "believers". The investigations made by the Inquisition were seriously hindered by the unwillingness of the country people to help, and the Cathars could still count on influencing many noble families.

Such was the position up to the middle of the thirteenth century, but by the early seventies the picture was very different. The Inquisition was now much strengthened, and since the mid-fifties had been returned to the province of the Dominicans; they had succeeded in breaking up the network of contacts between the heretics and discouraging those who had formerly supported them. The aristocracy, coming more and more under French influence, no longer protected the "good people", who left their homeland in large numbers to seek refuge across the Alps.

Towards the end of the century the stand taken against the Inquisition by the towns of Languedoc encouraged some of the "perfect" to return home and take up their activity there again. However remarkable this late renaissance was, the times had changed and there was no hope that the sect could hold its own in southern France. The Inquisition again had everything well in hand, although there was still a great deal to be done to root out the heretical faith from among the peasants of the inaccessible valleys in the County of Foix.

Languedoc offers the best opportunity of tracing the development of Catharism. Here the sect had become part of the life of the people during the relatively long period of favourable conditions,[10] and it was not overcome until violent change affecting the whole country gradually robbed the heretics of their strongest backing in society.

A similar development can be discerned in Italy, but there the picture is a more complex one; the heretical movements were inextricably involved in all the antagonisms which rent the scattered world of the Italian cities. The term "heretic" served too readily to designate a political opponent, and the same seems to have been true (in so far as can be judged from the sources) for the towns of Dalmatia, on the other side of the Adriatic. Nevertheless the heretical movements were a real force in the Italian cities, nourished by the growing struggle between the secular power and the Church, and the consequent abnormal state of affairs. The Cathar sect took root here some time round the middle of the twelfth century; its ecclesiastical organization, taking form at a time of doubts and conflicts brought over from the East, covered the northern and central regions of the peninsula. It was a much more scattered affair than in Languedoc, where the territorial limits of each church were clearly laid down.

For the most part the Cathar bishops in Italy had their seat outside the towns; here, too, the Cathars found their safest refuge in the fortified homes of the aristocracy in the rural environs of the towns. This was the case in Florence, for example, where the most powerful families took the Cathars under their protection. The picture we are given by the records of the Inquisition in Florence is very like that we found in Languedoc: the "good people" had a profound influence in the life of the noble families.[11] The resistance of these great patrons could only be broken by force of arms, in order to rob the Cathar sect of its position in Florence for good.

In the North, in the region of Verona, the fall of the powerful Ezzelino III da Romano (1259) brought about a political change unfavourable to the heretics, the full consequences of which were only seen later. For a long time the Cathars still kept their "house" in Verona itself, but they felt most secure in the strong *castellum* on Lake Garda; at this time Sirmione was something of a Montségur, where the leaders of the church of Bagnolo gathered, as well as some important refugees from Languedoc. It took several months of siege by the French king's troops to reduce Montségur (1244), but a single stroke by the secular and ecclesiastical powers in Verona was enough to get the heretics out of reach of their protectors (1276). Two years later the tragedy was played out in the amphitheatre of Verona, as it was in the town of Bram after the fall of Montségur.

The burning of about two hundred Cathars as late as 1278 does not suggest that it was primarily spiritual weapons that overcame the sect. There can be no doubt that both Franciscan and Dominican monks were well equipped to work among the people, but these orders were formed when the Cathar movement was already deeply rooted and could only be destroyed in merciless battle. The real domain of the Dominicans was preaching, but from the outset it was combined with *inquisitio haereticae pravitatis*, and judgment on the heretics was soon mainly in their hands. The Fratres Praedicatores did not neglect the need for theoretical polemising with the enemy, armed as the heretics were with an excellent knowledge of the Bible. No learned theological arguments, however, could break the bonds of feeling between the Cathars and the society around them, in which they were firmly rooted. Contemporary reports reveal quite clearly that questions of dogma were of little importance compared with the force of the reverence evoked by the pure lives of the "good people".[12] Nor did their self-denying way of life fall at all short of the Franciscan ideal of poverty in the spirit of the Gospel. The times forced them to search for sources of income to maintain their church, and sometimes even to demand its dues harshly. They themselves, however, lived in simple poverty, as even their enemies admitted. Not even a sceptical modern scholar can fail to acknowledge that the exemplary life of most of the "perfect" effectively prevented the practice of repeated administering of the consolamentum (after mortal sin) from completely upsetting the order of the Cathar churches. In fact there is evidence of this practice from the earliest times of the sect in the West,[13] and there is no need to see it as evidence of a later process of ecçlesiasticalisation.[14]

In fact, it would seem to be difficult to find in the history of Catharism that tendency to become a "settled church", ready to accomodate to the world, that A.Borst attributes to it. Indeed the world itself did not give the sect time to accomodate.

The "ecclesia Dei" had to fight ever harder for its very existence, and the "good people" moved about the world like "sheep among the wolves". Over and over again, in times of great trouble and face to face with death, the "good people" gave proof of their moral strength, supported by the extreme self-sacrifice of their "believers". Their example was strong enough to draw others with them, even to the ultimate sacrifice, as in the last days of Montségur.

We are not concerned here to write an apology of this mediaeval dualist sect; it is the importance of this sect for the history of the western world which must be objectively assessed. The dualist doctrine itself, by its very nature, could not pass the limits of myth; it made no contribution to the philosophical or theological thought of the West. With its emotional and ethical values, however, Catharism was a vital force in the development of western society and answered to the needs of the time. In the end it was not defeated by the spiritual superiority of the West, but crushed by the material superiority of the Church and the secular power combined. The "living West", in the end, could only defeat the movement by calling up a "Holy War" and instituting the courts of the Inquisition.

The history of the sect in Bosnia was curious; the failure of the crusades against it and of the Inquisition allowed the Patarene faith to survive for over two centuries longer. A handful of Franciscan monks were powerless against the native heretical church, securely protected by the Bosnian nobles. During this long uninterrupted development the "Bosnian Church" became an integral part of social life and even fulfilled some purely secular functions in it. Here the mingling of the "ecclesia Dei" with the world did indeed leave its dark mark on the life of leading Christians, as can be clearly seen in the remarkable career of *gost* Radin. Radin, however, is the only figure in the Bosnian church which stands out at all clearly, and this one case cannot be generalized. The few references in the sources show that the social structure of the "Church of Bosnia" was not homogeneous; *Gost* Milutin moved in aristocratic circles at home and abroad, and was buried in precious brocade, under an ornamented stone tomb.[15] On the other hand, the *krst'janin* Radoslav was a simple man whose hand, "heavy from hard work", was not used to the pen.

The "Patarene houses" had privileges of their own, and a contemporary accused their inmates, perhaps rightly, of receiving only the wealthy with honour, and driving away the poor. The great mass of ordinary *krst'jane*, however, lived as simple peasants - as indeed we have seen in Languedoc, too.[16] In Bosnia these people kept their religious identity under Turkish rule, after the fall of the Bosnian state.

In the last decades of Bosnian independence the external situation became unfavourable for the heretical church, but the backing it had among the nobles held off its destruction until the eve of the fall of

Bosnia. In spite of all the curious features of the growth of the "Patarene church", its end, too, was due not to internal decline but to external factors. The organization of the Patarene church was in fact destroyed only in the domain ruled by the King of Bosnia; the leading men found a safe refuge in the domains of Stephen Vukčić Kosača, whose foreign policy was never as unequivocally directed towards Rome as was that of the last kings of Bosnia.[17] *Gost* Radin remained the highest authority in the hertseg's suite, and witnessed his last hours.

In conclusion it can be said that this mediaeval dualist "sect of many names" reveals a remarkable continuity in its basic forms, common to all branches in the West and in the East. The names Bogomil, Cathar, Patarene and many others, are merely different ways of designating the same heretical society which, in spite of features peculiar to one or another region, and in spite of conflicts between individual groups, never lost its awareness of unity and a single origin.

Yet this defines only one fundamental aspect of this historical phenomenon. The sect never lived in isolation, but took part in the growth of society in the country concerned. Historical research must consider which elements in the sect's doctrines had the greatest effect on the world around it, ensuring that its influence was not limited to a narrow circle of believers. Where there is sufficient material the social basis of the movement should be studied, as well as the interplay of historical factors determining the rise and the fall of the sect.

It is from this standpoint that the development of this mediaeval dualist heresy has been studied in this book, and it is our belief that further research should follow the same lines.

[1] See *Borst*, 67 ff., 90, 95 ff., 98, 101 f., 104 ff., 109, 115, 119 f., 120 ff., 128 ff., 135, 142, 223 ff.

[2] See *I.G.Klinčarov*, Pop Bogomil i negovoto vreme, Sofia 1927.

[3] See Chapter IV above, p. 56.

[4] *D.Obolensky*, Bogomilism in the Byzantine Empire, in: Actes du VIe Congrès Intern. d'Études Byz., t.I, Paris 1950, 293.

[5] *Puech*, Catharisme médiéval, 78, 79.

[6] *Dondaine*, Hiérarchie II, 267, 269.

[7] See *Griffe*, Aventure, 169 f.

[8] Ibid., 159.

[9] See *Lacger*, art. cit. (see Chapter XIV sbove, n.19), 899.

[10] *E.Griffe* does not accept earlier suggestions that the "southern civilisation"

played the decisive role in the successful spread of Catharism in Languedoc (*Griffe*, Aventure 168 ff.). Nevertheless we consider that this phenomenon did owe something to the cultural climate of the South, so different from that of Northern France. (See the apt definition of this difference in *G.Duby* - *R.Mandrou*, Histoire de la civilisation française au moyen-âge, Paris 1958, 131 ff.). It is worth noting that it was in this Occitan society, where the lyrical poetry of the troubadours had given such powerful expression to a new attitude to women (to the women of the aristocracy), that women took such an important part in the heretical movement. Griffe is right in stressing that Catharism did not affect by far the whole of Occitan territory. Yet even the anti-clericalism of the rural nobility, which the author regards as the most significant factor, was a phenomenon discernible over a much wider area and in itself does not explain why Catharism took such firm root precisely in some particular regions. *Y.Dossat* had already pointed out that heresy found it harder to penetrate where the bishops had preserved their temporal power. (Besides the conditions obtaining in the diocese of Albi, he referred to the different situation in the City and Bourg of Narbonne. See the article by Dossat quoted in Chapter XIV above, n.19). Griffe points out that "la contagion cathare, partie de la région située entre Toulouse et Albi, a gagné, de préférence, les pays qui dépendaient des vicomtes de Béziers ... Évidemment, les frontières féodales n'ont pas constitué les barrages hermétiquement fermés, mais il semble bien qu'elles ont joué un rôle assez important et qu'elles expliquent, au moins en partie, pourquoi le cheminement de l'hérésie cathare est allé de l'ouest vers l'est et du nord vers le sud". (*Griffe*, Aventure, 177). It is certainly true that local factors must be carefully considered in dealing the geography of heretical movements in what is now Southern France.

[11] See the documents published by *F.Tocco*, Quel che non c'è nella Divina commedia o Dante e l'eresia, Bologna 1899.

[12] See the words of a nobleman of Languedoc, recorded by Guill.Podiol., t.19, 200. Cf. *Tocco*, op.cit., No.13, p.49,(testimony of Renaldus de Pulce): "...et dixit quod vita eorum placebat sibi, et eos bonos homines tenebat, ..."

[13] The consolamentum lost its efficacy in consequence of mortal sin, committed either by the person who had received it, or by the one who had administered it. From the point of view of dualistic doctrine there was no real difference.

[14] *Borst's* conclusions, that Catholic rites and customs were adopted by Cathar churches in the West, are not very convincing. We should remember, for example, that the sect could have adopted the system of three forty-day fasts while still in the East. The existence of these fasts is attested as early as the beginning of the ninth century in the Byzantine Church. See *K.Holl*, Gesammelte Aufsätze zur Kirchengeschichte II, Der Osten, Tübingen 1928, 197 (quoted by Borst himself).

On the other hand the author wrongly attributes to the influence of the western *heretical* environment, that the Cathars would not take any oath; this is attested from the earliest days of Catharism in the West (see the letter from the clergy of Liège to Pope Lucius II, PL 179, col. 937, and the Acta of the Council of Lombers, Mansi XXII, col. 166). This principle, however, like that of refusing to kill any human being, is attested in Zigabenus for the Bogomils (Euth.Zig., ed. *Ficker*, 104, 10-13 = PG 130, col. 1324A).

[15] See the article by *V.Skarić*, quoted in Chapter XIX above, n.78.

[16] See Chapter XIX, n.132.

[17] See Chapter XIX, p.316.

Index

GPSR Compliance

The European Union's (EU) General Product Safety Regulation (GPSR) is a set of rules that requires consumer products to be safe and our obligations to ensure this.

If you have any concerns about our products, you can contact us on ProductSafety@springernature.com

In case Publisher is established outside the EU, the EU authorized representative is:

Springer Nature Customer Service Center GmbH
Europaplatz 3
69115 Heidelberg, Germany

Batch number: 10824476

Printed by Printforce, the Netherlands